CT SUITE

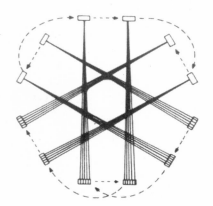
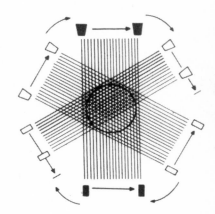

BODY, COMMODITY, TEXT

Studies of Objectifying Practice

A series edited by
Arjun Appadurai,
Jean Comaroff, and
Judith Farquhar

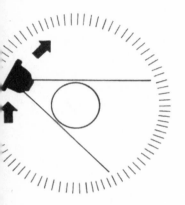

BARRY F. SAUNDERS

CT Suite

The Work of Diagnosis in the Age

of Noninvasive Cutting

DUKE UNIVERSITY PRESS

Durham and London

2008

Printed in the United States
of America on acid-free paper ∞
Designed by Amy Ruth Buchanan
Typeset in Minion and Gotham
by Achorn International
Library of Congress Cataloging-
in-Publication Data appear on the
last printed page of this book.

*Duke University Press gratefully
acknowledges the support of the
University Research Council at
the University of North Carolina,
Chapel Hill, which provided funds
toward the production of this book.*

CONTENTS

ACKNOWLEDGMENTS

This book would not exist without the generous attentions and critical tractions of professors, colleagues, students, and friends over fifteen years or so.

Special thanks are due Judith Farquhar—for her marvelous examples of discernment and clarity of explanation, and much more, over many years—and for her initial welcome of this book as part of the Duke University Press series *Body, Commodity, Text*. I also thank Ruel Tyson for contributions to my scholarly vocation and formation at so many crucial junctures, including many parts of this book. And I thank Larry Churchill for welcoming me to the interdisciplinary conversations at UNC Social Medicine, which has become such a fine intellectual home.

Among colleagues at the University of North Carolina, Chapel Hill, who shaped the book most directly—by commenting over the years on writing, offering bibliographic recommendations, thinking with me about biomedicine-as-culture, and more—are Keith Wailoo, Tomoko Masuzawa, Carol Mavor (all of whom critiqued the dissertation that preceded the book), Kevin Parker, Richard Clark, Victor Braitberg, Fletcher Linder, Della Pollock, Mark Olson, Keith Cochran, Stephen Pemberton, Joel Elliott, Roper Marks, and Larry Russell.

I am indebted to my hosts at the University Hospital Department of Radiology where I conducted my fieldwork in the late 1990s. The chair and vice-chairs and the directors of CT services were welcoming and cooperative. Members of resident and attending staff, many technologists, several nurses and administrators, and various clerks, schedulers, and couriers were helpful in answering questions, allowing themselves to be recorded and photographed, and generally tolerating my intrusions into busy routines. This book is thin tribute to many performances of skill and care that I witnessed. Any misrepresentations of radiological views and activities in this book are my own. I am also grateful to Dr. Melissa Rosado de Christenson and colleagues for my welcome in a fall 2000 visit to the AFIP Department of Radiologic Pathology.

The book has benefited from feedback on various presentations in scholarly colloquia and meetings: Duke Sawyer Seminar on Cultures and Medicine; University of Chicago Department of Anthropology; European Association for Social Anthropologists (Vienna); Association for Social Anthropology (Manchester); Society for the Social History of Medicine (Manchester); Society for Literature and Science (Los Angeles); Tanner Humanities Institute (Salt Lake City). I am particularly grateful for the perspicuity and help of several host colleagues in these settings: Priscilla Wald, Cristina Grasseni, Simon Cohn, Jim Bono. Helpful commentaries included those of Joe Dumit, Jean Comaroff, Barbara Stafford, Brian Rotman, Timothy Lenoir, and Andreas Roepstorff—among many others.

Two anonymous readers of the book manuscript at Duke University Press were meticulous and generous in their interventions. Ken Wissoker and Courtney Berger deserve thanks for securing their help—and also for their own patience and prudence throughout manuscript revisions.

Over a longer durée, Kathryn Montgomery's insights into literary methods in medicine—and friendship—have been very important. Another friend, Liza Wieland, lent her poet's ear to this project when my own editorial judgment was at low ebb.

At the University of North Carolina, I am fortunate to have recurring teaching roles, and fine colleagues, in both Anthropology and Religious Studies. Students in several graduate seminars have offered helpful critical commentaries on chapter drafts. In Social Medicine, I remain deeply grateful for my teaching appointment (and for one leave!) while this book project was in the works, and I am particularly appreciative of warm collegial support I enjoyed from Sue Estroff, Nancy King, Gail Henderson, Bill Lachicotte, Jon Oberlander, Don Madison, Kendrick Prewitt, Bill Kerwin, W. D. White, Terry Holt, Des Runyan, Alan Cross, Rebecca Walker, Judy Benoit, Lisa Perry, and others. In Medicine and Family Medicine, I thank Fred Sparling, Tim Carey, Byron Hoffman, Todd Granger, Mark Gwynne, and Warren Newton for recognition of my scholarly work. At Chatham Hospital, erstwhile home away from home, I thank Jim

Adams, Annette Willett, Cam Austin, Bea Morehead, deb Garner, and John Dykers for the same, and for their friendship.

Two able assistants supported specific phases of this project: Rachel Winters helped with bibliographic work; and Lisa Smith transcribed tapes of radiological shoptalk. Several archivists provided special help beyond reproduction of images: Michael Rhode (National Museum of Health and Medicine, AFIP) and Eileen Mathias (Academy of Natural Sciences). A grant from the University of North Carolina's University Research Council helped defray publication costs.

Finally: I express my heartfelt thanks to my family and friends—especially my wife, Susan—who gave me the support and freedom to think and write and kept their impatience for the book's finish mostly under wraps.

INTRODUCTION

For the tasks which face the human apparatus of perception at the turning points of history cannot be solved by optical means, that is, by contemplation, alone. They are mastered gradually by habit, under the guidance of tactile appropriation.

• • • **WALTER BENJAMIN,** "The Work of Art in the Age of Mechanical Reproduction"

In hospital-speak, the "CT Suite" is a series of rooms: scanner room, waiting room, reading room, file room. And closets. All dedicated to Computed Tomography, or CT scanning—a mode of diagnostic imaging.[1] So, in part, this is a book about space.

But rooms—architectural forms—are just one connotation of "suite." There are various linked *practices* distributed in and among these rooms. As in a musical suite (classically, a sequence of dance pieces: *allemande, courante* . . .). Cutting (*tomo*) and writing (*graphein*) are tomography's eponymic practices— but they are linked with many others. This is a book about practices.

A third kind of suite is one of *attendants* (the first "suites" were courtly entourages: King, Queen, Minister . . .): the retinue of persons who attend to, and wait on, diagnostic images. This book is about social roles that make up an economy of diagnostic attentions.

And all these suites are arranged around the CT scanner, its viewbox, and other pieces of apparatus—a suite of furniture. This book is about how high-tech machines fit into an institution and its ethos.

These suites all constitute *followings* (*suivre*): rooms linked to rooms; writings following cuttings, and more; clients and trainees following radiologists, in pursuit of expertise; viewboxes linked to scanners. All following slices of culprit disease. All these have been followed in turn, and re-presented here, by a diligent ethnographer.

. . .

Diseases have long had their distributions in populations, their vectors—in households, herds, villages. Within bodies, however, it is only during the last two centuries that Western medicine has understood diseases to assume particular shapes, to distribute themselves spatially in tissue. In the famous formulation of Giovanni Battista Morgagni, we came to conceive of diseases in bodies as devolving each from some "seat and cause." Diseases have been understood through visible, material, sited abnormalities—characteristic *lesions*—since the early nineteenth century. That was when the ancient imbalances of humors, the sequences of symptoms, the "botanical" constellations of superficial signs were reclassified, in the Paris Clinic, in view of the "bright light of the corpse."[2]

Cadaver upon the anatomy table: hub of a ritual alignment of attention, scalpel cuts, and memory (of antemortem events). It was this platform of the lesion-made-visible—a new mode of "spatialization of the pathological"[3]— that made diseases classifiable according to morphological types. Derangements of "tissue" came to serve as the gold standard for diagnosis.

Today, the hospital still honors the legacy of the visible lesion—alignment of diagnostic attention around a culprit morphology, a site. But things have become more complicated. Dissection of the corpse, the founding ritual of a lesion-centered nosology, was soon followed by other procedures of examination which produced seats and causes from *living* bodies. X-ray photography has been one of these procedures since 1896. The laboratory has been a nexus of many such procedures. Moreover, lesions once fixed, macroscopic, and re-trieved from the cadaver—tumor, abscess, tubercle, infarction, clot—became microscopic, molecular, fluid. Now we have coagulation cascades, vagal tones:

diseases once again of humoral balance and temporal succession, of sympathies and flows, that evade simple morphological characterization. But even if very small, the fixable, visible lesion remains important.

Why, after founding a mode and ethos of morbid spatialization, did the cadaver seem to slip out of service? No doubt there are disincentives to examine it—odors, stigmas, religious values. But also there are so many ways now to work our morphological magic, conduct examinations, without turning to the cadaver.[4] Our capacity to open bodies and plumb their depths "noninvasively" is a commonplace, and point of pride, in contemporary medicine—especially in diagnostic imaging. The diminished role of the cadaver today implies a transfer of epistemologic authority to practices and spaces outside the morgue—and indeed, a different status of death in the hospital.

• • •

A substantial measure of transferred authority now resides with a figure we will refer to as "doc-at-the-box." This figure, or scene, is found throughout popular and professional media: a physician, or members of a health-care team, in front of a viewbox, before grids of images—often CT scans.

The visual cliché of professional-before-image-grid is a reincarnation of doc-with-stethoscope or doc-with-head-mirror. Medical expertise remained cartoonable throughout the twentieth century as man-with-special-prosthesis. But the defining prosthesis these days may not be something worn or carried. The viewbox of images and the scanner which made them are institutional fixtures, objects which experts gather around. Like other exam tools, they organize regimes of attention, exchanges among specialists, and relations between doctor and patient.

CT is one of the key technologies of medicine's famed "imaging revolution": it naturalized the image-grid to the shop floor of the hospital and made it possible for this pictorial convention of doc-before-viewbox to stand for medicine in general. Yet oddly, this very scene, doc@box, is one from which patients have largely been excluded. Consumers of this scene do not know much about the practices that it crystallizes, or the ways they structure power and truth. The scene is something of a fetish, whose preciousness and authority we cannot quite explain.

• • •

Americans are enamored of CT scanning, and of medical imaging more broadly: believers in its faithful accuracy, in its place in the triumphal march

1. Docs-at-the-box

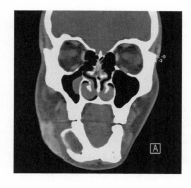

2. Cystic thing in Rohan's jaw

of progress. Scanning is such a standard part of healthcare—and seeing is so far toward believing—that we have come to value medical images as Evidence in its quintessence.[5]

But this wasn't so with some friends of mine a year ago. They are naturopaths, and they were wishing mightily that their son Rohan could avoid a CT scan. He had a nasty cystic growth in his jawbone—but they were reluctant to engage the "Babylon system" of the hospital. They knew that a scan would be, in addition to a year's supply of radiation, a staging test for the scalpel. They held out as long as they could, with turmeric poultices.

Rohan and his family eventually agreed to a scan of his jaw, after much deliberation. They were given an appointment, then found their way to University Hospital's second-floor CT receptionist and waiting room. Rohan was escorted by a tech to the scanner room, strapped to the scanner table, told to hold still while the table slid him into the donut ring of the scanner. Afterward, on his request, he was given a disk to take home. It booted up on his home PC. Rohan could scroll through a stack of 3-millimeter slices and contemplate the "thing" in his mandible. It took him days to do this: at first it was too horrific, too depressing. Meanwhile the images were viewed promptly at the hospital, and a report was dictated and printed for Rohan's record. The images were later reviewed by a craniofacial surgery team. Rohan's family did not receive the formal report, though the scan became a prop in theatrics of clinical persuasion.

Naturopathy gave in. The surgery took place, the thing has been excised. Rohan recovered well. CT scanning was instrumentally useful. Biomedical intervention trumped another competing religion.

Patients do not usually inspect their own images as Rohan did. (Though this is happening more frequently as we exit the age of film.) Patients have historically been excluded from scenes of CT interpretation.

Rohan's looking his lesion in the face—anxiously, courageously—seemed important. (Did it see him too? Or was its presumption blind—this encysted Other?) Fortunately it could be excised. We hope it is not a recidivist sort.

. . .

Rohan's tale could be misconstrued at this juncture: this book is not about illness experience or patient subjectivity—though these are at its edges, in its background, in its suites of cultural entailment. Nor is the book about medical images that escape the diagnostic enterprise and find their ways to patients' home computers or refrigerators (like so many prenatal ultrasounds). It is about "cultic" uses of CT scans—their primary circulations in the hospital's diagnostic economy and their place in the constitution of medical expertise. It is more about interactions among medical professionals than it is about caregiver-patient interactions.

And if Rohan's tale presents a coherent narrative—something we all appreciate in an ethnography—alas, this book, in general, does not. My representational modes are not primarily about coherence. Rohan's tale is a gesture toward normative arcs of institutional process. But throughout this book I have emphasized instead the contingency, the interruptions, the juxtapositions, that characterize—indeed, are constitutive of—a full suite of tomographic practice.[6]

In this I acknowledge—as in the book's subtitle—a debt to the cultural theorist Walter Benjamin. My fascination with medical imaging has been a bit like his with the Paris Arcades—novel architectures of visual absorption and commodification. After following a few of Benjamin's footsteps through the arcades of medical imaging and experimenting with modes of description, I have also come to realize how much I owe to his philosophy of history, and to his explorations of representational modes of montage.[7]

As an ethnographic document, this book represents events, interviews, and shop-floor exchanges in a University Hospital CT Suite in 1996–97. Names, numbers, and some genders are altered to preserve anonymity of informants and subjects. Most field description is rendered in a present tense—faithful to the ethnographic present of its witnessing—notwithstanding our present historical distance from those scenes. In many places I have reproduced words of my subjects in transcripts of recorded dialogue.[8]

These conversational exchanges are integral to the suite of ethnographic modalities that make up this book. They are not merely illustrative of claims developed elsewhere; nor do they simply mark empirical foundations. Quotations and transcribed conversations, at the viewbox and elsewhere, are meant as exempla: anecdotes that serve as morals, models, dramatizations in their own right. Most of these conversations' implications far exceed what I offer in commentary. I have retained a few long transcripts at places in the book where the reader's ear may be tuned to their weave of implications. Where it seems important, I include in these quoted exchanges some nonverbal noises of discursive practice—saliently, "filmflap" and "boardbuzz"—sounds of work at the mechanical film viewbox.

This summons a historical note. Since my initial fieldwork, the mechanical viewbox—source of filmflap and boardbuzz—is gone, and film nearly gone, from University Hospital. Now images are distributed through the hospital by means of a PACS, a picture archiving and communication system, a digital network. Thus a major reason for clinicians' visits to the reading room—to review films for themselves—has all but disappeared. Visits by clinicians to the reading room have fallen off considerably—and with them, the conversations and interspecialty performances that once accompanied film review and are analyzed in this book. Over the same period, the radiological caseload has more than doubled. There are compensatory mechanisms for communication among radiologists and their clinical clientele: the phone; a few new consultative sessions "on the wards"; improved formats of official reportage. But face-to-face interspecialty dialogue on the radiological shop floor has thinned. Radiologists do not agree on what this means. Their professional literature has focused on measurable utilities of digital tools and fewer interruptions: lower cost, higher throughput, fewer errors.[9] Toward the end of this book I will suggest some further implications. But throughout the book I allow that ethnographic moment—with film still central, and practitioner understandings of the digital transition still wobbly—to remain as it was, largely unencumbered by anticipation of film's imminent disappearance.[10]

The setting of my fieldwork—an academic medical center—specifies some of my interests. First, teaching and learning: this relationship makes conspicuous aspects of craft practices that might otherwise remain invisible or inarticulate. The craft I have investigated most closely is that of radiological reading: in many ways this is a book about learning to read. In describing radiological learning I distinguish several levels of training: attending (professor); fellow (advanced trainee with supervisory responsibilities), resident (trainee with an MD), student (trainee without an MD).

The setting of the teaching hospital also specifies the importance of disciplines—rigor, specialization, divisions of labor. In this book I consider a range of hospital service roles: not just doctors, but nurses, techs, clerks. I discuss how occupants of these roles work in institutional settings (not their lives at home). Among doctors, I distinguish several broad tribes: radiologists, clinicians, pathologists. (Surgeons are included in the tribe of clinicians—though they may differ from medical clinicians in many ways.) Thus I allude throughout to three primary forms of inter-disciplinary "correlation": rad-clin, rad-path, clin-path. In transcripts, I refer to myself as "investigator."

My own history of formation as a physician was an important condition of my work, not only of my capacity to comprehend radiologic diagnostic practice, but indeed of my welcome to study radiologists at all. My ethnographic study was accepted with misgivings and after much discussion by my radiologist hosts. With careful stipulations about confidentiality and consent, I was allowed to audiotape and photograph radiological viewbox work, but not to make video recordings. (Medicolegal risk was the express concern.)

Though it helped to be a physician, in this project it was crucial that I was, am, not a radiologist. This has helped me cultivate what Simmel called "the objectivity of the stranger": "distance and nearness, indifference and involvement."[11] And to avoid charges of impiety.

• • •

This book was conceived as an anthropological study of *diagnostic rituals*. I approach the hospital as a historically religious institution that reshapes practices of death and suffering.[12] I analyze dramaturgical aspects of medical work—roles, rhetorics, theatric effects. Yet hospital rituals often comprise quite ordinary practices: examining, pointing, pushing buttons, reading and writing, shuffling papers, speaking into phones and recorders. I describe these to restore to the cultic authority and refined purity of biomedical knowledge some of the humble, practical conditions of its making and custodianship.[13] In this I have deemphasized existential issues like belief and intention.[14] Apropos of religion, I do not subscribe to most of its received distinctions from science (along lines of spirit/matter, or superstition/reason, or denomination/state). The diverse practices of science have never been adequately summarized by the rubric of "method."[15]

From this book's inception I have engaged a broad range of work in humanities and social sciences—particularly in medical anthropology, and in interdisciplinary fields of media studies, science and technology studies (STS), and "medical humanities."[16] I have integrated much of what I have learned from insightful cultural analyses of visuality and representation, embodiment, performance, and narrative, in and out of biomedicine.[17] It is indeed difficult to specify key texts from these fields, but here are a few whose influences may be evident to some readers: Stafford's historical considerations of the rise of particular representational conventions; Reiser's writings on new technologies' effects on social relations in medicine; Montgomery's work on clinical thinking and its narrative structures.[18]

In my research I found that certain historical resources, largely pertaining to the nineteenth century, illuminated present ethnographic settings. But most of my historical forays are more philosophical than they are narrative—"to seize hold of a memory as it flashes up at a moment of danger":[19] I do not always connect historical circumstances causally to present practices. If "danger" seems dramatic, at least it also registers skepticism about conventional histories of triumphal progress—as they continue to shape effects of biomedicine around the globe, many of which are far from salutary.

My concerns in this document have been to describe ways in which a diagnostic technology—a marvelous apparatus like the CT scanner—works to shape social relations, cultural formations, and intellectual activities—even as it produces clinical evidence. Images unto themselves are not yet knowledge or evidence. Particular rituals of registration and display—especially the complex of practices known as "reading"—are requisite to the making of diagnostic evidence—and thence to the making of diseases. Diseases are made from *cases,* which comprise folders of films, discursive exchanges at the viewbox, correlations with nonradiological information, structures of thought. The term *case* signifies both ways of knowing and ways of showing. It is multivalent, referencing in turn a particular patient (a sad case), an instance of disease (a beautiful case), an institutional assemblage (a case series).

Apropos of evidence: I aim in this book for agnosticism about truth-claims of the radiological priesthood: as an anthropologist and scholar of religion I seek instead to represent one cultural nexus of truth making, at a particular historical juncture.

• • •

Three principal threads are woven in this book.

The first is *practice.* This is a variegated thread, as I refer to practice under many headings: gesture; rhetoric and discourse; technique, production; performance, theater, ritual. I do not have a typology or master schema of practice; rather, I assume practices are better approached in situated, incomplete, multiple—even messy—ways.[20] At times I address media upon which practices operate. And I address consolidations of practice in machines, protocols, and bodies. As the epigraph to this introduction suggests, I am particularly interested in tacit and unformalized aspects of practices—habitus, mimesis, skill—where they may refuse determination by concepts, rules, institutional logics, even intentions.[21] This shades into concern with aesthetics (in its broadest sense, one addressing feeling) and with figuration (agency and tropisms of language and media).[22] I observe how practices are connected in suites and systems—like economies. But again, not for the sake of metrology: an abiding impulse is to consider how aspects of practices can be elided, or reified, congealed in their products—and thus the products fetishized: as, for instance, the true reading, the unequivocal evidence, the unimpeachable expert. The arrangement of chapters names a general suite of practices: reading and writing, diagnosing, curating, cutting, testifying and teaching, exposition. (More on the book's structure below.) In places I have juxtaposed contemporary practices with historical scenes that they quote, resemble, or conjure up: film filing practices and the police prefecture; viewbox practices and the camera obscura, observation tower, and museum; teaching conference practices and the courtroom witness-box.

A second thread in the book is *death*. I address death in some places as a kind of haunting and in other places as a source of authority. Mostly I consider it in particular aspects—like interruption, or fixity. I am concerned with where and how CT diagnosis continues to trade in cadaverous substrates, notwithstanding CT's pretensions to being a "noninvasive" modality. How a filmic puzzle replaces mortal narratives and fixed tissue—or quotes them, or imitates them, or exchanges with them: these are central concerns of this book. Death's status and function is subject to historical change[23]—and this is perhaps of greater significance in the "imaging revolution" than the concerns usually cited—for example, unprecedented transparency, computational speed, graphic resolution. CT is a potent death-changing device.[24]

The book's third thread is that of *intrigue*. I owe this to Edgar Poe, who, along with Benjamin, has been a tutor of my ethnographic sensibilities. By intrigue, I mean primarily conditions of puzzlement or curiosity—in relation to the form of the diagnostic "case."[25] The reading room viewbox remains a locus of intrigue. This is not always a simple psychological matter of curiosity in the face of Nature's puzzle—a version of *intrigue* which the *OED* suggests is a "modern Gallicism." The viewbox turns out to be a social nexus, a place where images are rapidly invested with, and divested of, supporting information— where pedagogical interrogations or contradictory testimonies can be as productive of intrigue as images themselves. Intrigue has powerful affinities with technical and social forms of discontinuity, interruption, juxtaposition— named wishfully, but perhaps simplistically, by the radiological term *correla-*

tion. This is closer to older connotations of intrigue, having to do with plots, dissimulations—"intrications." Things one does with threads.

• • •

The chapters of this book name broad domains of practice.

Chapter 1, "Reading and Writing," is about modes and aspects of attention at the viewbox—the diagnostic "gaze." The chapter introduces social conditions of viewbox work and the materials on which the gaze operates. It also introduces some practices of writing that are most closely linked to reading. (Writing is a category of practice that extends throughout the CT Suite.)

Chapter 2, "Cutting," concerns the scanner itself, the kinds of slicing it accomplishes, and ways it structures activities of technologists and other staff. This chapter is about making images (rather than reading them); and in this it concerns itself with relations between images and fleshy bodies of patients.

Chapter 3, "Diagnosing," situates identification of culprit disease as a particular way of knowing. It provides historical background for my claim that the CT Suite is haunted by the nineteenth century. The chapter introduces the notion of diagnostic intrigue, describes some of its historical contexts, and sketches its logical and rhetorical contours.

Chapter 4, "Curating," describes archival practices that maintain collections of images and places these in broad historical contexts. It also describes practices that manage bodies in their passage through the hospital, as well as practices that assemble texts, for clinical, teaching, and research purposes.

Chapter 5, "Testifying and Teaching," returns to practices of reading, but within the particular frames of pedagogy (learning to read) and witnessing under pressure. It considers ritual contexts of teaching conferences, as well as consultative interactions around the reading room viewbox, and further develops the notion of diagnostic intrigue in showing how radiological expertise is constituted.

Chapter 6, "Exposition," departs from local institutional settings of University Hospital's CT Suite to consider two crucial external contexts. One is the Armed Forces Institute of Pathology (AFIP), America's preeminent custodian of cases of disease. Among the Institute's key missions are maintenance of a public medical museum (since the nineteenth century) and training of radiologists in how to correlate radiological findings with tissue. The other external Expo is the annual RSNA (Radiological Society of North America) meeting, the world's largest combined scientific conference and trade show. The scale of exhibitionary projects in these settings underwrites the cultural potency of medical images in and beyond their local cultic contexts.

The concluding section of the book, "Impression," is brief. It is named for the final, broad-brush, summary segment of a radiologist's x-ray report. Though also among its echoes is the Enlightenment philosophers' name for a momentary play of light upon the retina.

Thus the structure of the book. Glimpses into a series of linked practices, presented as an interrupted series. A suite.

By way of one last overview—and for readers who do not follow the aforementioned "threads" through the whole book—I distil this handful of theses. In a nutshell:

CT is about cutting; but it does more than cut up bodies, in its famously noninvasive/virtual way. It has for decades been cutting up hospital culture. It has been introducing, as so many technologies do, new forms of discontinuity and exchange within systems of expert labor, viewing practices, rhetorics of testimony, training regimes, modes of experience.

By producing discontinuity, CT revises prospects for narrative connections in the hospital. It cuts off some forms of story and rearranges others. It trades in surfaces (slices), and thereby complicates the hermeneutics of hidden depths. Its artistic affinities are more with cubism than with realism.

But CT is a technology of locating culprit lesions—and so is not as "postmodern" as we might presume—or as the rhetorics of the imaging revolution might have it. It is *haunted* by nineteenth-century projects of comparing, interpreting, classifying morphological specimens and residues—including, ultimately, the residue of the cadaver. Even as it endeavors to replace the cadaver. In the CT Suite, to invert a famous formulation, we have not ceased to be modern.[26]

The manifest cultural potency of medical images, which engage needs, anxieties, and desires—moral and aesthetic purposes, as well as epistemological ones—devolves from these images' primary cultic contexts: diagnostic work. One register of diagnosticians' moral and aesthetic involvements in medical images, which may sometimes have surprisingly little to do with utilities or outcomes, is their engagement with intrigue.

Apart from other virtues of intrigue—restoring forms of narrativity to cut-up hospital experience—the theatrics and rhetorics of intrigue are crucial to teaching modes of thought that escape determination by rules, protocols, metrological structures—especially conjecture and judgment. The status of "clinical judgment" is much debated in this era of ascendance of "evidence-based" medicine.[27] Whether and how such forms of thought thrive in evolving imaging practices is important to their larger situations in powerful medical cultures of the future.

1 READING AND WRITING

"If it is any point requiring reflection," observed Dupin, as he forebore to enkindle the wick, "we shall examine it to better purpose in the dark."
• • • **EDGAR A. POE,** "The Purloined Letter"

In the Body CT Reading Room, overhead lights are off. There is a museal glow from the viewbox, dappled blue by rows of images. The room has six empty chairs: no body is here—save the ethnographer's. Two double-tiered viewboxes and several computer monitors stretch along the wall opposite the door. On the viewboxes, rows of backlit ovals look like . . . what? Modernist photomontages? Cinema filmstrips? Eyes in the mantle portrait? The door, opened inward, displays names and pager numbers of the attending, fellow, and resident on duty. In these first morning hours, it still shows yesterday's names.

I am here to observe a tomographic workday. Oddly, this unpopulated room feels less empty than expectant: the viewbox is replete already with quiet filmic presences. I set up my tape recorder and sit.

The attending radiologist du jour arrives, having hung her coat in her office on another hall. She knows about my study, acknowledges my greeting. She has already met with the night-duty resident (elsewhere) to review the evening's emergency CT cases. It is 8:15.

En route to her seat, the attending taps a wall phone just inside the door-frame. "The DND button—it's my favorite one." She elaborates: "In here, we are doing physical examination. You don't walk in on a rectal exam. It's sacred."

Eventually I understand: DND is "do not disturb." I am surprised there exists such a button—that a mere device enforces this room's decorum. I had assumed the subdued ambience of this privileged space was conventional, tacit—not a technical production. The CT reading room is like a library chamber: rapt readers gathered around small pools of light, silent or in hushed conversation—chastened by the exigencies and nobility of reading. True, the CT reading room is, in University Hospital, also a kind of commons: door usually open, no warning light, no access code: strangers enter unbidden, without knocking. However, most visitors wait quietly, in deference to radiological concentration. Groups in discussion tone down. The quiet ambience of radiologic attention seems to need no extra enforcement, no librarian's shushing.

Though the DND button surprises me, I feel privileged to be within its cone of silence—to be counted as welcome within this space that is so mannered and protected. I am also reminded that my presence is welcome so long as I do not disturb. While this attending's touch of a button may be part of a morning routine, her explanation serves as a gentle warning shot across my bow, my ship from foreign waters.

I consider the attending's assertion that reading is sacred. Was the religious idiom chosen for me as ethnographer? Possibly—though her claim could be conventional, an echo of ancient Hippocratic pieties.[1] I am glad to consider CT reading as priestcraft, but I wonder: how is CT reading like physical examination? Materials inspected at the viewbox are silver-salted films, inked paper, phosphor screens. CT and other x-ray images *substitute* for bodies. Like ordinary photographs, they are fixed; but they do not trade on resemblance. One may observe similarities in silhouette between body and film—yet a patient would probably not recognize himself in his CT scan (or even, for that matter, in his x-ray). Relation between film and flesh derives from an event, a *registration*. CT images relate to bodies as do fingerprints to fingers, dossiers to subjects: they are traces, signs of passage. The substitutability of an x-ray image for a particular body relates to a historical moment of alignment—body, beam, film—during which shadows, morphological residues, were captured. If there is sacrality in CT reading deriving from bodiliness as such, it is transmitted to filmic image in that brief, regulated pose.

Perhaps associations between filmic exam and physical exam would be better staked not on similarities between objects (films and rectums), or on the modesty of human subjects (who are elsewhere),[2] but on similarities among gestures and practices.

The practices of "physical examination" vary across specialties, institutions, and periods: inspecting; palpating; percussing; auscultating; sniffing—even tasting.[3] These practices share an existential authority derived from *proximity*—the copresence (arm's length or so) of examiner and examinee.[4] They implicate bodies of examiners. These practices are focused, deliberate, and often absorbing (for the examiner); they constitute forms of intensely "personal knowledge." They are also styled, emulated, scripted practices, constituting forms of social, institutional knowledge.[5] They are among forms of attention on which the role of "attending" is staked. What sacrality they claim pertains to venerability of traditions—crafts of knowing, honed over years, taught in apprenticeship, by worthy teachers—as much as to vulnerability and trust of patients.[6]

In exam practices at the viewbox, *seeing* prevails.[7] Experts look closely, search for legible signs of disease, say what they find. Like physical exam, reading CTs engenders absorption. Perhaps close viewing—deliberate, intense, arm's length—accounts for part of CT reading's sacrality. Yet other professionals and craftspersons look closely at objects, from arm's length and closer, without claiming that their work is, *ipso facto*, sacred. What distinguishes attentiveness of the diagnostician from that of the welder?

Again, there are venerable traditions of seeing and knowing that inform work at the viewbox. Sacrality here, as in physical examination, trades on more than intimacy of inspection, or priestly deliberation, or traces of malign agencies of disease. The quiet, dark reading room is a social space with a complex history. Exercises of CT reading are ritualized, informed by doctrines and canonical texts, organized around anxieties that are social as well as personal.

This chapter is about viewbox practices of reading. It is about scenes and contexts of reading, about conventions and structures and habits of reading, about conversations informed by reading. And it is also about writing. Tomographic writing is not just Nature's writing, tracings of bodily norm and anomaly.[8] Writing in CT is, like reading, a suite of practices.

Proximity and Autopsis

At the Body CT viewbox:

Investigator: Dr. Bynum, have you always been able to read a study from six feet away?

CT attending:	Ah, yah.
CT fellow:	He usually does it from the back of the room.
Attending:	Yah.
Fellow:	He prides himself on that.

[*brief pause*]

Attending:	You know what, I don't know whether this is right. Is the patient bleeding? I don't feel like this is—is this an artifact or is this real?
Investigator:	What are you looking at?
Attending:	This patient's got anasarca with subcutaneous edema.
Investigator:	Mm-hmm.
Attending:	This is a sickie. This is a sick patient.

This virtuoso radiologist positions himself as close as he needs in order to see—no closer. Even as he redirects conversation to images before us, he leans back. He habitually sits farther from the viewbox than do most of his colleagues (two to four feet is average), and this distance is emphasized by occasional use of a pointer. The acuity of his vision (and, by extension, his diagnoses) is emphasized by such remove.

On this occasion, the attending preempts further remarks about his "sharp eyes" by addressing the images. Look here, look with me. See what is salient: see comprehensively, urgently: "this is a sick patient."

Seeing for oneself is more central to the sacred task of each CT radiologist than, indeed, feeling sick is to being a CT patient. Likewise this is a goal of most visitors to the reading room: to view things on the viewbox *for themselves*. The dramaturgical center of the reading room is the lookout's role, the observer's role—not Parsons's famous sick role.[9] The visual empiricism which unfolds in the reading room is a drama of proximity to the image and, above all, *autopsis.*

"Autopsy" is wrested here from common usage—opening a corpse, as in the coroner's inquest[10]—and restored to broader connotations. *Autopsis* is a term from ancient rhetoric which pertains to testimony of the eyewitness.[11] It is about seeing for oneself and *saying* what one has seen.

| Neuroimaging fellow: | [*on phone with attending:*] Are you coming down soon? [*pause*] Oh, OK. 'Cause Terrance [a neurosurgeon] called again. He wanted you to look at this. |

	OK, well, I'll tell you what it looks like [*laughing*]. Tell me your diagnosis. There's a nonenhancing, no mass effect, high signal T2, low signal T1 lesion in the white—periventricular white matter kind of going up in the centrum semiovale. Looks like PML, but he's fifty-nine years old and it's the wrong location, so it's something demyelinating. And it was read outside as a tumor but there's nothing there. [*laughing*] OK. [*pause*] OK, I'll tell Terrance. Thanks. [*laughing*]
Investigator:	What's the word?
Fellow:	I guess he didn't trust my, uh, description.
Investigator:	So he's gonna come down.
Fellow:	In half an hour . . . [*laughing*][12]

Autopsis comprises the fragility of seeing, the perils of poor seeing. As mode of personal knowledge, autopsis connotes existential weight—a sense of immediacy for the subject whose faculties are enrolled in an act of attention. Yet ocular witnessing in a teaching hospital is often a social event as well—embracing both object of viewing and context for bearing witness—audience, dramatic conventions, scripts. The term conjures a larger representational and rhetorical economy within which some modes of viewing and modes of testifying are superior, convincing. If autopsis connotes self-possessed direct observation, it also connotes a problematic of persuasion and reception, a performative milieu in which the veracity of the viewer is not so much presumed as it is at stake.

Native Vision

Body CT attending:	Today I haven't been doing any work other than just using my eyes and my few neurons.
Body CT fellow:	That's quite all right.
Attending:	I haven't done any physical work.
Fellow:	You don't get paid to do physical work.
Attending:	That's right.
Fellow:	You get paid for the mental / visual.
Attending:	Actually, I did do some physical work yesterday. I did a couple of GI studies. I even did a couple of walkie-talkie upper GIs yesterday.

Notwithstanding this attending's distinctions between seeing and doing, it would be a mistake to apprehend the viewbox "gaze" as a simple, coherent, or merely visual experience. Looking comprises a multiplicity of gestures, especially pointing. Even without a patient's body at hand, the visual/visible is intertwined with the speakable/audible and the tactile/touchable.[13]

"Did you see that chest wall process, Don? There's both a intraparenchymal component as well as a component out in the chest wall. Um—Barry, can you see? Need to pull up a chair . . ." Summons to closer engagement, and specific postures. Radiographic reading is usually done sitting. It is not that images cannot be reviewed standing: they often are, especially by visitors. Viewing images involves various postures—sliding back to see more wholly, leaning forward for minute details. But for radiologists, sitting is a posture of method—and of being at home at the viewbox. Legs tucked under the writing surface, feet on the floor. (In some reading rooms, feet work film transport controls; in some they work dictation controls; in some they work an auxiliary "hot light." At University Hospital's CT viewboxes they are often more solidly planted.)

A senior attending recounts his visit to a distinguished New England hospital with a large radiology department, during which he was struck by the battered chairs in a reading room—mismatched, "all taped up and everything." In contrast, chairs in his CT reading rooms match and are in good shape—adjustable heights, smooth casters, hinged backs—and comfortable.

Once seated, radiological vision uses few prostheses. Occasionally one sees a reader of mammograms holding a magnifying glass. And on computer monitors, magnification and contrast can be adjusted. But diagnostic film viewing, including CT reading, is mostly macroscopic: it employs a "native" vision, a repertoire of squinting and scanning and gazing, a few feet from the image surface. This is especially true of CT as compared to "plain" radiography (e.g., chest x-ray): CT scans are relatively coarse-grained.

Unlike "life-size" plain radiographs, CT images are typically displayed an order of magnitude or so smaller than the specimen they reference, with many images on one sheet of film. Film size is standard, but the "matrix" of slices on each sheet—"3 by 4" (columns by rows)—is variable, subject to differing conventions, even to ad hoc specification by readers.[14] The smallest CT findings represent millimeters of tissue. One CT fellow quotes an attending—"My eyes are calibrated to two millimeters"—and then quips that, for his part, he is accurate to "plus or minus five millimeters." Conventions of CT display calibrate an arm's-length gaze, with respect to film, to the scale of arms-length handwork, with respect to the body. CT findings represent objects that might be held between the fingers.

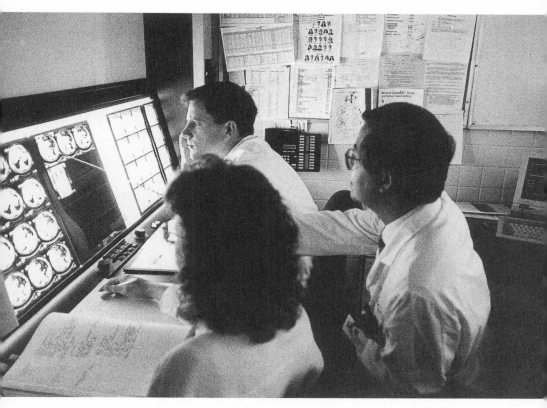

3. Attending pointing

One indication of how scale and size relate to CT reading is a recurring phrase: "too small to characterize"—as in, "little low density area in the kidney, too small to characterize." This has a concrete meaning: "[If] we put cursors on that we will not get a valid [Hounsfield] number." CT readers often note darker or lighter flecks too small to be registered to density scales (whereby liquid is distinguished from gas or solid, fat from muscle). Findings too small to characterize are not below the threshold of the significant—but are unquantifiably ambiguous.

"Cursor on the lesion" highlights practices of *pointing*. The cursor is a prosthetic extension from hand to screen, a gesture. Radiologists do a lot of pointing.

As a fellow at the board moves images with right hand on the toggle, she points with her left hand at specific locations. A clinical attending beside her points and asks: "Is this compatible to a met?" Pointing is collegial: it orients the gaze of another to what one is inspecting. It is a feature of conjoint autoptic

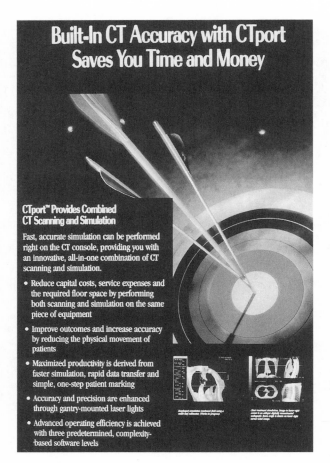

4. Indexing, targeting: Toshiba CT Exhibit, RSNA meeting, 1996

practice: seeing for oneself, together.[15] Pointing is more common in CT than in some other kinds of collegial radiologic viewing. This is in part due to the multiplicity of viewable locations—many slices, much detail, at small scale. Verbal denotation becomes clumsy: "look at slice 33" or "below the level of the renal arteries."

A corollary function of pointing is more impersonal: indexing, targeting. The index connects a regime of search to a locus of suspicion. Index>Target is a staple of the radiological imagination: arrow in the bullseye, crosshairs on the enemy: accuracy in service of conquest.[16]

Body CT resident: [*deleting patients from MagicView archive*] Hey, Amy, what am I doing here? . . . OK, see if I click anywhere in

	here nothing happens, but if I click on one of these little dots right here, that happens. If I click over here on this dot, that happens—
Fellow:	It must have something to do with this—this is like a magnifying thing. I don't know what that one does.
Resident:	I better stop before I screw something up.
Fellow:	That suggests that something's open still, though.
Resident:	But you know—
Fellow:	Select that one and open it and see if it's—
Resident:	Uh.
Fellow:	Now press the . . . little arrow thing.
Resident:	Oh, OK . . . I like this. Now this has 512, wait. Now, when I'm doing this all I'm doing is I'm using more information from the CT scan, right?
Fellow:	I think so, yeah.
Resident:	I think that—
Fellow:	This is awesome. That's cool.
Resident:	That is real visual.

Saccadic Vision[17]

One might imagine that the usual visual mode at the viewbox is the rapt stare: if not fixation on the target, then perhaps a stare akin to that whose "modernist vocation" has been charted by Rosalind Krauss: "the stare's relation to pattern."[18] However, viewbox looking is, like much ordinary looking, often discontinuous, interrupted—by checks and hesitations, twitches and winks,[19] constant shuttle among images, labels, and other nonpictorial data. "Scan" captures not only the dynamism of the scanner apparatus but also the roving gaze of the radiologist. Radiologists look back and forth between films and folders, films and texts; they look alternately at contiguous slices to infer continuities of structures crossing them, or to ascertain if an anomaly on one, missing from the next, could be an artifact. They compare left and right sides. They compare current scans to former scans, slice by slice.

Twining of radiological vision and thought is irregularly staccato; this is often evident in preliminary reviews of studies on the board, when images move in rapid, halting succession.

[*salvo of viewbox film movements: two seconds, three seconds, three seconds, two brief ones*]

CT attending:	[*chewing*] Did you see that flash by you?
CT fellow:	[*chuckling*] I did.
CT attending:	I think she has lung stuff.

Trajectories of the radiological gaze have been charted meticulously by researchers concerned with efficiency. From an abstract of one study:

> An eyetracker, a device for recording eye and head movement, was used to determine the scan patterns during the interpretation of single and multiple computed tomographic (CT) examinations presented on a four-over-four viewbox. CT examinations . . . represent complex viewing situations . . . Radiologists viewed patient folders containing . . . CT chest examinations and dictated a report. Eye movement was recorded with an eyetracker and video camera. After mounting the films in order, radiologists generally started with a sequential scan through the entire examination, followed by careful viewing of two to four clusters of three to six images, followed by dictation.[20]

This account was oriented specifically toward workstation ergonomics—a Taylorist agenda.[21] It described how radiologists divide a CT study into manageable parts. Other eyetracker studies have mapped finer aspects of image viewing—like "pause" or "dwell" locations in the play of a viewer's eye movements, sites where the gaze rested temporarily—sites presumably seen as suspicious, but perhaps only unconsciously so. Making such locations known to the viewer, through computer assistance, might reduce missed findings.[22]

One could map encounters between a gaze and any kind of image. But different images summon pauses with different affective dimensions.[23] In analysis of radiological viewing, the play of the gaze is presumed to signal unconscious suspicions—part of a radiological "unconscious optics."

This is a term formulated by Walter Benjamin about photography—especially cinematography—which also exposes, and anatomizes, phenomenologies of viewing: "With the close-up, space expands; with slow-motion, movement is extended . . . The act of reaching for a lighter or a spoon is familiar routine, yet we hardly know what really goes on between hand and metal, not to mention how this fluctuates with our moods. Here the camera intervenes with the resources of its lowerings and liftings, its interruptions and isolations. The camera introduces us to unconscious optics as does psychoanalysis to unconscious impulses."[24] For Benjamin, the camera's association with the unconscious relates to its shifting among magnifications, proximities, and speeds—camera relative to object, movement of film relative to movement of the Other. Benja-

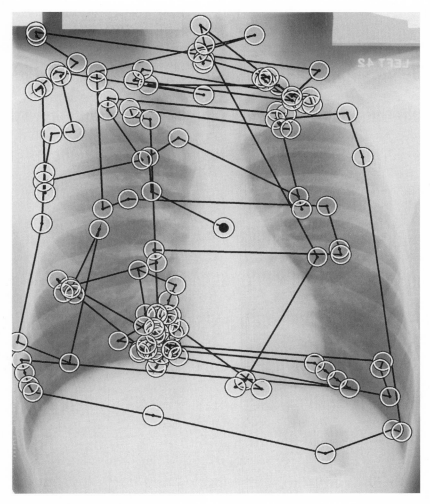

5. Gaze tracking on a chest x-ray

min likens the "boldness of the cameraman" to surgical intervention: "a debate of steel with nearly fluid tissue."[25]

Things are different with CT. What is brought to consciousness is not some unknown aspect of a subject's gesture, but a hidden bodily structure. There are kinships between a scanner's circumferential movements and the motions of a movie camera.[26] But with CT, the arc the beam describes around the specimen is fixed, unlike a handled scalpel. From scan to scan, there are fewer opportunities for paraintentionalities and tacit knowledges of techs or tomographic subjects to inform the images.

There is, though, a great deal of paraintentionality in play for radiologists inspecting images on the viewbox.

Apparatus and Confidence

CT attending:	That's why I say—is that an artifact that we're looking at? I—usually my eye can tell the difference—usually the attenuation is at least ten plus units. What is that?
CT fellow:	Sixteen.
Attending:	Is that right? My glasses, I need a new pair of glasses. So you're trying to tell me this thing and this thing is the same? Huh.

The self-certainty that attends autopsis—"usually my eye can tell the difference"—is familiar, and yet, in light of subliminal suspicion, a bit puzzling. Self-certainty is a Cartesian legacy. Yet it was Descartes's famous doubt that led him to the *cogito*. Along the way he dissected an eye of a cow—and there found, among other things, a space in which to invert the world. Vision became the model sense for Descartes, who confirmed autonomous faculties of the perceiving subject for the Enlightenment and beyond: centered on an interior *camera,* into which light penetrated via a hole, a window—the pupil.[27]

Confidence in the visual habitus of the CT viewbox is indebted to this post-Enlightenment legacy of the truthfulness of sight.

The Cartesian bovine eye resembled a structure known to early modern Europe: the camera obscura, with its hole and, opposite, surface for the inversion of the scene outside. This dark room of travelers, artists, and literati helped produce, and naturalize, a mode of faithful representation—faithful to linear light rays, to passive perception, to separation of observer from view. The camera obscura, by materializing certain optical principles, exhibited both the reproducibility of the scene outside and the universality of ocular anatomy, the dark rooms in every person's head containing screens for projection.

Another early modern device anchoring the truthfulness of sight was the gridded frame of perspectivism—a tool for inscribing coordinates of a scene, bodies receding toward a vanishing point, on a two-dimensional surface.[28] A visual field bound by Euclidean or Albertian coordinates valorized the monocular vantage point and the logic of the steady gaze—privileged it over the "glance" and, indeed, the "scan."[29]

6. Athanasius Kircher, Camera obscura, 1646

One more device of modern visual confidence—not unrelated to the camera obscura and the window-grid—is the lookout. The Enlightened gaze sought positions of power: overlooks, reconnaissance towers—"birds-eye" views. Lookout posts served colonial administrations. They also served European polities: dioramas, panoramas, maps—representational devices locating viewers in central, apical positions, with scenes and objects of empire arrayed before them. Timothy Mitchell refers to these as tactics of "the world-as-exhibition": "a place where the artificial, the model, and the plan were employed to generate an unprecedented effect of order and certainty."[30]

The hole which admits light to the dark room demarcates a threshold, and the modern observing subject is assigned a place behind it, at the screen. This subject-position is juridical.[31] Modern theories of vision often held the spectator, whether in a room or on a tower, firmly separate from a world of things and others.

Enlightenment valorization of vision, and scientizing of representational codes, followed on the Reformation's deprecation of vision. According to Martin Jay, separation of the visual from the textual in Reformation thought was "crucial in preparation of the scientific worldview."[32] Protestant iconoclasm traded on availabilities of printed texts in national languages. Images were liberated from sacred tasks: "Protestantism no longer really desired the assistance of visual aids in teaching the mysteries of faith."[33]

If camerae obscurae (later, photographic cameras) helped naturalize sightlines of realism, they also naturalized the two-dimensional representational surface. Confident seeing valued the reproducibility of the flat image obedient to realist codes.[34] John Berger describes this issue of reproducibility (speaking to European oil painting): "Its model is not so much a framed window open onto the world as a safe let into the wall, a safe in which the visible has been deposited."[35] The commodification of images and the reification of an object-world are bound up historically with these developments; realist codes are now firmly embedded in our experience.[36]

Are images on the CT viewbox like an exotic panorama? Do they conform to realist codes? Is a legacy of the camera obscura lived out in scenes of reading room autopsis? There is in CT a stipulation that the scene being imaged—the patient's body—remain still enough for its spatial coordinates to be assembled reproducibly. CT scan images are held to be faithful to their anatomic refer-ents—though not according to perspectivist geometries. The reading room is darkened. Faculties of radiological perception are receptive, juridical—before a screen. A reader sits at the viewbox awaiting the next case in the docket. At-tending, *attendre*. And readings are thoroughly commodified.

But the reading room viewbox in University Hospital is different from an observation tower, a *camera obscura,* or another photographic device. It is brighter, often more social, and does not array spatial depths in relation to a vanishing point. Instead of one cyclopean viewpoint, one perspectivist scene, the reader at the CT viewbox faces a wide display of tiled windows, contingent scenes, none of which conveys, by itself, spatial depth. CT vision is not a per-spectivist gaze—as is, for instance, the peephole view of endoscopy. If the two-dimensional surface of the CT image remains a site of faithful representation, it is not according to tenets of realism.

One nonrealist aspect of CT images relates to their summation of mov-ing viewpoints. A historical precedent is the "wandering optical axis" used in some comparative anatomic illustrations to eliminate distortions of parallax and perspective. Mobile viewpoints were also developed by northern Euro-pean landscape painters, in part following what Svetlana Alpers has called a "mapping impulse."[37] Production of topologically rigorous images through integration of mobile, *non*perspectival viewpoints is of defining significance for CT, and all tomography.

Another aspect of CT's removal from presumptions of realism is exempli-fied in work of the Cubists—who were extremely interested in x-ray images from their inception—in part for their elimination of receding depths.[38]

Some features of CT viewing belong to radiographic viewing in general. These features are not necessarily medical: in 1896, x-rays were greeted as extensions of ordinary photography.[39] And though bodies were prominent

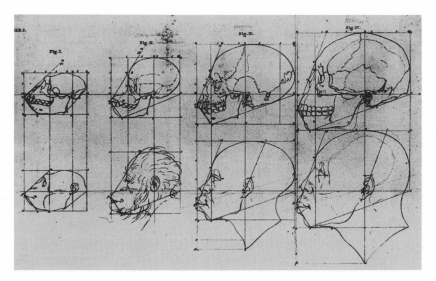

7. Petrus Camper, Craniological specimens drawn with wandering optical axis, 1791

objects of early radiography—like the *hand mit ringen* of Frau Röntgen—they were objects of sexual interest or morbid concern more than medical investigation.[40]

Yet it was apparent that x-rays were not conventional photographs. They were "skiagrams"—shadow-figures. As objects of public marvel, their clinical reception was framed by caution, even skepticism. The new ray collapsed depth, could not be refracted or focused.[41] By impacting shadows on a surface, radiography created new forms of ambiguous density. Problems of viewing were sometimes formulated in terms of duplicity or peril: understanding superimposed shadows required thinking about the relative positions of film, x-ray tube, and object.[42] Notwithstanding the interest of Cubists: x-rays, rather than eliminating metaphysics of depth, intensified cognitive reconstructions of depth.

Visual practices of CT reading modulate those of general x-ray interpretation. Slices make volumes into surfaces: they trade in a kind of superficiality. Compared to conventional x-rays, notes one attending, "CT makes it too damn easy for you . . . You can see things sitting right there." In his view, CT's sparing of the viewer from deciphering overlying shadows makes trainees lazy.

CT images are widely apprehended as truthful. However, their truth is not one for which untutored vision is prepared. Patients remark on feeling handicapped by lack of training. The complicated appearance of CT images breaks

8. Marcel Duchamp, *Nude Descending a Staircase No. 2*, 1912

commonsense links between autopsis and confidence. The "obviousness" of CT lamented above by the attending presupposes a certain level of skill, an expertise which must be learned.[43]

Often seeing at the viewbox is a collective enterprise—multiple aligned gazes, each autonomous, but calibrating its sensitivity, adjusting its focus, in accommodation to others.'

CT fellow:	Here's another cyst there, or a little—yeah, it's a cyst. You see that one? [*boardbuzz omitted*]
CT resident (first-year):	Yah, I didn't see that one, actually.
Fellow:	It was there before.
Resident:	Where'd you see that before? On the old film—oh yeah, right there.
Fellow:	They were both there.
Fellow:	OK, that's—that's the old one, good.
Resident:	Where's pancreas?
Fellow:	There was ah—on the old one—pancreas is here, see it?

Rhetorics of interactive seeing can veer from explanation to contradiction. The performative scaffoldings of teaching also serve agendas of self-convincing, self-persuading.

Body CT attending:	Maybe that's a hemangioma up at the dome of the liver, it ah—just looks too—
Body CT fellow:	I know what you mean—
Attending:	—regular. It's just not the ah—
Fellow:	It's right at that area where you don't trust what you see.
Attending:	Mm-hmm.
Fellow:	'Cause you're coming into the dome, into the ah— curved surface of the liver.
Attending:	Mmm . . .

[*pause: twenty seconds of silent looking*]

At the viewbox, faithful seeing means, at times, shared *mistrust*—mistrust not so much of what is seen per se but of the capacity of the seen to be faithful to the "real."

Findings

"Findings" is the term heading the descriptive section of a formal x-ray report. It implies searching; it affiliates with discovery.[44] Findings are bigger than silver grains, smaller than scans; they can be shown to a colleague, marked, described. Findings are referents of a summary radiological "Impression" (concluding section of a formal report).[45] They may or not be lesions—there are normal findings—but they have been discovered, against some background.

The coming-to-notice of a finding is an originary event. A disease announces itself; prey enters the scopic field: residues of origin henceforth cling to this finding. Even in CT, wherein an "original" image is already reconstructed from a digital dataset—and "originals" are indistinguishable from copies—"findings" preserve a whiff of origin.[46]

How some visible enters the realm of the found is not to be understood in merely optical or cognitive terms; it is inflected with aesthetic valences, which radiologists are sometimes at pains to restrict. For example, "satisfaction of search":

Radiology attending:	If you're satisfied with what you find, you miss everything else. Because you're happy that you've come up with this . . . lesion in the lung, you missed a clavicle that's busted or something like that. Satisfaction of search, it's another issue.
Investigator:	It's a term in the moral education of radiologists.
Attending:	Yeah, satisfaction of search, and it goes along with things like tunnel vision.[47]

And counterweight of the finding: the missed finding. Misses are caught through serial readings, by different readers.

Body CT resident:	[*hanging up phone*] So did I miss something big on this?
Body CT fellow:	[*at viewbox*] I see some bowel wall thickening.
Body CT attending:	I see—yeah, good—that's a good pickup. I think that's real.
Fellow:	Yup.

A "pickup" by the fellow becomes a "missed finding" for the resident. The resident's miss is not grave: indeed, praise for the fellow, along with the attending's subtle equivocation—"I think that's real"—mitigate its sting.

"Look at this—that looks awfully real to me," says a Body CT attending. "I think we need to call the doctor." Here "awfully real" connotes shock, urgency.

Not all findings are compelling. Reality of findings is often a matter of view-box equivocation. Ambiguities may be amenable to technical resolution. CT attending: "I am unsure as to whether that is in fact real. I tell you what I'd do. She's an inpatient, right? You ought to bring her back for thin cuts. You don't have to give her any contrast—just do 2 by 2 millimeter cuts through that area."

What is not real is often "artifact." Artifacts are held to represent flaws in imaging technique or in computer reconstruction, rather than aspects of bodily substrates being imaged.[48] They constitute a backdrop against which any new finding must be considered.

One common artifact is "volume averaging." It occurs when a CT slice is tangential to a border between different densities, so that a portion of the border, on film, has an intermediate density.

Body CT attending:	What do you think of that nodule there, Don, that left upper lobe nodule?
Body CT resident (first-year):	Mmm. It appears very dense. I mean like—um, am I looking at—seeing the same thing?
Attending:	Yeah, this thing here?
Body CT fellow:	I didn't—just to play devil's advocate, could that be the bottom of . . .
Resident:	. . . it's volume averaging—
Attending:	It's volume averaging of that first rib costochondral junction.

To avoid being tricked by volume averaging, one can discount findings that do not extend across contiguous slices—and one can make thinner slices.

Then there is "motion artifact" (from breathing, peristalsis, shifting on the table). This is mitigated now with faster scanners; in past it required inert patients, and sometimes rescanning. In MRI, one kind of motion artifact, said to be due to laminar flow in vessels, is sometimes referred to as "ghosting" artifact.

New, unprecedented artifacts arise. One day the Body CT fellow is puzzled by lines on several films, and asks his attending: "What do you think that artifact is? We keep getting that every once in awhile."

Attending:	This thing?
Fellow:	Yeah.
Attending:	It's only on one scan out of the, ah—in the film?
Fellow:	Mm-hmm. All the time, it's only on one.
Resident:	And it's only like on every other, I mean on every—
Attending:	Have you shown it to the technologist?
Fellow:	No, we should, though.

Artifacts call attention to agency and apparatus, to techniques of data collection, in a way that real findings do not. Though the imaging appearance of a "normal kidney" is just as mediated, its presumed reality outweighs the technical artifices of its production.[49]

In general, only the real is worthy of official commentary.

Body CT resident (first-year):	[*dictating:*] There is a low density area seen in the posterior right lobe of the liver period. This appears to be motion artifact and may represent some volume averaging period . . . [*portion omitted*]
	New paragraph. Impression colon: One period. [*to fellow:*] Hey, on that—on Mr. Schmidt, I mentioned that low density area in the liver that's consistent with motion artifact. In Impression, I need to say it—What do I say? This is—just "a low density area"?
Body CT fellow:	I wouldn't even put it in your Impression.

Artifacts are a pre-occupation at the viewbox. They confound recognition of the real. As one attending puts it, discussing a "follow up lymphoma" study: "Don't want to give disease where it's not due, we don't want to ignore it when it's real." Artifacts summon cautionary viewbox tales:

Neuro CT fellow:	Did you see what I was looking for? On the reformations?
Neuro CT resident:	Just at the tip of the odontoid where it looked kind of irregular.
Fellow:	Mm-hmm. That's why Otis always teaches, if you're going to do the reformation of the odontoid, always do it in two planes. Because my favorite story is—last year,

	one of my—a co-fellow read it in the coronal plane, called an odontoid fracture, so they put burr holes in . . .
Resident:	Ohh . . .
Fellow:	. . . and put the patient in a halo, and got the films back a day later and said, "Oh, we were only kidding about that." It was artifact.

A colleague's mishap—an error resulting in bolting of hardware to a patient's skull—is surely only a "favorite" tale because of its teaching value. Vexed relations with the real are part of radiology's moral and pedagogical fabric.

Norms and Names

"Normal's got to be second-nature . . . Normal is your friend . . . Normal anatomy, in terms of a reading . . . You will see pathology for the rest of your life. If you know normal, you know a lot." This advice, from attending to first-year resident, criticizes a particular interpretive performance and conveys advice of general import. "Make friends with the normal." A widely used student text says this: "You will find that you build your idea of the expected normal appearance of various structures in various ways. You are informed by instructors that a given x-ray film is normal. You compare what you see with what you know about normal anatomy. You compare a problem film with one you know to be normal. And . . . you can often use the opposite side of the same patient as a norm standard."[50]

Friends should be chosen carefully. One Neuroimaging attending objects to colleagues' willingness to call brain atrophy "normal" in elderly patients. "See, I maintain this is normal. I feel atrophy is not normal. Let's see how old she is."

Norms are not merely objectified sets of limit criteria—though they are often presented thus in textbooks.[51] The normal becomes a structure of the expectant sensorium—an aesthetic device of viewbox practice. One attending relates norms of expectation to pleasure: "Now, if the images are not pleasing to the eye, we will generally call the tech back and say, for example, 'The images are way too contrasty'—an expression that I traditionally use is they 'hurt my eyes' . . . They're so contrasty they hurt my eyes and they need to be refilmed . . . And if it's a special case, we will go down there and say, 'This is the way I want the images, so let me look at the images before you film them,' or 'Let's do this,' or 'Let's vary these parameters and see if we can see what I expect to see.'" The normal translates into suites of attention. With an abdominal CT, one may inspect organs in accustomed order: trace the bowel, look for fluid in dependent recesses. Each step is informed by expectation: sizes of organs, homogeneity in liver, adequate bowel contrast. Marking regular progress through a stepwise inspection is a series of fulfilled expectations—nods to familiars, to friends. Such inspections proceed in silence for a solo reader, but in pedagogical settings are often punctuated with commentary.

> Fellow: [*to resident:*] . . . All right. Lungs, no nodules. Liver, kidney, spleen, adrenals, panc, normal . . . abdominal, pelvic, retroperitoneal, adenopathy and fluid . . . No . . . change. Good for this man.

The litany of the checklist offers comfort—proper reconnaissance, summation of friends well-met. It recurs in viewbox conversations and in some dictated reports. Trainees articulate series of normal findings, whereas attendings abbreviate: "normal study"; or, "abdominal organs all within normal limits." The standard explanation is that trainees are insecure and need to make their efforts explicit until routines are sedimented in habit.[52] "Security" is further testimony to affective, expectant components of radiographic attention, dependent on the comforting familiarity of the normal.

Checklists raise issues of naming.[53] Naming findings can be a courtesy extended to friends, or a way of controlling enemies. Naming is one of the rhetorical reflexes of seeing for oneself, and a prerogative of one who testifies. When the Body CT fellow demonstrates a normal, though infrequently seen, cardiac finding to the resident—a "moderator band"—the resident conveys deference and admiration: "I've never heard of that in my life."

Naming and quizzing are commonplaces at the viewbox. Normal may not mean common: a friend may not have been encountered in some time. Sometimes the naming exercise is summoned by clear demonstrations of "normal anatomy."

Body CT resident:	That's the uh, celiac?
Body CT fellow:	Mm-hmm.
Resident:	And this is your SMA.
Fellow:	Yup.
Resident:	Where's SMV, right here?
Fellow:	Yup. That's the portal-splenic confluence right there, so the SMV is going to come down out of it.

Resident:	What is this, the splenic vein?
Fellow:	Yup.
Resident:	Splenic vein. I see a splenic artery.
Fellow:	Yes you do.
Resident:	I do.

When radiologists are unsure what a finding should be called, they often say "could be." "Could be a fibroid or—with the ovary, it might be the left ovary." This rhetorical form, the present conditional—could be, might be, may be—happens with many scans each day. These locutions convey wobble, ambiguity, and equivocation in interpretation—some of which is finessed at later stages of film reading, some of which is irreducible.

An exchange at the Neuroimaging viewbox begins unequivocally, at the tip of the fellow's finger:

Neuroimaging fellow:	His pituitary gland looks normal. And there is a pituitary gland, more importantly. His optic chiasm is there. Um . . . [*pause*] Anyhow, um, his vents [ventricles] look good. I don't see anything to suggest—
Neuro resident:	That could be consistent with pseudotumor?
Fellow:	Always could be.
Resident:	Always could be.
Fellow:	Yeah. [*laughs*]

Laughter celebrates this bloom of equivocation about names and norms.

In formal dictation, the locution "may represent" is the usual formula for such equivocation. "Low density area in the IVC just above the bifurcation which could represent flow artifact or thrombus." "Represent" is rarely used without qualification, but it is used often with "may" or "could."

Knowing the normal establishes modes of attentiveness, rhythms of confident and comforting naming, in relation to which the abnormal may, eventually will, intrude. One's friendship with normality is a kind of alliance against an as-yet-hidden enemy. The diagnostic gaze is inflected with worry, its incantatory reference to normality sometimes producing a kind of suspense,[54] not blithely neutral but suspicious, on guard. When a lesion appears, it can be as a malign agency, disruptive of the alliance with the normal. On such occasions the image may look back.

Agency of the Lesion

The Neuroimaging fellow describes something catching her eye: "It drives me crazy all the time. OK, say you're looking at this patient and they come in with 'question meningitis' [on requisition] or something like that. See all this enhancement in here. See, all that. You'll be sitting here going 'Jeez, I don't know' . . . and on and on. I've kinda learned to blow it off, but it always catches my eye and bothers me." The expression is common enough: "caught my eye." Shrugging off a worrisome appearance is a skill developed over time: a capacity to repress something that intrudes. Voicings of "worry," feeling "disturbed" by some lesion, being "concerned," are common—though some radiologists seem more disposed than others to rhetorics that attribute moral depth, or malign agency, to filmic findings.

The aspect of an image that "pricks" an observer is referred to by Roland Barthes as its *punctum*. The punctum reaches out to "arrest my gaze": it "overwhelms the entirety of my reading; it is an intense mutation of my interest, a fulguration."[55] It would not do to understand the punctum as a specific locus, necessarily: while often a detail, it can also be a more diffused feature—or an aspect which escapes immediate notice but crops up later—or an aspect of premonition, of warning. "It is what I add to the photograph *and what is nonetheless already there*."[56] And like a dim star, the punctum does not accommodate scrutiny: it ceases to have its pricking, traumatizing effect when it is studied.

On the viewbox, a lesion often plays this role of punctum. So, too, can an entire study. If a grid of images is a forest where some beast is hiding, then the very potential of discovering this creature—or being discovered by it—can be the source of a prick effected by medical images. This prick marks personal investment in the image: for someone taking up the position of the patient, the worry about a potential predator; for the stranger physician, the search for prey.

There are other formulations of the capacity of the object, or the image, to look back. Benjamin explores this in reference to visual exchanges in early urban crowds, where the encounter with the gaze of the Other was sometimes shocking—for example, in suddenly meeting the gaze of a prostitute.[57] For Lacan, in the "scopic field," it is not so much that the spectator is unsuspectingly watched, but rather that the gaze she solicits cannot see her.[58] Such relations between images and viewers could be difficult to discern in viewbox practice; on the other hand, something like them is called into play whenever images on the viewbox cease to be inert objects and call upon, press back upon, the desires, worries, suspicions of their viewers. But this inverts the issue: in diagnosis, neutral objecthood is the derivative state: the finding begins as friend or foe.

Grids, Slices, Depths

A patient queries conventions of representation in CT:

Patient: Is the CAT scan like you have—you know when we were children, we had those little books and you flip them and you make the dog look like it was running?

Investigator: Yes—that's really a very interesting question. I think for radiologists, the CAT scan—it's as if you took a bread loaf, sliced it up and put each slice of bread out on a grid—and there's a particular order that they show on the grid, but—and you're asking the radiologist to imagine the whole loaf by looking at this grid of slices. And I think that talented radiologists, some point far along in their training, can think in depth by looking at these slices. They can say, here's a bubble in the bread, because they see a small circle in one slice, a larger circle in the next slice, big circle in the next slice, then it gets smaller again. And I think they—in their mind's eye—can see the bubble. But there are fancy computer programs now that will allow you to kind of flip through the—they even have some back here where they can look at them—almost as if it's a cinematic thing, moving through the body.

It is already possible—indeed it is now the norm, on computer displays—to view CT images the way this patient imagines—as if a stack of animation cards is being flipped.

So one representational convention to consider, one belonging to the era of film, is the *grid*—radiographers call it the "matrix"—for example, "three by four" (columns by rows). The matrix lays out a series: left to right, one row to the next, top down—like a comic book: ordering slices on a surface.

Grids are ubiquitous in biomedicine—and in sciences, and throughout bureaucracies. Much of the grid's prominence on the radiology viewbox is explained by pedestrian utility. But the CT grid is also a bearer of aesthetic effects. How to understand these? Rosalind Krauss has remarked that the grid in twentieth-century art "emblematizes the Modern."[59] Krauss remarks on a paucity of grids in nineteenth-century art, with two exceptions: works on optical physiology (Goethe, Helmholtz) and symbolist paintings of latticed windows. Both kinds of works were charged by dualisms of materialist and spiritualist concerns. Subsequently, in the twentieth century, grids were pressed into myriad and "schizophrenic" forms—signifying total comprehension (all cells accounted for) or extreme partiality (gridlines extending infinitely). Grids like Mondrian's function both as frames for the visible and as objects of vision. Krauss suggests the grid derives its cultural "success" from its capacity to re-present, in a symptomatic fashion, the twin embarrassments of religiosity and material positivism of the nineteenth century.[60]

On the CT viewbox, the grid is, likewise, both infinite and partial: it establishes an endlessly extensible structure, but the specimen-slices it makes comparable are particular. Proliferating grids accommodate ever-thinner slices. CT's capacity to divide *space* endlessly across the grid reminds us of an analogous grid-function intimated by the patient above: the infinite divisibility of time and motion—as by the frame-sequence of a cinematic filmstrip.

Within the filmic grid is the "*slice*"—the most important representational convention in CT display. "Cut" is the common term in informal settings: "It's just on that one cut and you don't see it above or below." "Slices" show up in formal dictations: "The scans were formatted into eight millimeter collimated slices." One also encounters "slice" in discussions of technique and image quality—as in "slice thickness." The term "section" is rare on the CT shop floor (though common in pathology).

Slices are definitional of tomographic representation of any sort—conventional or computed. At the same time, the slice has become a common form of representation across many biomedical disciplines—not just radiology. Sectional displays are used in anatomy textbooks, in cell biology, in biochemistry; they are standard fare in pharmaceutical ads and on covers of "throwaway" journals. Outside medicine, sectional images are deployed across a range of nonbiological disciplines, including architecture, engineering, and geology.

The representational device of the section came to prominence only recently. In anatomic representation, layered dissections—from ecorchés to "surgical" anatomies to posed skeletons, in perspectivist view—have been far more common since the Renaissance.[61] Sectional images of bodies or body parts were rare in the eighteenth century and uncommon in the nineteenth, across a range of media, from painting to lithography to wax modeling, and across a range of presentational contexts.

Before considering sectional images in biology, it is instructive to consider development of sectional representation in geology, especially well docu-

9. Piet Mondrian, *Composition 1916*

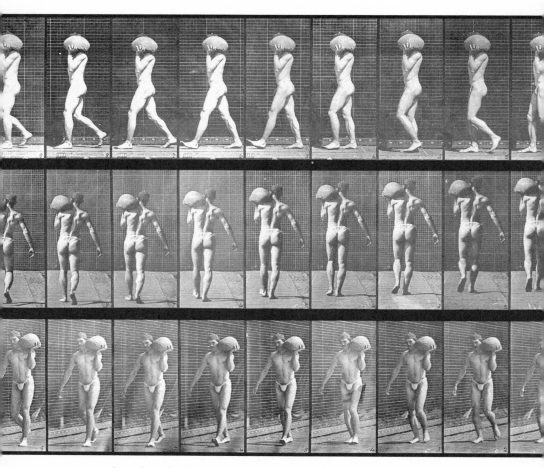

10. Eadweard Muybridge, *Animal Locomotion*, 1896

mented by the historian Martin Rudwick. Geological sections are (apart from cliffs and canyons) largely not matters of direct, comprehensive visualization: "A modern geological section is a highly theoretical construct. It is a kind of thought-experiment, in which a tract of country is imagined as it would appear if it were sliced vertically along some particular traverse of the topography, and opened along that slice in a kind of cutting or artificial cliff."[62] Geological sections were rare in published work before 1811. The oldest sectional illustrations of the earth, from the late seventeenth century, were "cosmogonical"—to illustrate claims about formation or history[63]—and "theoretical" in design, that is, lines ruled and diagrammatic. Early sectional images served polemical purposes and did not conceal their conjectural nature.

Mining activities in the late eighteenth century contributed new empirical materials (core samples) and utilitarian purposes to geologic diagrams.[64] Geologic sections were still conjectural, but modes of representation became more "realistic." "By the beginning of the nineteenth century, straightforward realistic sections based on underground evidence had probably become a standard form of record-keeping in mining concerns, although few were published, probably for reasons of industrial secrecy or state security."[65]

Though it is not clear that any realism is straightforward (*pace* Rudwick), in the case of geologic publication, sectional representation became complex—in part influenced by national traditions. The British emulated "engineering" traditions of schematic drawing.[66] In contrast, the *Essai sur la géographie minéralogique* produced in 1811 by Cuvier and Brongniart included sections drawn freehand and more expressively. French illustrative conventions were continuous with Cuvier's important work in comparative anatomy (addressed in chapter 3, "Diagnosing," and chapter 4, "Curating").[67]

Neither of these modes of sectional representation began as naturalist; there was the artifice of the ruler on the one hand, and exaggerated scale and tentative extrapolation on the other. The naturalist equivalent of the geologic section was the cliff view—favorite of travelers and Romantics. There were complex interplays among depictions of cliff views and those of geological "traverse" sections. As traverse sections became more common—largely around mining interests—they spawned another generation of "theoretical" sections, supporting inferences about geologic processes—sedimentation, mountain building, and so on.[68] Gentlemanly scientific society was the matrix of integration of these different traditions through the nineteenth century.

What of sectional images in other scientific and craft traditions besides geology? There was in the fine arts, by the eighteenth century, a complex tradition of "dissective" illustration which made use of various cutaway views—of bodies, buildings, cameo brooches. The art historian Barbara Stafford discusses cutaway views from architecture, stonecutting, and anatomy, claiming continuities and a general Enlightenment interest in "study of depths." However, her analysis suggests that surgical and anatomic traditions of dissection—involving limited portals of exposure, and removal of one layer at a time—provided dominant representational devices.[69]

In anatomic representation, full sectional images, transecting a body, remained highly exceptional in early modernity. Leonardo da Vinci depicted median sagittal sections of male and female bodies in the early sixteenth century; and a handful of anatomists depicted sections of brain, eye, and sexual organs in the seventeenth century. In the eighteenth century, sectional images became

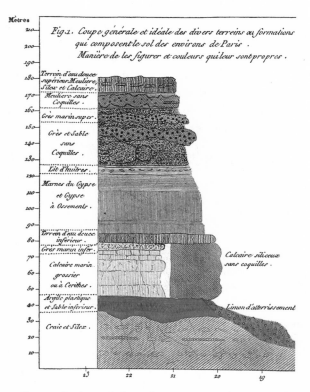

11. Georges Cuvier and Alexandre Brongniart, Slice of terrain, 1811

more common but remained supplemental to more conventional dissective representations.[70] Around this time (1780) the first microtomes were invented, for cutting thin sections for microscopy—but initially these were used primarily on plants. Microtomes did not become staple instruments in biological research until the 1870s. Then they were particularly useful, according to the historian Nick Hopwood, in exposing complex suites of tissual change and involution in an emerging discipline of embryology—for which microtomic work was often twined with craft-practices of wax modeling.[71] Microtomes also became important tools of neurologists, concerned with continuities of long structures (nerve tracts) traversing complex organisms. And they were indispensable for histologists and pathologists concerned with microscopic structure of organs in health and disease, such as J. J. Woodward at the Army Medical Museum (see figure 12).[72]

One technical obstacle to sectional depiction in gross anatomy—to macrotomic, as distinct from microtomic, work—was the inability to stiffen the body,

12. Joseph Woodward, Photomicrograph of tumor section, Philadelphia Exposition, 1876

to keep parts together along the section plane. In the early nineteenth century, a Dutch anatomist, De Riemer, published an atlas of transverse sections of frozen bodies. Freezing techniques were rediscovered by other nineteenth-century anatomists—notably Nicolas Pirogoff, who published a five-volume atlas of body sections in the 1850s.[73] The first publication of serial sections of the *same* body was Dwight's *Frozen Sections of a Child,* in 1881.[74]

Even with frozen cadavers, it was difficult for anatomists to produce detailed studies, because of tissue disruption by the saw. Operators developed various ways of polishing a cut surface, for example, with pumice powder and revolving wheels. Around 1903, the first sections of formalin-hardened cadavers were made.

In anatomic representation, sectional views were not widely used until the nineteenth century. They were not materialized in anatomic models until late in the century,[75] and they were not published to support surgical craft "while the patient is on the table" until 1905.[76] This development roughly parallels that sketched by Rudwick in geology. In both disciplines there were significant technical problems in opening substrates to viewing. Both disciplines' most "realistic" sections were bound up with "invasive" agendas: core samples and mining; saws and surgery.

In CT, "section" can signify a mode of "data acquisition" as well as a mode of image display. Display is the concern here. CT images, like geologic sections, are "reconstructions." This term has specific meaning in computing and the mathematics of "filtered back-projection." It alludes to complex manipulations which turn a circumferential collection of transmission measurements into a matrix of X- and Y-plotted local densities. Even though a displayed image slice might correspond to a plane intersecting the body, it is understood—and signaled in the term "reconstruction"—that this slice is not a direct representation of a cross-section of the body but, rather, is a mathematized projection.

Reconstruction of CT data need not follow the geometry of the slice. For example, the "Digital Human" datasets, compiled from photographed cross-sections of bodies, and from cross-sectional CT and MR images, are often deployed in presentational modes of traditional surgical anatomy. In one Digital Human software program, a user can view a muscle perspectively, "remove" it (make it transparent) with a mouse click to show what lies behind it—and never perceive that the volume of the muscle is assembled from stacked slices.

The choice of slices to display information about a volume is neither essential nor, in itself, "real." Slices convey particular relations among structures and spatial domains—but they obscure others, and require feats of imaginative connection. One radiologist reflects: "Many of us have developed the ability

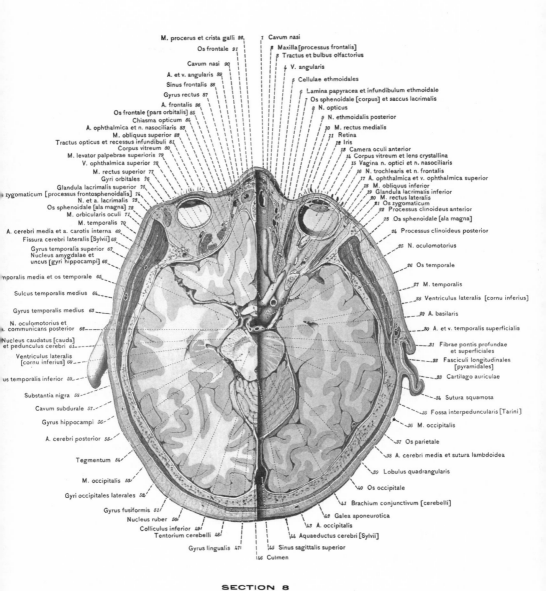

M. procerus et crista galli *92*
Os frontale *91*
Cavum nasi *90*
A. et v. angularis *89*
Sinus frontalis *88*
Gyrus rectus *87*
A. frontalis *86*
Os frontale [pars orbitalis] *85*
Chiasma opticum *84*
A. ophthalmica et n. nasociliaris *83*
M. obliquus superior *82*
Tractus opticus et recessus infundibuli *81*
Corpus vitreum *80*
M. levator palpebrae superioris *79*
V. ophthalmica superior *78*
M. rectus superior *77*
Gyri orbitales *76*
Glandula lacrimalis superior *75*
zygomaticum [processus frontosphenoidalis] *74*
N. et a. lacrimalis *73*
Os sphenoidale [ala magna] *72*
M. orbicularis oculi *71*
M. temporalis *70*
A. cerebri media et a. carotis interna *69*
Fissura cerebri lateralis [Sylvii] *68*
Gyrus temporalis superior *67*
Nucleus amygdalae et uncus [gyri hippocampi] *66*
mporalis media et os temporale *65*
Sulcus temporalis medius *64*
Gyrus temporalis medius *63*
N. oculomotorius et a. communicans posterior *68*
Nucleus caudatus [cauda] et pedunculus cerebri *61*
Ventriculus lateralis [cornu inferius] *60*
us temporalis inferior *59*
Substantia nigra *58*
Cavum subdurale *57*
Gyrus hippocampi *56*
A. cerebri posterior *55*
Tegmentum *54*
M. occipitalis *53*
Gyri occipitales laterales *52*
Gyrus fusiformis *51*
Nucleus ruber *50*
Colliculus inferior *49*
Tentorium cerebelli *48*
Gyrus lingualis *47*

Cavum nasi *1*
Maxilla [processus frontalis] *2*
Tractus et bulbus olfactorius *3*
V. angularis *4*
Cellulae ethmoidales *5*
Lamina papyracea et infundibulum ethmoidale *6*
Os sphenoidale [corpus] et saccus lacrimalis *7*
N. opticus *8*
N. ethmoidalis posterior *9*
M. rectus medialis *10*
Retina *11*
Iris *12*
Camera oculi anterior *13*
Corpus vitreum et lens crystallina *14*
Vagina n. optici et n. nasociliaris *15*
N. trochlearis et n. frontalis *16*
A. ophthalmica et v. ophthalmica superior *17*
M. obliquus inferior *18*
Glandula lacrimalis inferior *19*
M. rectus lateralis *20*
Os zygomaticum *21*
Processus clinoideus anterior *22*
Os sphenoidale [ala magna] *23*
Processus clinoideus posterior *24*
N. oculomotorius *25*
Os temporale *26*
M. temporalis *27*
Ventriculus lateralis [cornu inferius] *28*
A. basilaris *29*
A. et v. temporalis superficialis *30*
Fibrae pontis profundae et superficiales *31*
Fasciculi longitudinales [pyramidales] *32*
Cartilago auriculae *33*
Sutura squamosa *34*
Fossa interpeduncularis [Tarini] *35*
M. occipitalis *36*
Os parietale *37*
A. cerebri media et sutura lambdoidea *38*
Lobulus quadrangularis *39*
Os occipitale *40*
Brachium conjunctivum [cerebelli] *41*
Galea aponeurotica *42*
A. occipitalis *43*
Aquaeductus cerebri [Sylvii] *44*
Sinus sagittalis superior *45*
Culmen *46*

SECTION 8

13. Section from Albert Eycleshymer and Daniel Schoemaker, *A Cross-Section Anatomy,* 1911

to sense three dimensions by looking at a series of two-dimensional slices. Or even, I could look at that kidney and tell you what's in front of it and what's in back of it very nicely even though it's at two and a half [2.5mm slice]—by just imagining where the vessels are going—coming and going. I may not have ever seen those vessels, but I know where they're going because of my other experiences in other sections you can take too."[77] Radiologists who look at CT images on a daily basis and claim the ability to extrapolate spatial relations from thin slices are proud of their skills:

> That's probably the hardest part of learning radiology . . . Putting the anatomy in 3D and knowing the relationship of the structures to one another. Because you can't communicate with—for example, you can't communicate with a neurosurgeon if you don't know the relationship of a structure to another one . . . The surgeon says . . . when I move that vein, or something, what am I going to find under it, or what am I going to find behind it? . . . And maybe that is not, maybe that topology is not on the same slide as that structure that he's talking about, so you need to integrate it . . . And very few people can do it from the beginning. And I think, to tell you the truth—this is my feeling—that even despite the training, very few people obtain the capacity to integrate in the format you mentioned.

This is a recent problem. Sectional images have existed in radiology since World War I but were not often produced as a series until the propagation of CT in the 1970s. Until the 1960s, in fact, a much more common tool for representing depth was the stereoscopic viewer. This produces a quite different viewing experience, a different phenomenology of depth.[78] The average radiologist in the 1950s would have had far more experience in using a stereoscopic viewer than in reconstructing volumes from slices. One attending recalled his experiences making stereo angiograms of postmortem hearts:

Radiologist:	Then I would make some stereo images of those and you could actually see the heart . . . you could fuse the image and kind of see where the coronary arteries are—circumflex—and very, very effective.
Investigator:	Mm-hmm.
Radiologist:	The trouble with that technology is that you can't share it with anybody.
Investigator:	It's a fairly private—
Radiologist:	You have to do it yourself. If you can't do it, you're just sitting there looking at it, you know.

Investigator:	Yeah.
Radiologist:	And so it doesn't really catch on as a teaching—unless everybody wears glasses . . .
Investigator:	I suppose that's tough.
Radiologist:	And it takes time and—but for years, stereo chest x-rays were a very popular technology. There were stereo viewers. When I first came here, there was actually a big device that had the two chest x-rays with slightly different angle.

[*brief segment omitted*]

Investigator:	And this was as late as 1973?
Radiologist:	Yeah.
Investigator:	When CT was being born, you had an old stereo viewer!

Later this radiologist discusses new digital reconstruction technologies and the return of 3D viewing using special goggles, or holographic displays. Radiology, in conjunction with the University's Department of Computer Science, is active in developing these novel displays. But these have little impact as yet on viewbox practice, he asserts, and are likely to be embraced more avidly by surgeons than by radiologists.

Because stereoscopic apparatus and tomographic apparatus have distant kinship relations (chapter 2, "Cutting"), and because the virtues of both modes of presentation were debated by investigators from the 1930s to 1950s,[79] I pursue this line of questioning—about tensions between 3D binocularism and sectional displays—with another attending:

Investigator:	Now at RSNA, you can walk by and see these holograms . . . [80]
Attending:	Holography of the vessels inside the head.
Investigator:	Do you think that this sort of magical capacity that radiologists [have]—some radiologists—to be able to do this work of integrating slices—is a historical phenomenon that's going to be superseded by a different kind of representation down the pike?
Attending:	No, I think that all those things are just—they're not really important. I think we're . . .
Investigator:	We're going to have films with a three by five matrix of slices for a long time to come?
Attending:	I think so, I think so . . . from a strictly radiology point of view, it will not change. Perhaps from the clinical use it will

change. I mean, there will be programs that will integrate all the images into a three-dimensional image and calculate the position of a probe and show the surgeon—which a lot of people for endoscopic and nasal surgery are using . . . So I think that will change that way. But I don't think in radiology—I think we'll be reading slices basically for a long long time.

Aesthetics at the Viewbox

In Body CT, the fellow calls over the resident: "Here's a beautiful example of gastroesophageal varices." Later, the attending remarks: "Boy, that's beautiful corticomedullary differentiation in the kidneys, huh?" In Neuroimaging, the fellow comments: "Look how pretty the gradient echoes came out." What are the implications of these rhetorics of aesthetic involvement?[81]

> Well, there's no question that I mean—one thing that we look at—the films— the studies have to be pleasing to the eye . . . The quality of the filming has to be pleasing to the eye. The quality of the study itself, whether the patient moved or not . . . One thing that strikes me about how radiologists are picky about things that are pleasing to the eye is if you go to a meeting—whatever, X clini- cal meeting—and they're showing slides . . . the slides seem to be a hodgepodge with different colors and stuff like that, and old slides and new slides mixed . . . Perhaps they're a little bit blurry, it doesn't matter. But if you go to a radiol- ogy meeting, everybody has their slides, you know, fancy computer slides. If one slide is a little bit blurry on the screen, the first thing that people say is "please"—you know—"the slide is blurry, please sharpen it."[82]

"Image quality," in textbooks, is a technical, rationalized issue. At the viewbox, however, images are objects of connoisseurship, and summon aes- thetic terms. "I think we should just say that pelvic images are degraded by artifact . . ."

One hears of beauty, and also darker rhetorics of aesthetic response, in re- lation to other dimensions of images than their quality. A Body CT attending comments: "I think we've got to at least raise the specter of invasion of the left side of the seminal vesicle. And it would fit with a carcinoma, as big, ugly, necrotic-looking, somewhat enhancing mass. Right?" "Ugly" here character- izes a lesion which marks its bearer as unfortunate: there are features of ma- lignancy—irregular margins or contrast uptake—or tissue death. It suggests recoil of the reader—tinged by fear, sympathy, perhaps morbid fascination.

It is surprising, therefore, that CT radiologists do not speak more readily about aesthetic or affective dimensions of reading. They forefront "intellectual" engagements of their craft. This senior radiologist is an exception:

> I can still remember the day that I decided to go into radiology. I was a medical student on a rotation at Hopkins, looking at a lateral cervical spine . . . of a young woman I believe it was. It maybe sounds a little bit corny but . . . there's a certain beauty . . . of the curvature of a cervical spine—lateral view cervical spine—of a long-necked individual. Not that they have more vertebral bodies, just that you see them better . . . The beauty of the articulation and the patterned repetition of the body and it just—it just to me was so exciting I wanted to spend my life doing that sort of thing . . . And even today, when I look at chest films—I aesthetically get much more of a kick looking at a chest x-ray or a bone, or a full bone film, than I do sorting through fifty images, slices and so forth . . . It has to do with training and what you like, but I'm an aesthetic person, I enjoy images . . . My office is full of images, as opposed to other people who don't have images in their office, but have lots of diplomas.

The same radiologist goes on to note that his enjoyment of radiological images is not merely spectatorial but also comprises pride and pleasure of craftwork. "And I think in many respects, my images [for research] . . . that I develop with my own hands, are in a way a work of art . . . And I've had people want me to use these things as Christmas cards, and they borrowed images to do this and that and so forth . . . And I've pictures that I've loaned people as gifts and so on . . . And they consider them works of art that transcend the pathology and structure and function that they also demonstrate . . . And that to me is important. I wouldn't be in radiology if I didn't feel very strongly about that."[83] Further remarks articulate a divide between two viewing attitudes. There is the tradition of the radiological generalist, engaged by holistic, sometimes sensual, images on "plain film." Another, newer attitude, that of the sectional imaging (CT/MR) specialist, involves taste for more stark, "quick and easy" display. The radiologist associates this latter attitude with the rising generation: "I have neuroradiologists [who] never look at a skull film any more." This question of "disenchantment" will be taken up again in chapter 3, "Diagnosing."

To understand aesthetic aspects of CT diagnosis—to understand *reading*—we must consider practices beyond the ocular. These practices intensify diagnostic gazes, link with them, differentiate them—as well as interrupt them and divert them. Such "paravisual" practices include handling films, answering the phone, dictating. Distributions of these practices reveal familiar social cleavages—especially specialization and rank. Specialist visitors make for uncertainty and disagreement; teaching contributes discontinuity. Reading CTs in University Hospital turns out to be interrupted reading, adjusted reading, contiguous reading, dictating, rereading. These issues are crucial to the question of disenchantment in CT diagnosis—and historical fates of intrigue in diagnostic judgment.

A Viewbox Consult

A Body CT attending and resident, in neckties, engage tasks across three workstations and two viewboxes.

Members of the Pulmonary Consult team enter: attending and fellow, two medicine residents, a student. One resident and the student are women; all but the student wear long white coats. They gather around the left viewbox and quietly discuss a chest CT (already discussed by radiologists). The Body CT attending finishes dictating at the other viewbox.

Pulmonary attending:	Hello, good Doctor Baumgartner. [*booming, mock formality*]
CT attending:	Distinguished pulmonary colleague, how are you? [*similar tones, stands up*]
Pulmonary attending:	I don't know how I am, sir. I haven't had my body CT for today.
CT attending:	You're looking at a chest; I suppose that's close enough to the body. [*His expertise: abdomen. He moves to left viewbox.*]
Pulmonary attending:	Pull up a seat.
CT attending:	Pull up a seat . . . why stand?

But the CT attending remains standing, and the pulmonary attending sits. Residents are right and left, the man squatting, the woman standing; the fellow stands behind; none takes the empty chair.

The attendings discuss "patchy opacities" in the lungs. They point. They move images on the board. They consider diseases that patchy opacities might signify.

The CT attending expresses interest in the patient's past treatment with methotrexate. The pulmonary attending disputes the relevance of this—on the basis of imaging criteria. "I think with methotrexate you can get a much more infiltrative, less interstitial picture." The tone of their dialogue is speculative; neither advocates a particular diagnosis. Another disease is mentioned by the pulmonary attending: "Eosinophilic pneumonitis . . . classed as a hypersensitivity pneumonitis." The CT attending asks if this disease "shows a restrictive clinical picture." (Earlier, the patient was said to have restrictive lung disease.)

The student offers recent pulmonary function test results. These are recited unselectively: normal and abnormal, some irrelevant. She mentions another imaging test the patient underwent yesterday: "a VQ scan that was low probability."

The CT attending then comments, not on imaging matters (the VQ scan) but on a nonradiological datum from the pulmonary function tests: "That's a pretty low DL_{CO}." This provokes a question from the pulmonary attending. "Could this be pneumocystis?" CT attending: "Could be pneumocystis . . . we've certainly seen this . . . ground-glass opacity."

So far, radiologist and clinician have each offered questions and comments that would seem properly to fall within the other's expert purview. The CT attending has queried medications and gas-exchange (clinical matters); the pulmonologist has commented on radiographic appearances. As this cross-disciplinary exchange has unfolded, each has taken the standpoint of the other—articulating testimony that in a sense belongs to the other. Such criss-crossing of specialty perspectives is not uncommon in fact-finding phases of case development. Here it seems enhanced by collegial comfort, by histories of prior consultations. But soon the colleagues revert to conventional roles—radiologist speaking to image-appearance, pulmonologist to clinical issues: [84]

Pulmonary attending:	Describe just what you're seeing?
CT attending:	[*authoritative tone:*] . . . ill-defined patchy ground-glass opacity,[85] primarily in mid lung fields . . . *this* is relatively uninvolved lung [*pointing*], this is relatively more involved lung [*pointing*]."
Pulmonary attending:	If I was to bronchoscope her, would middle lobe be a good place to go? . . . superior segment of left lower lobe . . . harder . . . we could try . . .
CT fellow:	. . . You've already talked about eosinophilic pneumonia?"
Pulmonary attending:	What's not good for that is . . . What is good is . . . Here the infiltrate is respecting the fissure. Eosinophilic pneumonia doesn't respect fissure lines . . . 1 to 1.5cm nodes . . . My inclination is to biopsy her . . . That's why I asked her if she ate lunch.
Student:	[*recites what patient ate*]

CT attending:	Did she get this scan before or after lunch? Because at this time she had a fairly empty stomach. [*chuckling*]
Pulmonary attending:	[*chuckles*]
CT attending:	Well, let us know.

The pulmonary team leaves. The CT attending says to the resident, "We're finding out enough about this patient to go ahead and dictate her." (Previously they planned to wait for old films.) "By the time we get done she'll have already been bronchoscoped, and we'll know the answer."

The attending wants to enter radiological testimony into the record before it is rendered moot by a tissue diagnosis. The logistical consideration is not utility—the clinical team has already been advised—and, indeed, a fully informed reading might still require waiting for old films. But the chronology should show radiology's place in the causal chain of diagnosis. Cogs of the hospital mechanism are turning. If the record showed a bronchoscopic diagnosis was made before the CT scan was read, this would belie the strategic role of the scan in staging the procedure.

A minute passes: the same pulmonary team returns, carrying a chest x-ray of the same patient. Surprisingly, upper lobe infiltrates are not apparent. If anything, lower lungs look "busier." This dissonant finding prompts the pulmonary team to reconsider the CT to verify a bronchoscopy target.

One of the pulmonary team phones the bronchoscopy suite. Meanwhile a medicine resident suggests discussing the case in an upcoming "Resident's Report" conference. "A good case for Monday. Let's just say it's PCP [pneumocystis carinii pneumonia]. Even if it isn't." Chuckles all around.

The other resident suggests: "Or, you could talk about methotrexate." The first resident replies, "Then I'd have to go looking up some papers on methotrexate lung." CT attending: "Share 'em [papers] with us when you get 'em?!" The cheerful directness of this request makes it unclear if he missed the cynicism of the residents' exchange, or if this is a tactical intervention, as if to say: don't cut corners, please *do* look up methotrexate: don't betray the diagnostic equipoise we have just worked out.

This exchange exemplifies several social features of viewbox interaction. One is host/guest relations: greetings, jocular exchanges—manners and customs of reading-room sociality. Another is interspecialty relation: differentiation and engagement of expert testimonies. A third feature is logistics, of contribution to actualities of patient care, and to the record. A fourth feature is pedagogical hierarchy—including, here, delegation of library work to trainees. In the sections that follow I will address each of these four features, toward an unpacking of social dimensions of reading-room practice.

Guest/Host Relations

On weekdays, when there is a consistent Radiology Department presence in the reading rooms, visiting clinicians are guests, and radiologists hosts. Patients are conspicuously absent. If patients review their films, they tend to do so in clinical, not radiological, settings.

When a clinician enters the reading room, she is usually greeted by one of its occupants. There may be a delay—depending on time of day and engagement of radiologists. If a visiting clinician knows what she wants—goes to the unattended viewbox, finds a particular study, inspects it, and leaves quickly— she might elude greeting. But on most occasions, a visitor will be offered assistance.

Residents are the reading room's steadiest occupants on weekdays. One is assigned to Body CT, and one to Neuroimaging, each month. The resident is responsible for many workday tasks, and tends to defer privileges of greeting to superiors.

The fellow is the other steady weekday occupant of the reading room. Fellows are also assigned to CT (Body or Neuro) for a month at a time.[86] They are much more experienced, more fit to play host/consultant roles. The fellow is likely to greet a visitor if he is the senior person on duty.

The attending is often a more intermittent occupant of the reading room, even on weekdays. Though the attending has a supervisory role and is accountable for the work of the service, it is possible for her to remain hands-off— especially if she has confidence in the fellow and resident. Attendings have other offices. An attending is assigned to a CT service (Body or Neuro) for a day or two here and there. Some attendings cover other services at the same time; thus they may need to migrate between reading rooms. On days like these, the attending expects the resident and fellow to handle viewbox busywork. But when the attending is present, she is as likely as the fellow to greet a visitor and offer assistance.

Radiologists' host manners reflect departmental ownership of the reading room and traditions of consultative service: theirs is a "shopkeeper" habitus, one which acknowledges that professional strangers are key clients. Some clients become familiar, and guest-host relations can be genuinely warm. Yet radiologists' manners are not necessarily deferential: they are inflected with the power of the shopkeeper as well.[87]

Divisions of Diagnostic Labor

A CT reading room is a watering hole in the savanna: different beasts come and drink, perhaps mingle, look warily at competitors, move on. Clinician visitors come to look at films and "pick the brains" of radiologists. These clinicians represent various sections and services from the hospital: Renal Consult, Pulmonary, Medical ICU, the Liver Transplant Team, and so on. All these formations, and more—constituted around clinical disciplines and service regimes—send emissaries to CT viewboxes.[88]

For some services the pilgrimage to the viewbox is routine. On Pulmonary Consult service it is a rare day with no chest CT to review; Neurology likewise travels to the Neuroimaging reading room at least daily. Some services give greater weight to collegial interaction, some to direct viewing of images.[89]

The radiological division of labor that influences CT viewbox discourse the most is that of Neuro from Body. The Neuroimaging reading room and the Body CT reading room are separate domains, with distinguishable caseloads—at least Monday through Friday. This separation of facilities derives from divisions of faculty expertise (reading) rather than tech expertise (image production).

The Chair reflects on specialization in his Radiology Department—a "medium-sized" one—over the last decade. First there was General/Diagnostic Radiology, Pediatric Radiology, Nuclear Medicine, and some Vascular/Interventional and Neuroradiology. New modalities (ultrasound, CT) were combined in an Imaging Division—a "technology-based" division—typical of a phase when institutions are first discovering "the capability, the limitations" of technologies. On his arrival in the early 1990s, he divided Diagnostic Radiology into organ systems and made individual modalities, including CT, into "services"—as larger radiology departments were already doing. "Like an x-ray machine. Gastrointestinal radiologists use CT, neuroradiologists use CT."

University Hospital's organization of radiology services and specialties emulates that of "big places." Within official specialty divisions there are recognizable subspecialties. One of the neuroradiologists is particularly well versed in ENT (ears, nose, throat). Another is the acknowledged cervical spine expert. Another conducts vascular neuroinvasive procedures such as embolizations of aneurysms. On a given day, a clinical team might discover a favorable match between a problem they bring and the expertise of the CT attending du jour.

Or the match might not be ideal—the team may be instructed to "show these images to Dr. So-and-so, and see what she says."

Though medical specialization did not develop in hospitals—indeed has stronger historical associations with universities[90]—certain specialties have been hospital-based from the outset: radiology, pathology, anesthesia.[91] Radiology has developed, like pathology, as a consultative discipline: its clients have historically been clinicians rather than patients.

The CT reading room is a special crossroads, between sensory-cognitive faculties of individuals and the disciplinary faculties of a school of medicine. It is not only a place for observation of films on the viewbox; it is also a place for observation of one specialty by another, of consultants by clients, and vice versa. Clinical specialists come to viewboxes to study images, to solicit the advice of radiologic specialists, and to perform expertise.[92]

Trafficking in Images

Logistics of intellectual work in CT depend on movement of images. In the late 1990s, film is still the crucial image medium.[93] CT images are "acquired" by technologists, then printed on film, and delivered to the reading room for review. The tech usually makes the drop.

Techs are not always greeted upon entry to the reading room; they are not guests as clinician visitors are. Techs interrupt reading when they need something. "New study: OK to send this lady home?"

A tech enters the Body CT reading room and lays a film stack at the fellow's elbow. Without shifting his gaze from the viewbox, the fellow registers the drop with a quiet groan. Tech, with mock solicitousness: "Such heavy sighs!" No response: the tech leaves, unoffended. The fellow, rapt at the board, silently nudges the film stack with his elbow, toward the resident. The resident takes the stack in his lap and rolls his chair (backward, long leg-strokes) to the other viewbox. From his seat, the resident lifts the stack of film in both hands, leans forward, and slips it behind a nylon retaining line on the board. The line snaps on the pile. With both hands he spreads individual films out behind the line. His gestures have a taut economy. Delivery of this film stack from scanner to viewbox is made with minimal verbal accompaniment: these are oft-repeated routines, firm hierarchies.[94] The tech's film drop has the air of a servant's attendance at table; the resident's response to his fellow's nudge is as snappy as a hard salute.

Timely film flow, once a rationale for having the reading room close by the scanner room, relies on hands and feet.

Body CT resident:	[*on phone with clinician:*] She gets done down in the Emergency Room, but we need to have a hard copy to read . . . [*pause*] They'll wait for like three or four scans to get done before they bring them all at once . . . [*pause*] I have not seen a hard copy on her at all—I mean, I've got your message, so as soon as she comes through I'm going to page you with the results—I'm—I mean I'm going as fast . . . [*pause*] Yeah . . . [*pause*] I got you, so—I mean, I realize it's important. Well, I'll call down to the ER and see if they could bring up some scans. I mean, she's on the monitor, I know she's done, but I just have to see a copy we can put them in and go. So. I will um—9726, that's your pager, yeah—I will let you know as soon as we get her off the press. You bet. Bye. [*hanging up*] [*laughing*] Man, you gotta love it, man. Buy, sell, sell, buy!

For new studies, techs do most film carrying. Old films are handled by Film Management staff. Each day a pile of old film folders is delivered to the reading room, from a list prepared by the resident the day before. Yet each day there are missing jackets, as well as "add-on" studies.

Body CT fellow:	Nine add-ons? Well, we had nine add-ons yesterday, right?
Body CT resident:	Yeah.
Fellow:	I think it's going to be that way all the time with the order entry.
Resident:	The problem is that . . .
Fellow:	They'll just keep coming—they'll just keep rolling in and then we'll just—
Resident:	But you're never going to be prepared for them, that's the problem. You're not going to have anything pulled up on the person.

The new computerized Order Entry system, allowing inpatient ward clerks to bypass Central Scheduling, has made the CT team's workload unpredictable and has delayed access to old films.

Here the Body CT resident has paged the courier to request old films on two of nine studies on the morning schedule: "How are you doing this morning? Good. Um, couple films—ready? Um. Marcus Porter, medical record number

1039828. He had a C—Porter, mm-hmm—he had a CT on 7–12–96 and . . . Kiefer, Samuel Kiefer I believe, with a K, K-I-E-F-E-R, but it's 298435—he had a CT done 8–3 of '95, of the abdomen. And I think that's it so far." Unfortunately, the courier is not always available. Late one afternoon in Body CT, the fellow has gone to retrieve newly printed films from the ER/Neuro scanner (across the hospital). The attending wonders why. The resident explains: "Our courier leaves at four, so we called the ER people and said, 'Well, who's supposed to bring us films now that our courier's gone?' and they said, 'Well, the Film Management people are,' so I called the Film Management people and they said, 'Well, the ER people are.' So I called the ER people back and he said, 'Well, I'm the only one here. I have to stay and answer phones.' I guess Dale finally said, 'I'll go get 'em myself then.'"

Notwithstanding the centrality of film in the late 1990s, images also arrive in the reading room independent of film. For a year or so, the computerized image-handling system, MagicView, has been in use.[95] This allows radiologists to review images before they are printed.

Attending to tech, on phone: "Skip? This is Dr. Bynum. Need a favor from you. Lincoln? And Denton? Those. Yeah. Can you ship the images, ah, back to the MagicView? Thanks. Bye." The MagicView system moves images when there is some urgency—or when the tech needs to know if a study is complete, before letting a patient leave. The system is particularly helpful in allowing Body CT radiologists to review scans from the Emergency/Neuro scanner, across the hospital—and whenever a printer is down. "Shipping" of images through the MagicView system happens on the initiative of the tech or on request of reading room staff; the system does not (yet) routinely convey all cases.

A student at the MagicView in the Neuroimaging reading room, deleting cases already reviewed, remarks: "Wow! At the same time as we erased Smothers, Forscher showed up. And as one disappeared and the other reappeared, the computer started to make a lot of frightening noise." The attending chuckles.[96] The MagicView's idiosyncrasies are still new and sometimes subvert the ideal of effortless, spontaneous image transport.

Near the end of one long day in Body CT, the film printer has been down an hour, and the reading room team is impatient to receive a few "reconstructions" via the MagicView, so they can finish dictating a case.

| Attending: | [*to tech in doorway:*] Weren't you going to send us some images? |
| CT tech: | Yeah, but—I don't even think it [MagicView] is sendin' 'em now. |

Fellow:	Can't even do that now?
Attending:	Well, can we come in there [control room] and look?
CT tech:	In just a minute. Farley's looking at her case we just did.
Attending:	Oh, OK.

[*brief segment omitted*]

CT tech:	It won't send down to the ER scanner at the moment either. [*tech exits*]

[*one-minute conversation*] [*phone rings*]

Resident:	Body CT. Hey. Um—[*to colleagues:*] Do we want to look at the recuts on the monitor? [*to phone:*] Yes.
Attending:	Since we can't look at them here. Should we take our book and take notes in there?
Fellow:	Yeah, maybe we'll bring this thing [old study] with us.
Attending:	We're mobile, we go anywhere where there's images.

[*filmflap*] [*all leave for control room*]

Hi-ho, hi-ho. The expert gaze can rove. But this chippy declaration of independence from the reading room—"anywhere there's images"—is a bit disingenuous. To chase images upstream, to the source, the scanner console itself, is to leave a customary and comfortable spot on the assembly line. The team is eager to close its workday.

A few minutes later, on return from the control room, the same team tidies up studies on the viewbox and MagicView—and anticipates a scan that is not done yet, which the on-call resident will read overnight.

[*boardbuzz*]

Resident:	I'm going to slide these guys down. Was there just one or was there two cases that you wanted to take to that Imaging conference tomorrow?
Attending:	I can't remember which cases those are over there now. I think there was just one down there. [*studying MagicView:*] Taylor we can send back because I talked to the doctor on that one.
Resident:	OK. Hanlon was the—
Attending:	Which one?
Resident:	—duplicated IVC.

Attending:	OK, we wanted to hold on to that one. And how about the neurofibroma?
Resident:	OK. Do you want me to just go ahead and take it down or do you want me to just—
Attending:	No, just leave it. Remember tomorrow just to hold it for me. [*boardbuzz*]
Resident:	OK.
Attending:	And what are the others in your hand there?
Resident:	Cornwallis, the one who didn't show up, I'm going to hold on to him—
Attending:	OK.
Resident:	And ah Holt, pancreatitis/sepsis, that can go back. They never came by to look at the—
Attending:	Yeah, we can send that back.
Resident:	And then Taylor can go back.
Attending:	Yeah. All right, so I think we're done. We'll just have to make sure—you'll have to make sure there's films on these tomorrow. Maybe if you're lucky you won't have any films tomorrow. Wouldn't that be terrible.
Fellow:	Oh, that would be a mess.
Resident:	What about Imma?
Attending:	I'm leaving those there instead of erasing them just because in case you don't get films or something, at least you've still got that. Let me set this book here. So we'll just leave these here for now.
Resident:	Even the disaster [trauma patient]?
Attending:	Well we could eliminate that one. Uh oh, what happened? [*unfamiliar screen*]
Resident:	[*instructing attending:*] "Select patient." So we can get rid of the disaster.

[*brief segment omitted*]

Attending:	All right. Would you have believed we'd be through at this hour?
Resident:	No.
Attending:	I wouldn't either.
Resident:	We did twelve, um, on this side.
Attending:	We didn't do that many really. How many did we do? [*filmflap*] [*counts in other logbook*] Eight.

| Resident: | Twenty. |
| Attending: | So it's only twenty, it seemed like a lot more than that. |

One by one the day's cases are sorted: send back, delete, hold on to him. This collective remembering treats cases with remarkable economy: surname, disease, thin descriptor. Most images leave the reading room with less ceremony than they entered: dumped from system memory, returned to a folder, into the courier's pile.

CT teams in 1997 are still getting used to the docket the MagicView presents. Unlike a schedule, a pile of folders, or a page in the log, the list of cases on the MagicView bears no handwritten annotation. And things have been hectic all day: members of the CT team have been distributing attentions differently, with respect to images and to reading room clientele. They need to regroup. Days often end with such salvos of recollection, around some form of the docket—recapitulations of a day's distributed thinking.[97]

Radiologists prefer not to let films go until they have dictated a reading. Here, an exception is made for a senior surgeon, because of clinical urgencies and his VIP status:

Chest surgeon:	So what do you really make of these little liver things all over . . . ?
CT resident:	Look like mets.
CT fellow:	Dr. Bynum was pretty worried about them and thought, in this particular case, if it makes a difference in your staging, MR might help . . .
Chest surgeon:	All right. [*pause*] Can somehow this make its way to the ah, noontime thing—noontime conference? How would we expedite that?
Fellow:	Um, is there like a noon—is there a Chest Conference?
Chest surgeon:	Yeah, this afternoon.
Resident:	Any chance they could make it back here after?
Chest surgeon:	Sure.
Resident:	Um, sure, as long as they get back here afterwards then sure, you can borrow them for an hour or two hours, whatever.

[*segment omitted*]

| Chest surgeon: | Yes. OK—I'll send Sam up—our coordinator. |

Fellow:	OK.
Chest surgeon:	I don't want to be held responsible for this.
	[*chuckles*] [*leaves*]
Fellow:	[*chuckles*]
Resident:	OK.

[*surgeon reenters with one of his entourage*]

Chest surgeon:	OK, we have a volunteer.
Fellow:	We have a volunteer.
Surgical team member:	I'll be responsible.

A logistical goal of the reading room—timely reporting—is at odds with surgical priorities here—to get the case to Chest conference. A procedural exception is made as a professional courtesy, and for the good of the patient. Yet the surgical attending admits he would be a weak link in the chain of custody. Fortunately, diligence of assistants can redeem the unreliability of chiefs.

Yet chiefs also presume on assistants. No mercy is shown to one dutiful student searching for old studies on the RadCare workstation (part of the radiology information system):

| Medical student: | What is this screen? |
| Neuroimaging attending: | You're about to delete the whole Radiology database . . . You're about to flunk your Neuroradiology rotation. |

[*laughs all around*]

Hierarchy

Time to clear yesterday's cases. The attending turns to the resident (first-year): "Don, my good man, when you've got a minute . . . How good are you at pulling films down from the board?" The mock formal tones are patronizing—yet the resident betrays no indignation: "I'll do the best I can."

Hierarchy is enacted in viewbox pedagogy and informs teaching opportunities in cases—from normal anatomy for students, to fine points of differential diagnosis for experienced residents and fellows.

| Neuro CT attending: | Where's the central sulcus? |

[*student on right points*]

| Attending: | Where's the frontal lobe? |

[*student on left points*]

| Attending: | Good. |

Relative rank is often legible in positions before the viewbox. A senior person sits centered before the images, with junior members of the team at her elbows—or looking over a shoulder. On occasion, the senior person may sit back, to allow closer scrutiny of images by trainees.

A case in Body CT: Norris. The fellow narrates past history and summarizes old radiology reports, while the attending points out imaging features with a pointer. The attending then turns to the resident. "When we get up old films on Norris, I want to ask you a few questions." The resident says, playfully: "And I'll put my hands out, and when I answer wrong you can rap them with your pointer."

At times, rank seems less about pedagogical agendas than about dividing work. Carrying and mounting films are tasks for trainees. Then there is the phone. Incoming calls interrupt reading of images as well as viewbox conversations: they are usually handled by residents. This duty is sometimes enforced in quiet but dramatic ways: an attending may allow the phone to ring unanswered—or may request that someone else answer it. "You get the phone; I'm chewing." Or an attending seated next to the ringing phone may pick up the handset and simply pass it over to the resident.

Readers in Transit

As images flow in and out of the reading room, there is also ebb and flow of readerly agency. Members of the on-duty team are summoned to other settings and engage other tasks which are not, strictly speaking, those of readership.

One Neuroimaging attending is renowned for being unable to sit still. He reads studies quickly and summarily, and often leaves the viewbox to intercept studies as they are printed, in the control rooms of Neuro CT or MRI. He says he likes to check in with techs and to feel more connected with patients and their circumstances. "I just read 'em better if I'm a part of it." He also feels that reading films sooner, "timely service," is good—even as he acknowledges that awaiting old films for comparison may make for "fewer mistakes." Reading studies apart from the rest of the team also means these studies are less likely to be teaching cases.

14. Resident on the phone, Body CT reading room

The shuttle from the Neuroimaging reading room to MRI is short. But when Body CT attendings have MRI duty on the same day, they must traverse the entire hospital to get to the MRI suite. Body CT attendings also sometimes cover Ultrasound:

CT attending:	I'll probably have to run down to Ultrasound a minute for this biopsy . . . [*on phone:*] Hello? So what's happening? So he scanned the patient? OK. All right. So, I'll give it a few more minutes before I come down. So he's already consented him and everything else. OK, thanks. Bye-bye. [*hangs up*]
CT resident:	[*dictating*]
Attending:	I'm covering Ultrasound and CT today, that's why I'm going back and forth.
Investigator:	Oh, I see. That's quite a day.
Attending:	Well, I have fellows both places, so—without fellows I wouldn't be able to do it.

Having a good fellow, or resident, on duty can also enable attendings to combine research or administrative activities with reading room supervision. "Dr. Githens [a resident], you are here today to make my life so much easier . . . make me able to go to my office today . . . [so I] don't have to do any scutwork."

The CT team may leave as an ensemble to attend noon conference. If one must stay behind to maintain a viewbox presence, or to keep work from piling up, it is often the fellow. Her educational needs are less acute.

CT fellow:	I don't see anything—now you don't have to dictate these now. You can go to conference now.
CT resident (first-year):	Yeah, I'd like to . . .
Fellow:	And then we'll dictate them when we get back.
Resident:	But actually, let me just go ahead and just dictate 'em. I mean . . .
Fellow:	'Cause Dr. Baumgartner has a conference for you guys. Slides.
Resident:	Oh, well then I'm going. Sorry.

Tasks in the scanner room can also summon CT readers from viewbox posts:[98]

CT tech:	Mr. Gupta refuses to drink his oral contrast.
CT fellow:	He does?
Attending:	Dr. Alpers [a resident], I'm sure, is [a] very persuasive physician.
CT resident (third-year):	Yes. [*reading requisition:*] "Rule out pancreatic pseudocyst."
Tech:	In times before, we haven't gotten him to drink it.
Fellow:	Is that right?
Tech:	Yeah.
Attending:	We'll see how persuasive Alpers really is.
Resident:	I can be very persuasive.
Attending:	Dr. Alpers will be using his charm there.
Investigator:	I'm going to go watch, watch you use your charm.

Mr. Gupta becomes a significant logistical problem this day, and a distraction for Dr. Alpers. He never does drink contrast—and his refusal provokes much review of old records and films. Balking in the stream of contrast flow and patient flow leads to intensified archival interest. Eventually he gets his scan.

CT readers may have to confront more serious problems with patients. The major reason for keeping the reading room close to the scanner room is the possibility of allergic reactions to intravenous contrast. CT radiologists are notified when a patient with a history of prior reactions must be given contrast (usually with premedication). Also, some patients brought to the scanner are quite ill. Sudden emergencies are uncommon, but they occur—and when they do, the closest physicians are often the CT reading room team.

Handling Film

Handling film is (in the 1990s) part of radiological craft—an aspect of expertise as valuable as card handling to a poker player. Dexterity serves work flow—and displays competence.

Film-handling skills involve familiarity with labeling: when a radiologist holds a film to the light, she knows where to look for date, whether a CT study is contrasted or not, slice numbers . . .

Likewise, radiologists are familiar with conventions for "hanging films" on the viewbox. Some of these conventions are widely distributed, some are local. Because CT studies comprise many sheets of film, these conventions become important. The viewbox is approached from the top down and left to right. First contrast, then noncontrast images. First soft-tissue windows, then bone or lung windows. When possible, images from the prior study are placed directly alongside comparable current images. Some attendings have unique preferences, which they make known to residents and fellows working with them.

In Body CT, the resident, a first-year, has just painstakingly mounted sixteen or so films on the left viewbox. The fellow and attending decide they are not properly placed to make good correlations. They revise the arrangement. This "biphasic" contrast study is relatively new for this service; its mounting conventions are still underdeveloped.

In the MR reading room, a fellow has shuffled films for ten minutes to put them in order. He finishes and sets out to deliver another study upstairs "for conference." He is stopped at the door by a nine-member Infectious Disease service. They wish to view the very study he has just finished shuffling. The ID attending is impressed with how quickly the fellow mounts the films. "I say this guy deserves a fireball," he says. "It's your reward." His entourage is eating fireball candies. "Thank you," responds the fellow. Later, when taking the films down and restacking them carefully, he tells me: "I used to not keep them in order. That only happened once. You never know when you will have to pull them out for a team." A few minutes later, telling a colleague on the phone what delayed his delivering conference films, he explains what he has in his mouth. "The team came down and gave me one [candy] . . . I put up about twenty thousand films in about thirty seconds and they gave me one." He embellishes: "There was about fifty thousand of them [the ID team]—a whole flock—I didn't think the room could hold them all." His intellectual contributions to a patient's diagnosis pale beside his having tamed hordes of films and clinicians, one swarming on the other, with his expert film handling.

Film stacks easily get out of order—especially MRI and CT studies, which comprise various techniques and projections, to be hung in sequence. A resident in Neuroimaging struggles to sort out a large MRI study: "This one has sections of special sequences—that's why it's so hard to hang . . . OK—so we have an enhanced coronal as well. The irritating part is they *could* keep them in order—but . . ." On another occasion, the Body MR fellow is sorting through a "follow-up bone tumor study" and lamenting the jumble of films. Too many images is "a necessary evil," but about the jumble he comments: "Now sometimes a tech will have these in order—it's always a welcome sight—lately it hasn't happened. I dunno whether they're pissed off . . ." Films are slippery: it is in their nature to become shuffled and misplaced. Theories of filmic disarray that involve vengeful techs are uncommon.[99]

Protocolling

Each CT scan is performed according to a standard protocol, authorized by a radiologist. Protocols are recipes for tomographic cutting, stipulations of "technique": how to align the gantry, how thin to slice, when to inject contrast.[100] In practice these details are abbreviated under tidy headings: Routine head; High-res chest; Adrenal protocol. Techs could choose, and modify, most scan protocols without help: the radiologist's "protocolling" is often superfluous. But protocolling ensures the imprint of radiological authority on image acquisition: the protocol has the force of a prescription.[101]

Residents and fellows review each day's requisitions in batches, usually the day before, and mark protocols for each. Techs bring more requests throughout the day. Every protocol means work for radiologists and techs—so protocolling can be an occasion for sympathizing—and for wondering about client clinicians' intentions:

Neuro CT tech: Hey. I need to get a few protocols from you.
Neuro CT fellow: OK.

[*brief segment omitted*]
[*tech hands stack of slips to fellow*]

Tech: This one, I think this is the tenth.
Fellow: Holy smoke.

Tech:	I know, they just will not quit from the ENT clinic.
Fellow:	One of my most hated things.
Tech:	And this one is: "if it's open"—so I guess we'll let you look at it or whatever . . .—then they're supposed to get an MR screening of the IACs, and if it's ossified . . .
Fellow:	If what's open?
Tech:	Cochlea, is what they're telling me, 'cause it's a cochlear implant candidate? That's what the lady called down and said . . .
Fellow:	Mm-hmm.
Tech:	From ENT clinic, she said . . .
Fellow:	[reading:] "If open, do screening MR of IACs. If ossified, do not do MR."
Tech:	I don't know, that's what she told me . . . So, it's just a Monday, we'll chalk it up to that . . . That one of course is a minus [noncontrast].
Fellow:	[shuffling slips] This one . . .
Tech:	One of your favorites again.
Fellow:	[whispering:] Oh my god!
Tech:	It's hell day. [chuckle]
Fellow:	OK . . . [shuffling slips, annotating one at a time]
Tech:	And another favorite of yours.
Fellow:	Oh, my gosh, I can't believe it!

[fellow returns slips to tech]

Tech:	Thank you.
Fellow:	Thanks.

[tech exits]

Investigator:	Do you have to approve all these? Or . . .
Fellow:	No, just protocol 'em. [door closes] And, you know, if it sounds like something stupid, or they sound like something . . . [chuckles] We wouldn't do any studies then [if we refused inappropriate ones].

[laughter]
[brief segment omitted]

Student:	Call up Dr. Farnsworth and say . . .
Fellow:	"Hey, why ya orderin' this? What, are ya stupid?"

Student:	Yeah.
Fellow:	Or if it's something that would be better—
Investigator:	Bite the hand.
Fellow:	[*chuckling*] Right—that would be better looked at with MR or something like that, you know.
Investigator:	Right.
Fellow:	Most of the time the techs are really good, and they know what's doable and what's not.

Assigning a protocol is often the first encounter between the reading room team and a CT study-in-the-works. This assignment can be mostly but not fully propositionalized. The scanner and its controls are predictable; the instant case is not. Assignment of a protocol is an application of rules to particulars: an exercise of discipline, classification—and sometimes improvisation. "Patient with icthyosis? I'll say just a routine chest."[102] Matching tomographic rules to clinical "indications" requires judgment.

Body CT resident (first-year):	[*reviewing stack of protocol slips*] If you're looking for bowel ischemia, there's no need to do a chest—
Body CT fellow:	I don't think so.
Resident:	They have "chest and abdomen." I'm just going to do an abdomen/pelvis. I mean—that's kind of—
Fellow:	Yeah, they probably just crossed the wrong . . .

Judgment is required in deciding whether to administer IV contrast to someone whose kidney function is poor, according to standard blood tests. Client clinicians occasionally request odd things, and protocolling may require the resident to exercise collegial oversight. Such exchanges constitute modest but important educational situations—contributing to the resident's repertoire of consultative interactions, and forming a professional persona within the wider hospital culture.

In Body CT, the resident queries the fellow about a protocol for a pre-transplant CT in a patient with cirrhosis.

| Fellow: | We usually don't do a CT for that. We usually do an ultrasound. Pre-transplant ultrasound, so they can look at all the vessels and . . . |
| Resident: | I wonder if I should . . . phew [*trails off*] |

The resident phones the clinical resident, whom he knows:

Resident: [*on phone:*] Bruce, Don. Hey, not much, man. I need to ask you a question on one of your patients. Um. Mr. Harlan Jenkins. This gentleman is an, uh—alcoholic cirrhosis, is it? Um, have you guys gotten an ultrasound done on him? [*pause*] OK, because um—I was talking to the Abdominal fellow, and he suggested that usually for, you know, for pre-liver transplant, we—you do uh, ultrasound. It, it gives you better vessel patency and—you know it's . . . I mean—

Fellow: They actually want to know the numbers before they go in there too. How wide the portal vein is and all that kind of thing.

Resident: Transplant . . . Are they concerned about a possible malignancy or something?

Fellow: Then he should get an MR.

[*Clerk, perfumed, enters with pile of folders. Leaves after one minute, without verbal exchange.*]

Resident: [*to phone:*] OK. [*to fellow:*] He's, he's—[*to phone:*] Hang on one second—[*to fellow:*] He's saying that they want—he said that the Transplant attending wanted a CT.

Fellow: Ah, then we'll do it. I mean they—he might have something specific that he wants. The person's probably already had an ultrasound, then.

Resident: Yah, that's true. [*to phone:*] OK, I just wanted to get a little more information on that—no problem, we're going to run it . . . [*portion omitted*] [*hangs up*]

Fellow: There's probably some kind of um, specific questions he wants answered.

Resident: [*to fellow:*] OK—is there any particular protocol you'd like to do on that?

Fellow: Yeah, I think you should do a biphasic liver. Which would be the best ah—the best we have for a hepatocellular carcinoma on CT. An MR would probably be a better examination for hepatocellular carcinoma if that's what they're looking for.

Resident: I don't know. He didn't even tell me if they had—if they were suspecting malignancy. He just said that the Transplant attending wanted a CT.

Liver transplant candidates get an ultrasound. MR is best, and biphasic CT next best, for finding liver cancer. Such are the facts, maxims, precedents, and rules of thumb that inform chores of protocolling. But no rules are absolute. And "Transplant attending wants it" trumps most objections.

Dictating

Each CT interpretation first takes form in a verbal report dictated to the RTAS—Radiology Transcription and Archiving System. The recorded dictation becomes available for telephone retrieval by clinician clientele.

The Chief of Body CT enters the reading room in administrator mode. He has a letter for the attending du jour pertaining to a residency applicant. He turns to the fellow, at the left viewbox, and compliments him on his dictations, which he has recently had occasion to review. The fellow replies modestly that from prior training he dictates according to "tried and true" format: "Technique; Findings; Impression."[103] The attending says he has noticed more than consistent structure: "You've obviously taken time to proofread it before it gets to me . . . You've given clinicians something to go on besides just 'there's a lesion in the liver.'"

Occasionally reading room staff receive feedback from client clinicians—usually complaints—about dictations. Neuroimaging fellow: "Let's listen to this 'confusing' dictation—I'm always interested in what people find confusing . . . Sometimes people who 'listen' to your report *don't*." The fellow cannot ascertain what might have seemed confusing, so she calls the clinician to find out. The attending du jour expresses annoyance that she is so solicitous.

Dictating takes concentration. Considerable pictorial information must be assembled in verbal form, in an order that accords with listeners' expectations. Imaging technique must be properly summarized. Abnormal findings must be described: location, size and shape, density. This summary may distil prior inspections and discussions of the images; it may be done from memory, or keyed to notes in the viewbox logbook; for experienced radiologists, it may be a real-time narration of primary image inspection. Beyond the abnormalities, a list of "normal" findings is often recited. This recitation may guide one last inspection of images—for example, according to standard anatomic series: "liver, gallbladder, spleen, kidneys, adrenals . . ."—to ensure nothing is missed.[104] After listing findings, the radiologist dictates a more summary and judgmental set of "Impressions"—what was once called "Interpretation." The arc of the entire dictation—from the circumstantiality of abnormal findings, through comprehensive recitation of normal ones, to summary impression—summons concentration.

In one diction, the Body CT fellow recites a formulaic list of normal organs, including "Gallbladder is not visualized." He then switches the dictaphone off, pauses, and says softly, to himself: "Well, it's pretty well-visualized." Rewind, play, pause; redictation: "Gallbladder is unremarkable." On another occasion, at the Neuroimaging viewbox, the resident catches herself early in dictation of technique: "Multiplanar"—rewinds—"Multiple five-millimeter axial images." Momentarily she forgot she was dictating a CT instead of an MRI and launched the wrong script.

Trainees receive pedagogical advice on how to dictate. Just Findings and Impression, says one attending. "Don't even talk about pertinent negatives." Much advice is particular to a study at hand. Attending: "Just say 'Soft tissue density in the right infrahilar region—unclear if it represents atelectasis, volume averaging, a small nodule, or unchanged . . . ' And note the patient is status post cholecystectomy. So don't say the gallbladder is present."

Questions arise mid-dictation. In Body CT, the fellow interrupts his dictation of a study already reviewed by the team. He carries a film to the attending to ask him about a small lytic area in one lumbar vertebra, previously unnoticed. "Good pickup," affirms the attending—and the fellow incorporates this finding into his dictation.

Dictation is often interrupted.[105] Some radiologists have impressive abilities to dictate in interrupted segments. A fellow in Neuroimaging, with a third-year student at each side, dictates after a brief episode of hilarity:

Neuroimaging fellow:	[*dictating:*] MRI of the brain and MRA of the Circle of Willis dated 10–20–96 period. New paragraph— History: Seventy-eight-year-old male, rule out infarct period. New paragraph—[*pause*]
Student 2:	[*leaning up and peering at board*] Oh dear.
Student 1:	What.
Student 2:	This is not this. [MRA *and adjoining* MRI *have different names*]
Fellow:	Well, that's a problem.
Student 2:	McAdams is not Ronnie . . .
Fellow:	Who am I reading?
Student 2:	You're reading—This is McAdams, this is Ronnie Fitch . . . [*boardbuzz x 3*] I don't know if Ronnie Fitch . . . [*boardbuzz x 3*]
Student 1:	That's Clarence, this is Clarence, all these are Clarence. So—

| Fellow: | This is very odd . . . OK, so now . . . [*boardbuzz*] We're reading McDougal, so this is McDougal's brain, that goes with this MRA, right? [*boardbuzz*] |

After clarifying the elements of the study to be read, the dictation is resumed—though justification for these elements is not evident from the requisition.

RTAS replay:	. . . infarct period. New paragraph—[RTAS *recording re-wound further*]
Fellow:	[*chuckles*] Wonder why they had to get a cervical spine to rule out an infarct [of the brain]?
RTAS:	Chapman on McDougal 75977, MRI of the brain and MRA of the Circle of Willis—
Fellow:	[*dictating:*] And cervical spine, dated 10–20–96 period. [*boardbuzz*] New paragraph— History: Seventy-eight- [*boardbuzz*] year-old male, uh rule out stroke period. [*boardbuzz*] New paragraph— Technique: Ah, multiplanar MR imaging of the brain was performed with gadolinium period. Additionally multiplanar imaging of the cervical spine was [*boardbuzz*] performed period. 3-D time-of-flight [*boardbuzz*] angiography of the Circle of Willis was performed [*boardbuzz*] period. New paragraph— Findings colon: Brain—[*long boardbuzz*] [*aside to student:*] I'm glad you noticed that. We would have been very confused when we got down here . . .
Student 1:	Yeah.
Fellow:	[*chuckles*] Uh, this ugly brain. [*dictating:*] Midline structures are intact period. There's a partially empty sella period. [*aside to students:*] Which is of no clinical significance at all, you just dictate it so they know you actually looked at the images.
Student 1:	On this, is this getting close to Chiari one [classification of cerebellar position]?
Fellow:	It's getting close. And the older you are the less leeway that you have. 'Cause if you're young your [cerebellar] tonsils can hang down five or six millimeters. If you're old, uh, you know, you don't want them to be hanging down more than

like three. Not sure why that is. You'd think they'd get . . .
hang down lower and lower as you got older. [*laughs*]

Student 1: Sag more and more, yeah, you know, gravity.

Fellow: [*singing:*] Do your tonsils hang low? [*boardbuzz*] [*dictating again:*] There is diffuse central and cortical atrophy with hydrocephalus ex vacuo period. [*aside to students:*] That always comes back on your dictation as some weirdo thing.

Student 1: Is that on the left? Is that what you're talking about?

Fellow: Well, just that the vents are big and it's, you know, it's . . . just hydrocephalus secondary to atrophy. [*boardbuzz*] Oooh. Yikes. [*dictating:*] There is a patchy high signal in the T2 . . . images throughout the centrum semiovale comma, periventricular white matter comma, and ah brain stem comma, mostly within the pons period. [*aside to students:*] So he's probably just really old and hypertensive or diabetic or something, you know, some kind of chronic stuff where he's got just demyelination . . . You know, he's got a couple old lacunar infarctions. [*boardbuzz*] And so we're looking for something acute on him. Sort of looking for cortical edema or enhancement, [*boardbuzz*] none of which I have seen yet. [*boardbuzz x 4*] Are we still on the right patient?

Student 2: Um, yeah.

[*boardbuzz x 2*]

Fellow: [*dictating:*] There is no evidence of intravascular or meningeal enhancement period. No cortical edema is appreciated period. No mass effect or intraaxial enhancing lesions are identified period. [*boardbuzz x2*] New paragraph—

MR angiography reveals [*pause*] a patent Circle of Willis period. The basilar comma, internal carotid comma, middle cerebral comma, and anterior cerebral arteries are patent period. No abrupt stenoses or occlusions are identified period. [*aside to students:*] See how they—he doesn't have—I mean he's got all his vessels but there's not like good arborization of any of the distal vessels. He probably has, you know, small vessel disease all over the place. [*boardbuzz*] We had one other MRA down here, didn't we? I could have sworn we

did—didn't we have two . . . Oh, OK. The one that's hanging out with Ronnie? There. [*humming*] [*dictating:*] New
paragraph—
　　Cervical spine colon: [*to students:*] You can already tell his
cervical spine's going to be tight. [*dictating:*] There is high T1
signal intensity throughout the vertebral bodies consistent
with ah fatty marrow replacement period.

The dictation unfolds in a dozen interrupted segments. Just as the fellow finishes, a clinical team enters. They wish to discuss McDougal—the same studies. The fellow summarizes her interpretations. The clinicians ask if a finding
from a year ago, a "syrinx" in the cervical spinal cord, is any better. The fellow
cannot say because a syrinx is not visible on current images (due to high signal
throughout the cord); she has not seen the prior study. The students search
for old films. They are nowhere to be found—even though Film Management
insists the jacket is signed out to Neuroimaging.

Pet phrases and stock locutions recur throughout formal CT reports. These
are beads on a string and aides-mémoire, allowing the radiologist to continue
after interruption, with the prompt of the last few seconds of dictation. With
such prompts, it is possible to interlace dictation of a report with pedagogical disquisitions, interrogations, measurement of lesions, inquiries by visitors,
song interludes, and phone conversations.

CT reports can be among the longest of radiological reports (along with
MRI reports). They address many images. In composing a report, brevity is de-
sirable, but not at the expense of potentially relevant detail. Radiologists have
long considered it their mission to provide comprehensive reports. Many other
specialties read CT scans for features germane to their practices—the neurologist for strokes; the neurosurgeon for vascular anatomy; the pulmonologist for
lungs—but the radiologist's reading is supposed to describe all findings, even
those outside the clinical specialist's purview.

Dictating is not always temporally associated with looking at images. Here
the fellow is looking at the jacket while dictating—absently, eyelids at half-
mast. She has not looked at images for a full minute. Her posture is relaxed:
leaned back slightly, body angled relative to board, legs crossed, left foot wig-
gling, left hand on the toggle.

Once CT reports have been transcribed as text on the RadCare system (see
chapter 4, "Curating"), the text must be "verified" by the dictating radiologist
(and attending) before a printed copy can be sent to the hospital chart or the
ordering physician. Verifying these reports is a chore. In Body CT, a fellow

speculates that every third report requires correction. He has a queue of 100 to sign. (Once he had 500.) The fellow admits that with these studies, some from last week, he has no recall of images; he is proofreading for form, not content.

During one lull in viewbox action, the Body CT attending slides his chair over to the RadCare terminal. "Maybe I'll sign some reports." There are eighteen, dictated yesterday. It takes roughly a minute to read through each report. At the bottom of the screen are "Action" choices: "Verify / Next Screen / Next Report." A single keystroke affixes his "signature" to each.

Backlogs in flow of reports concern all involved in their production. The department has made concerted efforts to have reports typed within a day. Once printed, the report enters the permanent record. The typed report is the ultimate product of the radiological work regime.[106] It is reviewed by a range of overseers: the clinician who ordered the study; consultants who peruse the chart; the CT attending; the chief of service; the chair. (Rarely by patients.) Dictation practice is embedded in layers of *dicta*—managerial, legal, academic.

Radiologists are self-conscious about clumsiness and verbosity of verbal reports—which they feel are nonetheless obligatory. One CT fellow reflects that comprehensive reports may be alienating to clients—and are perhaps suboptimal ways to convey what is important about a study. He believes digital systems will produce new reporting conventions that will render the descriptive text report obsolete. Future reports will include, he thinks, selected images from the CT study, with circles and arrows marking regions of interest.

But circles and arrows will never mark the sites of multiple origin, the polyphonic and dialogical dimensions, of "reading." The certified reading is signed to give it the stamp and authority of monologic authorship.

Writes of Passage

While many viewbox tasks are predictable—and I refer to them under the sign of ritual—neither the flow of images nor the rote aspects of practice quite fits the model of the assembly line. There is variation in staffing, scanner production, case mix—and complexity in reading. CT team members find it is helpful to mark stages of task completion.

Early one morning, the attending in Body CT takes papers from a clip on the right viewbox and begins to thumb through them. She explains she is reviewing the day's scheduled studies, the resident's system for marking, the annotations referencing old studies: "That's their job: they do it brilliantly . . . all the ducks are in a row." She speaks of "dovetailing" her plans. On the basis of her reconnaissances, she expects to attend the MRI conference at noon.

Not everyone tracks workload the same way. One Body CT fellow mounts paperwork on the viewbox along with films. The white PPF, the green requisition, and yellow sheets containing reports on old studies go behind the fishing line with the images. "This way when I come to this—like I don't have to remember anything about Wilma Sondheim, I just come here and it's all up." He has also marked the day's schedule, on the clipboard at the right viewbox, with green and yellow highlighting.

There is also writing on films. A Body CT fellow is measuring lesions with a caliper and marking numbers on the film with a red wax pencil, on the dark strip between slices. He slides his chair to the MagicView, toggles up an image on the same study, and places the cursor on the lesion to measure its density. He slides back to the viewbox and inscribes this measurement at the edge of the film. He does a fair bit of marking on this study, hunching forward. He also writes on the paperwork with his pen. He rubs out one wax pencil mark on the film and makes it again.

The wax pencil is an important tool of radiological film reading, especially in CT. It is used in spine studies (CT and MRI) to number vertebrae. In addition to level, size, and density, the wax pencil marks suspicious areas, findings to be discussed with the attending, lesions to be recalled at the time of dictation.

Body CT fellow:	Do you know what I desperately need? [*gets up, walks to other viewbox*]
Body CT resident:	Wax pen.
Fellow:	I can't read without a wax pen. I have to mark where I've been.
Investigator:	[*chuckles*]
Fellow:	Kinda, maybe it's a territorial thing.

On another occasion, the fellow notes that residents seem to have hoarded all the wax pencils in their study room: "They're like gold."

The mark of the wax pencil signifies the passage of the radiologic expert and cues those who follow. One clinician comments at the Body CT viewbox: "OK, there's the famous radiology 'circle sign'—I can't even figure out why it's there. What in the world is this?" The CT fellow responds, "A lipoma in anterior chest wall." Clinician: "It goes a long way up and down, doesn't it?" Erasable, but sufficiently durable to survive shufflings of film, marks of the wax pencil summon collegial attentions to territories of expert interest.

(opposite) **15.** Logbook, Body CT reading room

Wednesday

CTA/P
&c Pea CA
- ~~tea antiectasis~~ pharyngeal esp fatty liver
- Gastral wall thickening s/w known gastritis
- circumaortic ① LV.
- granuloma ① lung base

CTC/AP
sclerry CA
STD extending from ① peritracheal region @ level of ① subclavian ID cc arteries
inferiorly — surrounds aortic arch medially & laterally
subcarinal max AP dimension 3.0 cm & some around ① mainstem
bronchus   ① sub q nodule inf to ① scapula — NSC
① hilar nodes — NSC — 1cm
① apical density — NSC.   ① apical pl. based density — NSC
① pl thickening — NSC & LLL atx.
  RLL nodular pl. thickening — NSC
2cm nodule post to spleen (node vs. acc spleen) — NSC
② adrenal — 2cm LD lesion — adenoma vs not
Stable 2.5 cm STD post to ① kidney
  L/S — WLN in size

s/p bilat mastectomies
  Fat between chest wall & muscle on Rt
Tiny LDA Lt thyroid ; Lt gland > Rt
No mediast/hilar adenopath
No lung nodule
Linear area density Lt base NSC since 1994
Fatty liver
Fat ant to liver — different pattern (more than before
No lesion liver/adrenal/bud visualized/Spleen
No sign upper abdon adenopath

Lrge mass involving Lt breast, skin thickening
    5cm AP x 7cm trs x 6.4cm CC ; clips in area
      extends toward axilla ; to chest wall muscle
No lung nodules ; No signif chest adenopath
Fatty liver (entire liver not scanned
    Very Ltd views abdn
    Visual spleen OK

Tiny axillary nodes
Clips ant mediastinum + fat
Tiny prevasc. node
No lung nodule
& Gastric wall thickening is incomplete distension
Liver/spleen / Adrenals/ RK — WNL
Tiny LDA LL ~ 75 c
GB present

Another site of writing, throughout the CT reading workday, is the logbook. Each viewbox has one: a large bound volume, like an accountant's ledger. Each CT study is listed with the viewbox frames it occupies and salient findings. These are typically written when films are reviewed with a fellow or attending. The log provides a guide to the board for the day, a table of filmic contents. The list of findings for each study also constitutes a preliminary reading. Briefly, findings in the logbook may be the most definitive interpretation available; and they may be useful in dictating. The logbook summarizes a day's docket—and the cumulative caseload at a viewbox: a summary available for review by researchers and administrators for a variety of purposes.

Abbreviated findings in the log resemble what is written on the film jacket. Both inscriptions, logbook and jacket, are often made around the same time—though they are not identical, nor are they always made by the same person, or in the same order. Here the fellow is transcribing into the logbook information already written on the jacket: name and number, type of study, key findings. On another day, the attending is writing in the log, speaking as he writes, for the benefit of the resident writing simultaneously on the jacket, as they review images together.

Writing in the log and on the jacket also captures information conveyed by clinicians. A Body CT attending toggles the board: two panels rightward; three panels left; one more left. She leans back to take a bite of sandwich, then a sip of Coke. The fellow recalls some pertinent history:

Fellow: So the story I got from the clinicians on this guy. Oh, they didn't give me my yellow sheet with all my history on it, uh, uh . . . [*mock sob*] Well, I remember most of it, actually. He had—I don't know the dates . . .

Attending: Write it down before you forget it.

Fellow: He had left upper lobe lobectomy [*measured pace, writing*], dysphagia, tracheoesophageal fistula, at the level of the left mainstem, and they scoped him yesterday and biopsied this irregular looking mucosa in the esophagus. So their question—I mean, they know he's got a malignancy here—they're just wondering if it's a recurrence of his bronchogenic lung cancer, or if it's possibly esophageal—and they know we cannot make that determination, but they just wanted to get an overall view of what's happening here in the mediastinum.

Attending: [*after swallowing*] But he had bronchogenic [cancer]? [*writing as she speaks*]

The attending, like the fellow, speaks at the pace of her pen. The scribal function, writing to remember, shapes viewbox rhythms of observing and saying (and eating).

In the Neuroimaging reading room, one of two fourth-year students is keeper of the logbook, dutifully inscribing findings as the team clusters around the fellow, reviewing studies together. Suddenly the films on the board lurch to the right and keep moving—until the student realizes he has accidentally pushed the corner of the logbook into the toggle switch.

[*long boardbuzz*]

Student 1:	Oh, dear.
Student 2:	[*laughing*]
Neuro CT fellow:	[*laughing*] When I was a resident, that used to happen all the time and I used to feel like such a dork all the time for doing that.
Students:	[*laughing*]
Fellow:	Hoo hoo, huh, there goes the board, now everybody's lookin' at ya like, what are you doing? And [*laughing*] you're like, I don't have a clue. [*laughing*] And the board's just goin' by. [*laughing*] And you're just sittin' there with that stupid look on your face.
Student 1:	Dowee.
Student 2:	You're in charge of the book . . .
Fellow:	And your chairman's lookin' at ya like he can't believe he accepted you into the program. [*laughing*]

The fellow's anecdote is a kindness to the embarrassed student, but it also capitalizes on a slapstick moment: this logbook, tool of archival meticulousness, in the hands of a dutiful scribe, has toggled the assembly line into high-speed reverse, and temporarily derailed the ideal of faithful transmission of knowledge.

Now, at the Neuroimaging viewbox, the resident is rummaging for correction fluid to make a change in the logbook. "Just cross it out," says the attending, irritated. A moment later, the attending asks her, "Are you in the book right now? Could you write down brain 746927 negative?" The resident answers affirmatively, then queries the circumstances of the study. "Fell," says the attending. "What difference does it make? You don't write that anyway." The attending is enforcing brevity and expediency in approach to the log—resisting the resident's inclination toward excessive finesse. The log's role is provisional, even as it punctuates progress toward official closure.

The film jacket, like the log, is a prop for personal memory as well as institutional memory. Just prior to Neuroimaging conference, in informal conversation, the Neuroimaging resident hands her fellow a folder.

Resident:	Did you see this case?
Fellow:	This *is* my case. That's my handwriting.
Resident:	No, that's my handwriting. I just wrote that.
Attending:	You guys have the same handwriting?!
Fellow:	It looks exactly like mine—[*examining film folder*]
Resident:	This is the one—is this the one that . . .
Attending:	Where's your checkbook? [*chuckling*] . . . Clarissa, write me a check.

This amusing exchange overlays standard rationales for film folder annotation (to structure work flow, record a diagnosis) with another rationale: to record personal investment. Territoriality is exposed and accented here by comic confusion. The attending's rejoinder, invoking forgery, gently mocks these presumptions of radiological property rights.

Inscription—on folders, in logbooks, on film—marks progress through the workload of the CT viewbox. Notwithstanding reading's ocular and oral culture—the saccadic attentions of eyes, the intensity of dialogues, the vocalizing of definitive reports—pens and wax pencils and keyboards are crucial tools. Reading at the viewbox is guided by writing, interrupted by it, adjusted by it in its pace and rhythms. Could one say that seeing and saying are driven by writing? This is a difficult question. The primacy of the graph, the mark, is much contested.[107] Perhaps all that enters the field of the seen and the said has already been written on. Perhaps the lesion itself is already a writing, and all the radiologist's compulsive scribblings are reactive, derivative—part of an impossible struggle to catch up.

Looking Up

The Body CT resident comes to work excited about textbook reading he did last night:

Body CT resident (first-year):	It's cool . . . That little section last night just opened up a whole new—

Body CT fellow:	You mean the blue book?
Resident:	Yeah . . . I mean like wall thicknesses . . . mean I've just totally learned the good stuff about how to read those things.
Fellow:	I tell you, that book is just—
Resident:	It's the best—I mean for me, it's a struggle.
Fellow:	You know my opinion of it—I think it's a very important book to read.

Reading CT scans is not strictly an image-centered activity. The reading room is a space for correlating film with text. Textual resources structure film interpretation: radiological literature, patient history, laboratory data—forms of epidemiological and institutional precedent: *context*.

On shelves in both CT reading rooms are collections of books. Some are atlases demonstrating normal sectional anatomy. Others catalog CT appearances of various disease entities.

In the Body CT reading room, Webb's *High-Resolution CT of the Lung* is open on the counter. The fellow has unshelved it to aid in her differential diagnosis of "honeycombed" lung on a high-resolution chest CT. "I kind of use it as a picture—y'know, comparison thing—I just am not good at picking some of these." Here the image findings at the viewbox are submitted to a kind of field guide identification process: a collection of pictures of pathological entities is searched for things that look similar. The book provides a lineup of potential culprits (more on this in chapter 3, "Diagnosing").

An exchange at the Neuroimaging viewbox shows another way of using a textbook. Clinicians have provided a diagnosis—Wernicke's encephalopathy—to the fellow reviewing a brain MRI.

Neuroimaging fellow:	Well, his mammillary bodies are very small. [*boardbuzz*] Um, they're not high signal as we sometimes see, but they're small. [*boardbuzz*]
[*exchange omitted*]	
Fellow:	I don't see anything other than that. He's got a little bit of atrophy for 29. Um, although it's not as significant as I thought we might see in him. His temporal lobes look OK except his mammillary bodies are small. [*boardbuzz*] And he doesn't look like he has anything else going on to account for . . .

[*boardbuzz*] That's about it. He doesn't have any demyelinating lesions or anything. [*boardbuzz*]

[*exchange omitted*] [*clinicians exit*]

Left with her student, the fellow stands and pulls a book off the shelf.

Fellow: See, now that we told the clinicians what it is, now we need to look it up and see what's it supposed to have been.
Student: What are you looking up?
Fellow: Wernicke's. [*laughing*]
Student: Oh, Wernicke's?
Fellow: I just wanted to see if there's anything else. 764. They have this beautiful article in the *AJNR* not too long ago on Wernicke's. [*reading:*] "Nutritional thiamine deficiency . . . chronic alcoholic, classic triad, ophthalmoplegia, ataxia and confusion, blah, blah, blah, distribution involves both gray and white matter, periventricular regions, medical thalamic nuclei, mass, intermediate third, na, na, na, mammillary bodies are most frequently affected, the periaqueductal region, midbrain reticular formation and tectal plate are also commonly . . . imaging findings seen on T2 include hyperintense area around the third ventricle and aqueduct"—we didn't see that, did we?
Student: Nope.
Fellow: Do I believe you?
Student: You shouldn't . . . [*laughing*]
Fellow: [*laughing*] [*boardbuzz*] [*inspecting images*] I don't even see the aqueduct. OK. I see it. A little itty bitty tiny aqueduct. Now isn't that cute?

The inquisitive fellow wished to know from the text if she recognized in the instant images all that Wernicke's is "supposed to have been"—to refresh and supplement her recall. She perhaps wished to demonstrate to students a mode of engaged learning. The text on this occasion did not present a lineup of possible culprits; rather, it offered a list of possible appearances of a known entity—disguises, aliases, *modi operandi.* The instant case was subordinated to the text, to its authoritative description, its checklist of findings and typical images.

The textbook also prompted the fellow to reinspect the images—to discover an "itty bitty tiny aqueduct." Another fieldguide function: exposure of subtle features of the already identified specimen—prosthesis to native vision, a focusing knob.[108]

Texts can also help confirm a speculative diagnosis. Two members of a Pulmonary team come to look at the scan of patient Norris, with cholangio-carcinoma, booked for surgery. The CT attending points out widespread lung lesions: "Can't even count the nodules . . . Nodes . . . chunky here . . . pleural disease there . . . bronchi heavily calcified." The Pulmonary attending is interested: "Sounds like she needs a Pulmonary consult. Oldie but goodie . . . eighty-four [years old]?" He remarks that these "don't look like metastatic [cancerous] nodules . . . She could have MAI." Then the Pulmonary team leaves. When a Pulmonary fellow enters a few moments later, the CT attending points out the same nodules: "I think you've got to think some sort of TB." The CT fellow remembers a pattern of TB called "branching tree." CT attending: "What does that look like?" The CT fellow offers brief recollections. After the Pulmonary fellow departs, the CT attending and fellow undertake a final review of Norris's scan. The CT attending writes on the folder as he speaks: "Small bullae in here . . . [pointing] . . . superior segment of the right lower lobe . . . some adenopathy . . . small upper pretracheal . . . precarinal . . . coalescent infiltrates in the lingula and right middle lobe." Then the CT attending asks the fellow: "Anything else?" The fellow wishes he had a demonstration of the nodular pattern he recalls, involving the end of the "bronchovascular bundle." Texts at hand do not show this.

The CT attending begins dictating Norris's study—and the fellow leaves the room. The resident inspects images as dictation proceeds. As the attending finishes dictating, he remarks, "It'll be interesting to see what this turns out to be."

The fellow reenters with a book in hand and jaunty demeanor. The attending exclaims: "You found it!" The fellow is carrying Haaga's CT and MRI of the Entire Body, his finger on page 683. Images there show "bronchogenic spread of TB" looking much like Norris's CT. The attending remarks to the resident: "He's proud as a peacock." The resident is impressed: "You went to the library and got that, didn't you?" The fellow is pleased indeed: "Glad the old memory . . ." Shared pleasures of textual correlation.

The CT fellow remarks how fine a book this is. He recalls, however, that Haaga "wouldn't give us a copy." This is a bit of name-dropping. Haaga had been chair of Radiology where he did prior training—and has traveled "from one Body CT Chair to another."

In this case, the textbook lent support to diagnostic speculation about nodules on Norris's scan. It justified inclusion of TB/MAI in the list of diagnostic considerations; it would justify the Pulmonary team's pursuit of tissue, via bronchoscopy. TB and MAI are treatable, whereas biliary cancer in the lung could disqualify this patient from surgery. Retrieval of the textbook has also highlighted virtues of this fine fellow—diligence and good memory. Last, the text stimulated some professional "who's who" gossip, forefronting pedigrees and guild politics—and further enhancing the fellow's credibility.

Radiological literature also enters viewbox discourse in the form of the journal. Journals incorporate cutting-edge research that textbooks cannot.

Neuroimaging fellow:	Thanks. Bye. [*hangs up phone*] This is Francis [a Neuroimaging attending] with a request of the medical students. It's a bad one. He wants to know if somebody can go to the library and get this article. It's—let me write it down on something. It's impossible to find paper in here. Yeah [*laughing*]. That's the name of the article. *Archives of Ophthalmology.* I think he said—I think he said volume 111—anyhow it's 1993.
Resident:	Flip for it?
Medical student:	I'll go.
Fellow:	Yeah, I don't know . . . if it's good or bad to go the library.
Resident:	Is it raining out? I'll go. You go.

The reason for the request is a scan in the day's docket. The fellow peruses the images with the resident. When the student returns with the article, the team discusses the images again. Fellow and resident decide the images fit in categories 2 *and* 3 of a five-part classification—and joke about naming a new syndrome after the fellow.

An hour later, the attending who requested the article has joined the fellow and resident at the viewbox. The article is before them, but he does not pick it up.

Neuro CT attending:	So did you guys send medical students to fish out that um, article?
Neuro CT fellow:	Yeah.
Attending:	Does he say anything there?
Fellow:	Not really. It looks like—I mean I just read this part—but it looks like it's optic nerve hypoplasia with class three something or other posterior pituitary ec-

topia. I really do think that's top end of the pituitary. [*gesturing toward board*]

Summoning an article to the viewbox to refine diagnostic judgment is impressive. In the 1990s, MedLine search tools are in the reading room, but full-text journals are not yet available online—so such archival finesse unfolds rarely without efforts of students.

Writing Up

University Hospital imaging cases feed, as well as draw upon, the radiological literature. "Reportability" of cases is a matter of historical contingency. Usually publication requires a series of cases.

Oncology attending:	[*at Body CT viewbox*] You ought to talk to these guys, Staci, about a CT scan study—
Onc fellow:	That's right. I wanted to . . .
Body CT fellow:	I don't know if I'm the one to talk to—I'm actually the fellow who is leaving in a month.
Onc fellow:	Oh. [*laughs*] It's, it's possible. I'm actually looking at the retrospective data—that: There's a few small studies that have done CTs and chest x-rays in febrile neutropenic patients with prolonged fever. Um—chest CTs, and—the CT's clearly more sensitive, um, to pick up fungus.
CT fellow:	Um-hmm.
Onc fellow:	So, for this—I want to look a little bit more at retrospective data, but then maybe design a prospective trial to, at a certain point of febrile neutropenia, get chest CTs and chest x-rays.
CT fellow:	I mean, it sounds like an interesting idea, but how are you gonna pay for it?
Onc fellow:	[*laughing*]
Onc attending:	Yeah, that—that'd be another story . . . You—you—I mean, I think you could get it paid for. I don't think insurance is going to balk at getting a, a chest CT. I mean, I don't think they've ever—I've gotten three on this lady in three weeks and I don't think anyone's gonna call me and tell me that I'm not . . .

Onc fellow:	Yeah. I mean, it's like doing it at a point where it's reasonable.

[*brief segment omitted*]

Onc attending:	Who pays—who does—Sherri, who makes the IV contrast? Like, who—who makes money besides insurance companies off CT? Is it like Mallin—who's the . . . ?

[*brief segment omitted*]

CT fellow:	Mallinckrodt is one, but there's others besides. I think Mallinckrodt is the one that has—I don't know if they have Omnipaque or Optiray. There's different types. Do you know, Jim? Do you know if Mallinckrodt does Optiray or Omnipaque?
CT resident:	Uh, I thought they did both.
CT fellow:	Do they do both? [*laughing*]
CT resident:	I thought they did both.
Investigator:	Who paid for your dinner last night?
Onc attending:	Who paid for your dinner last night? Mallinckrodt?
CT resident:	No, that was not Mallinckrodt, that was um, Burlex.
CT fellow:	Oh, that was MR.
CT resident:	Yeah.
Onc attending:	I would—I mean if Bill and Hank go for it—you could talk to them too, Kayla—but I would see if you could perhaps get some people who would be interested in enhancing the number of CT scans done in the world to pay for this stuff.
Onc fellow:	OK, sure. Thanks.

[*team leaves*]

CT resident:	[*to investigator:*] How do you know somebody paid for my dinner last night?

The Oncology fellow is enthusiastic about an understudied question: how does chest CT compare with conventional x-ray in evaluating fever in cancer patients with suppressed immunity? The CT fellow understands the place of such a study. To "design a prospective trial" is a high ambition: rigorous enrollment criteria, randomization, careful counting. Of course the call of the unanswered question and the bright light of method are not their own sole rewards. Academic clinicians are expected to publish, to contribute to the fund of medical knowledge. They are expected to raise questions such as these—and, in answering them, to affect the fates of the immune-suppressed, control of fungus, decisions of Oncologists, flows of insurance dollars.

Crucial problem of research design: how to pay? In this case the problem is the expense of redundant imaging. Not all understudied questions are grant-worthy. Companies "interested in enhancing the number of CT scans done in the world" may help.

The other concern here—signaled by the resident's question, "How do you know somebody paid for my dinner?"—is that of research subjection: being identified and observed. This resident, more accustomed to watching than being watched, wonders why an ethnographer should remember his dinner plans last night.

Sometimes case series for University Hospital research are assembled retro-spectively. "In two or three months we've seen a couple of cases or three cases of this, and I say, 'Well, I haven't seen this in the literature' . . . Well, I usually give the residents—tell one of the residents—you know, 'Go, run a search' [for a few more cases]." Development of such research depends on recall of recent journals and gaps in the literature, and inquiries of institutional archives, in dialog with viewbox work.

Institutional case volumes can create conditions hospitable to research, or not:

Investigator:	What about the issue of [clinical] productivity, at the level of studies per day, in an academic institution, in tension with one of the things that you seem to have really done so well, which is to generate contributions to the literature?
CT attending:	Well I think it's a—I think maybe it's a delicate balance. You need one thing to do the other one . . . You know if you're not generating enough cases . . . you cannot do any clinical research, and if you're generating too many cases, then you don't have time to do it.
Investigator:	Yeah.
Attending:	So it's a very delicate balance. For example here, I think we're in the perfect situation, because we generate, you know, forty, forty-five cases a day—so that really, you know, gives you enough clinical work to make it fun, but enough free time to be able to pursue other things. I know that people

who are in busy places . . . Sandra Newlon, who was a fellow last year, she went to Vanderbilt, and I keep asking, you know, "When are we going to do a project together?" And she says, "No, I don't have time." It's way too much work on a daily basis. And at other institutions, many institutions' promotion and free time is based on . . . the money [patient revenue] that you bring into a department, no longer on your productivity as an academician.

The same attending also notes how research agendas generally fit well with radiological work, as distinct from more extended casework of some other clinicians:

CT attending: I mean [in radiology] you do a case, you look at it, you understand it, you dictate it and it's over. Which is very different, for example, from clinical care, where the problem seems ongoing and constantly changing . . .

Investigator: Right.

Attending: So I think that in a way, we are more similar to the surgeons than we are for example to the internal medicine people in the way that . . .

Investigator: Being able to sign with a flourish and . . .

Attending: It ends. The problem . . .

Investigator: Next paper, next case.

Attending: The problem ends, next paper, next case.

[brief segment omitted]

Attending: I mean here I have ten or fifteen cases or something and want to publish them.

Investigator: Right.

Attending: They're [tissue-]proven, or I'm going to do this type of sequence, I'm going to try to image this this way in twenty patients and see what happens—and I analyze it. It's much easier, I think.

Investigator: Establishing the bounds of your project.

Attending: Yes, of your project. And it's much easier to do it in radiology, and that allows you—I think that if you exploit that, that allows you to publish considerably.

The brevity and closure of the radiological case facilitates closure of the case series. Whereas the clinical case remains emplotted, difficult to contain, always hinged to another phase of experience or evaluation, the radiological case is isolable, countable. Radiological cases fit well with logics of inclusion, exclusion, and computation that govern much research. But research is still not easy.

CT attending:	I think it really takes . . . about ten years of hard work to get noticed and recognized [as a researcher] by the rest of the community. I mean you cannot do it in two or three years. It does really take—it's about a ten-year period. I mean that is when the majority of your productivity and promotions . . .
Investigator:	Starts to come to—
Attending:	Yes.
Investigator:	It takes several years just to learn a system and begin to get . . . the machine oiled?
Attending:	Yes, it does take about ten years. And people think that because one writes 150 papers, you can just sit there in the computer and just type the paper from the top of your head—and that is not the truth. I think that I myself, and many other people that I've seen that are very, very productive, struggle with the papers, struggle with the structure of the sentences, and it's—
Investigator:	Time consuming.
Attending:	It's a time-consuming situation. And it's not easy to write a paper. It's not easy to correct a paper that a resident has written, so . . . It's a lot of time, you know, but it's a very—it's a very rewarding issue.

Good research takes years and help. Papers require close reading and hard choices, of words and images.[109]

And interludes between episodes of film reading. Such interludes are not so much found as made—by trainees who maintain work flow and afford attendings the prerogatives of "overreading." The extent to which interludes favor research also depends on proximities, for example, of offices to reading rooms. Research papers are assembled in and among heterogeneous spaces.[110]

Apart from external conditions, academics and nonacademics differ with respect to enjoyment of research: "Some people get a lot of enjoyment of seeing the paper, and you know, doing the page or galley proofs—whatever they call it nowadays—and then seeing the paper published—and not everybody can understand that."[111]

One Body Imaging attending presents a paper in noon conference on MRI's superiority to CT in certain contexts. I ask why he called it "controversial." He responds: "People don't want to hear there might be something better than what they've built their careers on." Later that day in the CT reading room, he finds patients to enroll in a study. "This patient is eligible for a free MRI . . . [because of these] liver lesions." He will pass on their names to an assistant. His research subject recruitment is piggybacked on diagnostic services—but driven by preference for another diagnostic modality located elsewhere.

Recruiting patients for research is a familiar viewbox agenda—though viewbox readers need reminding. The MRI spectroscopist (a Ph.D.) is in the Neuroimaging reading room, enthusiastic about a new way "to get rid of retroorbital fat" on MRI. He wants to enroll a couple of students in a small study. "Done it with phantoms . . . but the proof of the pudding is to do it on humans." The students are pleased with an opportunity to be scanned. It is a novelty, and they will understand better what patients go through. Their images may find their way under the MRI viewbox, into one of the slots labeled "Research films—do not remove."

Diagnostic work at viewboxes in University Hospital is intercalated with research agendas. Film slots designated for diagnostic studies adjoin slots designated for research studies. From slots like these, cases flow out of the reading room, to be accumulated and processed, connecting bodies of patients to the corpus of radiological literature.

Reprise

The CT reading room is a visual space, a home and a school to the diagnostic gaze. The visual crafts practiced there are acutely presentist—absorption in, attention to, films and viewscreens at roughly arms'-length—yet are informed by old conventions and traditions.

Visuality in diagnostic practice is phenomenologically complex. Seeing for oneself, autopsis, is twined with hearing, pointing—and existential confidence is twined with dialogue and rhetorics of persuasion. The itinerary of a radiologist's gaze turns out to be saccadic and not always consciously controlled. The radiological gaze is not always confident: it is expectant, searching, somewhat anxious, reassured by friends (the normal, the nameable), alert to confusions between findings and artifacts. Though films on the viewbox are thoroughly reified as specimen objects, now and again something "catches the eye" of the radiologist. Radiologists are not the only agencies in the reading room. And looking, however intensively rationalized, is inflected with a range of aesthetic responses.

The famous transparency and two-dimensionality of the radiographic image are complex. Thin slices, which make some things "easy," have not quite eliminated depth. Even though CT disimpacts superimposed shadows, the radiologist is still challenged to reintegrate slices into a volume as part of "interpretation." So hermeneutics, the old practices of revelation, penetrating the signifier to reach hidden depths of significance, are not so much eliminated as they are inverted: the depths are already revealed, and if one wants back the old humanist signifier (the body) one can (with practice) reconstruct it.

Neither the grid nor the section, defining exhibitionary conventions of CT viewstations in the late twentieth century, is new. They are modes of representation which have been naturalized across a range of disciplines since the nineteenth century, some rather unexpected: geology, high modernist art. Aesthetic effects deriving from these devices are complex. But proliferation of grids of slices in popular media depictions of medicine suggests that there is a certain potential for commodification of scientism in this combination—apart from its diagnostic utility.

Amid visual and cognitive work of knowing and unknowing, a great deal of handwork is evident. Operating the viewbox, using the dictaphone, writing while speaking, shuffling films while chewing: these practices define chains of evidential custody, mark passages and territories among and across cases. Sometimes they enact hierarchies—as when the attending picks up the phone and hands it to the resident to answer. Many of these practices exemplify expertise. Some are described in scripts and protocols—how to reboot the MagicView. Others are emulated, imitated. Many of these practices are constitutive, not derivative, in the theater of knowledge making.

Despite the restricted situation of the reading room, much flows in and out. Images are "shipped" via the MagicView; films are carried in and out by couriers and others. Voices are mobilized via phone, paging, and dictation—and in the persons of visiting clinicians and techs, students and nurses. Textbooks and articles are carried in from the library. These movements are only partly predictable. Portions of these flows are intensely monitored for timing, thoroughness, accuracy—in logbooks and dockets, tables and printouts—by overseers of various kinds.

The reading room is, for all its insularity, a busy place. It is crisscrossed by instrumental imperatives of many kinds—production of readings and reports, retrieval and communication of old knowledge. But instrumental utility (and its blockages) is too limited a frame to account for the range of practices indigenous to the viewbox. The reading room is also a locus of guest-host manners, joke telling, gossip, singing, career advice, meal taking, vacation planning, and more. I have included much of this within the frame of my representation, if not always explicit analyses. This is partly a matter of "role distance" (next chapter): radiologists and residents are absorbed in their work, but they are not ascetics. These activities are sometimes superimposed, sometimes mutually interruptive. And there are the interjections of visiting clinicians, addressed further in chapter 5, "Testifying and Teaching." All these crisscrossed practices and agencies involve hierarchies and divisions of expertise, intensified by pedagogical agendas. They make reading and writing themselves suites of practice.

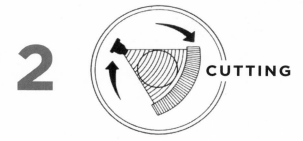

2 CUTTING

At length he approached the huge trunk, walked slowly around it, and examined it with minute attention. When he had completed his scrutiny, he merely said, "Yes, massa, Jup climb any tree he ebber see in he life."
• • • **EDGAR A. POE,** "The Gold-Bug"

CT scanning materializes a circumspection. The scanner's ring defines the path of a spinning x-ray tube. Yet the center of this orbit—a table—is not so much a center of gravity as a locus of substitutability. Specimens on it may be dense or frothy, flesh or phantom.[1] Any interposed specimen modifies the rotating beam. CT's "cutting edge" is not a hard one: the beam is "attenuated," absorbed.

"Nowadays getting a CT scan can be the easiest thing in the world. The scanners are quick—patients don't have to lie there that long—the beds are more comfortable—they [patients] are in and out!" A CT tech compares now to the old days. And surely, scanner tables of any sort are better than the autopsy table, site of postmortem cutting, the originary practice of lesion finding.

In other practices of spin cutting, the object is usually spun. With spiral slicing of ham, cutting by the lathe, at the potter's wheel: the object is spun and the cutting edge is fixed. This is inverted in CT.

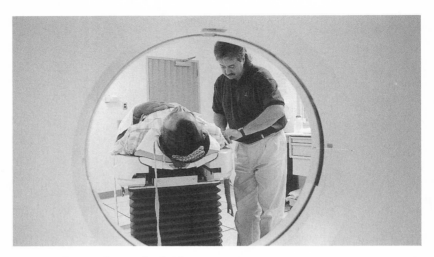

16. Gantry bore, patient, and tech: Neuro CT

Spin cutting has little gestural kinship to the probing of a knife or the chop of the guillotine. Nothing like "Brain Cutting"—weekly ritual slicing, upstairs in a different building, across from the Morgue—wherein pathology residents and faculty cut the week's neuropath cases—in plastic smocks, with broad knives, over stainless tables. Brains are cut linearly—no spinning—on metal pans. One hand of the pathologist cups the top of the specimen; the other draws the blade through it horizontally, parallel with the table. Hand-cut slices vary in thickness.

CT has no blade, but a ring—defining movement of two devices. Within the ring, x-ray *emitter* and *detector* both orbit, reciprocally and opposite one another, around their common center.[2] This reciprocity of motions within the CT circle materializes an embedded genealogy, a ghost in the machine, from the early history of tomography, when the x-ray beam was first put into motion.

Tomo-Technique: Black-Boxing Gestures

Poe would have been fascinated by a device for cutting space rather than mass—"stereotomy."[3]

In the years 1914–17, several radiologists, in independent contexts, undertook to move an x-ray tube during a single exposure. One was trying to blur rib shadows to improve images of the heart.[4] Another, an Italian military doctor, moved his fluoroscope to localize embedded projectiles.[5] These movements were pragmatic and relatively unsystematic—extensions of ordinary handwork of tube- and plate-placement. (One imagines contortions to manage these move-ments with two hands, or countdowns to synchronize with assistants.)

In 1921 a French physician, André-Edmund-Marie Bocage, submitted a patent application for a more precise system of motion imaging based on concepts he developed in military service. His innovation—"analytical radiog-raphy"—involved mechanically coordinated "translation" of radiographic tube and a sensitive plate, "such that their primary displacements are always syn-chronous, parallel, of opposite direction and in a ratio of constant magnitude. Under these conditions there exists between them, in space, a single fixed plane in which each point always has a corresponding image point on the plate; hence, only the organs contained in this plane are in focus. The other organs form only diffuse shadows."[6]

Bocage's machine of "*radiotomie*" was not operated until 1938 (for want of a manufacturer); others claimed temporal priority in the invention of similar focusing techniques. Analogous techniques were developed in several contexts independently, by workers in different languages, disciplines, and nations, largely unaware of each other. Their devices all reciprocally translated an x-ray source and plate during an exposure, to define a plane of focus. Several were developed for localization of foreign bodies, in military settings.[7]

The lapse between Bocage's patent and its materialization in an apparatus signals that radiotomie was more than an extension of ordinary handwork. It involved an elaborate machine for precise coordination and fixation that mere handholding could not afford.[8] In standard diagrams of classic tomographic principles, the motions of x-ray tube and film are linear—but they could also be arcing, elliptical, or spiral. In any case, positions and gestures were "black-boxed" (integrated into the machine).[9] Radiotomie and its cousins were tech-niques of *mechanical* reproduction.

Assimilation of such machines to common geometric principles obscures interesting differences. Among these are the *names* assigned to respective tech-niques. In Italy, Allesandro Vallebona—credited with the first device to pro-duce an actual sectional image, in 1930—termed his technique "stratigraphy."[10] A Dutch pioneer, Ziedses des Plantes, termed his method "planigraphy." Not until 1934 did a German manufacturer patent the "Tomograph" and therein coin the most enduring term. America's earliest machine (1938) was dubbed the "Laminagraph."[11]

Different names imply different conceptions of the machines' spatial-perceptual targets: strata, planes, laminae, slices. Differences addressed in his-tories of tomography tend to be mechanical. Practices are elided—what was

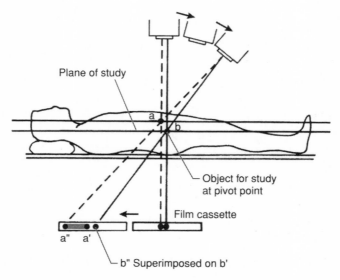

Plane of study

Object for study at pivot point

Film cassette

a" a'

b" Superimposed on b'

17. Conventional tomography: plane with *b* remains in focus while *a* is blurred

grasped, moved, transferred in use—issues of gesture and timing. Rarely are hands depicted in diagrams of tomographic "technique."

Consider the versatile tomographic apparatus patented by William Watson in 1936–38. This machine could make either a tomographic image (moved during an exposure) or a stereoscopic pair of images (exposures before and after movement).[12] Practical interventions—with the *same* machine—produced very different images, for different modes of spectatorship. This exemplifies limitations of merely "mechanical" historiography, without consideration of craft-practice and usage.

A fine supplement to standard genealogies of tomography is Lisa Cartwright and Brian Goldfarb's exploration of early photographic and protocinematic experiments.[13] They cite the French cinematographic pioneer Louis Lumière, who in 1920 (before Bocage's patent) described a new technique, "*photo-stéréo-synthèse.*" This involved moving a photographic camera lens and plate during an exposure, coordinated in relation to the object being photographed, such that the focal plane of its image was restricted. Only portions of the object in that plane remained in focus; portions behind and before were blurred. A series of exposures for a given object, with new focal length for each exposure, yielded a series of receding object-planes. Serial exposures could then be laminated as "a recreation of object-space"—an exploration of depth quite different from perspectivist views.[14]

Lumière's experiments forefront more than mechanism and geometry. Adding his work to the genealogy of early tomographic devices highlights again the importance of embodied (craft-practical) gestures along with machinic (engineered, black-boxed) movements—and of representations' effects on modes of viewing.

Coordinated movements of x-ray source and film around a pivot point, articulating a section: these features of early tomography remain definitional of CT. However, section planes in CT are geometrically different. In conventional tomography, the x-ray source and film lie outside the section; in CT, the x-ray source, detector, and pivot point are all within, and define, the cut.

Cross-Sectional Tomography: Orbits and Diffusions

Movements of tube and film in the earliest tomographic innovations were linear or arcing, not circular. Nor did they produce cross-sections ("transverse" or "trans-axial" sections).

In 1939, an Englishman (Watson, noted above) patented a device that could produce a cross-sectional radiograph of a rotating specimen.[15] Other contemporary devices made similar cross-sectional images. In 1940, a Hungarian patented a sectional imaging system involving optical reconstruction from projections—a method that anticipated some core mathematical operations of CT.[16] In 1945, a Japanese radiologist (Takahashi) developed devices which produced axial sectional images using optical back-projection.[17] By the 1950s, transverse sections were being produced in various parts of the world, by a range of different devices, some commercially available. All used some combination of rotational movements of specimen, x-ray source, and x-ray plate.[18] Development of these machines required corporate investment and more elaborate production facilities than did early tomographic devices.

The history of cross-sectional radiological tomography is, like that of plain tomography, one of contestations of priority. But one early experiment with 360-degree imaging bore no explicit relationship to x-rays—and is unmentioned in standard histories of CT. This is the motion-photographic experiment in 1925 (also recounted by Cartwright and Goldfarb) which the poet E. E. Cummings proposed in correspondence with the physician and filmmaker James Watson. "Seeingaround" entailed spinning an "objectsubject" in relation to a fixed camera, or putting a camera in orbital motion around a fixed objectsubject, or some combination. Cummings suggested that such reciprocal movement "actually means, or constitutes, 'solidity' on the part of the subject."[19] Though his camera movements were not intended to cut, per se,

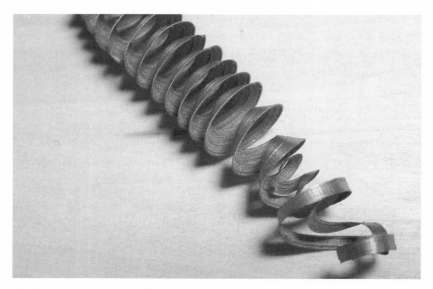

18. Metal shavings from a lathe

such motion-photography is, like tomography, an embrace of blurring in the service of imaging spatial volume. Like radiologists, Cummings was concerned with alternatives to perspectivist "behindNess."

Cartwright and Goldfarb assimilate interests of Cummings, Lumière, and early tomographers to a shared pursuit of "volumetry" and antagonism to "lenticular optics." But for present purposes this understates the importance of cutting—Poe's "stereotomy."

Recall the remarks of Benjamin (in chapter 1, "Reading and Writing") about the "optical unconscious"—referring to a spacetime beyond awareness, excavated by motions of the filmic apparatus. The mechanism that is for Benjamin most emblematic of modern "mechanical reproduction" is not the lens, camera, or film, but the *shutter*. The "click" of the shutter "gave the moment a posthumous shock."[20] (More on this in chapter 3, "Diagnosing.") Both the violence this figures and its unconscious communion with time-beyond-death are crucial to grasping the mechanico-gestural cutting of CT.

When Godfrey Hounsfield built a prototype scanner at EMI in 1970, the x-ray was secondary.[21] More basic were precise motions: rotation, and incremental translation of energy source and detector along the axis of rotation. For this Hounsfield used an old machinist's lathe. This cutting tool of the woodworker's, metalworker's, and lensgrinder's shops predates steam and required, well into the industrial era, continued adjustment of blade in contact with spinning object.

In Hounsfield's prototype, the lathe was not used for spin-shaving—but it did rotate a central object. A phantom (later, anatomic specimens) occupied the position of the shop worker's dowel. A gamma beam source occupied the position of the blade. There was no craftsman's varying of position and force of the blade: the gamma source remained at a fixed distance from the rotated object—and moved incrementally along the axis of rotation. No touching of cutting-tool to specimen: touch was specifically preempted, relegated to a standard distance.[22]

This standardization of distance remains paradigmatic for CT. It is black-boxed into the instrument. CT's pre-scribed, mechanical not-touching of the body (some scanners turn off if flesh contacts the gantry) is unlike, in comparison, medical ultrasound—which requires constant contact of transducer and flesh, frequent adjustment of position and pressure. The fixed CT beam/detector orbit materializes at best a *mimesis* of handling. There are honorable antecedent devices for not-touching in diagnosis. One is the stethoscope—which seemed to some of Laennec's contemporaries to disrupt older, intimate modes of auscultation. Many other diagnostic technologies obviate the need, or diminish the inclination, to touch patients.[23]

Radiologists and some historians might insist here that what is important about CT is neither geometries of orbiting nor Hounsfield's lathe. They would point out that crucial developments were mathematical—computational methods for "back-projection," or reconstruction, of an array of absorption coefficients from a series of line integrals. These were the accomplishments for which Hounsfield and Cormack shared the Nobel Prize for Medicine in 1979.[24] Indeed, mathematics of tomography have long transcended the clinic, x-rays, and dimensions of human specimens: they now link techniques that span many scientific disciplines and substrates—geological, astronomical, industrial, medical—and energy types—x-ray transmission, nuclear emission, magnetic resonance, optical transmission.

This brings us to one last form of motion which is often invoked to describe CT's situation: "diffusion."

CT's diffusion across clinical landscapes started in England, with Hounsfield's work at EMI and with neuroradiological collaborations at the Atkinson-Morley Hospital. From there CT was exported to North America and elsewhere. CT scanner installations across the United States mostly preceded evidence of clinical utility.[25] CT's costliness did not deter its entry into medical markets: mere availability of the technology produced desire and demand. The famous skepticism and rigor of medical men were enchanted by new images and their promise.[26] Indeed, the choice of the United States as the landscape for CT's diffusion was made in early marketing planning by EMI,[27] whose executives knew that the British Department of Health and Human Services was not going to fund its early use, but expected evidence of utility to follow once U.S. entrepreneurship put the machine to work.[28]

Through the early 1980s, there was alarm in the United States about spread of CT scanners. This coincided with steeply rising costs of healthcare. The National Health Planning and Resources Development Act of 1974 required states to implement Certificate of Need laws, restricting new scanner installations. This succeeded in curbing some scanner acquisitions, but accelerated others, and contributed to maldistributions of access.[29]

Diffusion implies passive drift, from higher to lower concentration until impermeable borders are reached. The term is analytically weak.[30] It elides the economics of desire and anxiety—public and professional—which have defined America's, and now the "global marketplace's," love affair with images. It elides significant maldistributions of scanners and scans by nation, ethnicity, class, and insurance status. Twenty-six-million CT scans were performed in the United States in one year of my fieldwork. The United States is second only to Japan in scanners per capita.[31] Distribution of scanners worldwide, particularly of new machines, remains uneven. And the pinnacle of "diffuse" CT usage—screening, or "preventive," scans for otherwise asymptomatic persons[32]—fiercely resisted by third-party payers—will serve only the worried affluent in the near term.

With respect to the history of CT scanners, which I have barely developed: while it would be foolish to deny the importance of Fourier mathematics, faster computers, and better x-ray tubes and detectors to the evolution of CT scanners, it is important to recognize the most radical innovations which these instruments materialize and rematerialize: the replacement of touch and handling by orbiting at a fixed distance, the replacement of the blade by the attenuated beam, and, after Benjamin, the "cut" itself—which, however noninvasive, still "gives the moment a posthumous shock."

Apparatus in Hospital Spaces

In most hospitals, a CT scanner has a room of its own. Scanners tend to be left in place for years. They are heavy machines (though some move among clinics and small hospitals on special trucks). Patients and radiologists come to them.

This is one sense in which CT scanners exert a kind of gravity in the hospital. Like the lights and table of the operating theater, or large archives—elements of modern medicine's "workshop complex"[33]—they structure circulations of materials, persons, and practices.

CT's special place entails little architectural fanfare. Patients find the CT check-in desk by reading signs on the wall, but they might not know the scanner room door even if standing beside it. At University Hospital, the Body CT scanner room has a door just like most of the others on a white hallway. There is a radioactivity label on the door, and a light fixture in the corridor ceiling above it that says "X-RAY IN USE"—but is never lit.

At University Hospital, there are two main scanners, with distinct tasks. There is a Head Scanner and a Body Scanner—formally, Neuro Scanner and Body Scanner.[34] Until 1988, two scanners in adjacent rooms served a single patient stream—a double-barreled setup. Then they were moved a few hundred feet apart and began to separate head cases from bodies. For a while the scanners remained on the same floor. In 1997 they are in separate buildings.

Radiology nurse:	This is Body and that's Neuro . . . And you don't send a head up here.
Investigator:	Right.
Nurse:	Those things don't happen. Even if they don't have—
Investigator:	Bodies'll go there after hours, and . . .
Nurse:	Oh, everything goes there after hours. But during the day, there is turf.

A longer tale of the head/body division of CT begins with the temporal priority of brain imaging: historically, the head was scanned first. The first CT apparatus was constrained to head applications on technical grounds: the head could be immobilized for the long exposure times required; also, the first experimenter with the EMI prototype apparatus was a neuroradiologist.[35] Several years passed before scanners were successfully applied below the neck. The brain has been constitutive of CT agendas.

In the early 1980s, when body CT scanning was "diffusing" across the United States, many radiology departments scanned bodies and heads in the same scanners and assigned reading of all CT scans to the same radiologists. Dedication of scanners and reading rooms to heads or bodies is (in 1997) barely a decade old.

At University Hospital, the Head Scanner is in the Neurosciences building. Because this building also houses the Emergency Department, this Neuro scanner handles emergency CTs as well—including bodies. The Neuro Scanner is the Emergency Scanner, the after-hours scanner—the "high-volume" scanner for the entire hospital. X-ray tubes have shorter lives there. The Radiology Administrator explains: "CT people still want to divide things up the way the radiologists read . . . The Neuro guys want to read all the head cases . . .—no, excuse me, spines and heads . . . And the Body people want to do everything else. Chests, abdomens . . . Any cavity parts. So what we try to do is keep the studies where the physicians are who are going to be doing the readings. We picked up the Neuro reading room when we opened Neurosciences [building] and moved it to Neurosciences." The Neuro/Body split follows divisions of readerly expertise. The Chief CT Technologist (who works for the Hospital rather than the University) confirms the scanners are the same, and techs rotate to both: "We were hoping they would put the two scanners side by side and then it would increase the throughput tremendously." So not all economies of effort and throughput accord with the division Head/Body: this division pertains to skills of faculty.

I ask the Chief of Body CT how his unit fits in the department. He explains that the Chair reorganized the department six years earlier along the lines of the Mallinckrodt Institute. This meant anatomically based "Sections"—Chest, Abdominal, and so forth. Body CT became a "Service" within the Abdominal Section—as is Ultrasound and Body MRI. Though the Chief of Body CT is a chest radiologist and a member of the Chest Section, he reads abdomen CTs and MRIs.

So Body CT is a service, enlisting faculty representing different sections (Chest, Abdomen). And each section comprises faculty who read films on other services besides CT (MRI, Ultrasound)—as well as some faculty who do not read CTs at all.[36] The Department's division by body part or organ system is crisscrossed by divisions of labor according to imaging modalities.[37] "Section" and "service" are twined units of departmental structure in which faculty claim multiple and differing citizenships. "Body" turns out to be a coarse designation for describing the expertise of CT radiologists.

Another issue that informs Neuro/Body specialization is that of hygiene, of purity and danger.[38]

| Investigator: | Logistically, things are crisscrossing between both scanners. Plenty of bodies flow through the head scanner. |
| Neuroimaging attending: | And to tell you the truth, we don't like that. I mean we would like for it to be the Neuro scanner. |

Investigator:	Just heads—
Attending:	Yes, just heads. For example, another issue is the Angio suite. That gets used by the Body people, but we don't like that. We would like for it to be the Neuroangio suite exclusively.
Investigator:	Because your work would be cleaner?
Attending:	Yes, yes . . . I don't like going in there and doing a cerebral angiogram when I thought that the patient before that had a drainage of an abscess on the same table.

Heads are cleaner than Bodies. In part this has to do with oral and rectal contrast administered for Body procedures. Neuro CT procedures—for example, CT myelograms, which involve placement of a catheter alongside the spinal cord—require a clean environment. So the head/body split in CT, historically related to the fixability of the skull and the immobility of the brain, widens in teaching hospitals because of divisions of medical school faculties, and issues of plumbing and cleanup.

Scanner Room Flows

Each scanner room has three doorways: one for "stretcher patients" from inpatient wards; another for "walkie-talkie" outpatients, from a waiting area down the hall; and a third, separating each scanner room from the adjacent "control room," for techs. The patient doors stay closed during scans and open only for ingress or egress of patients. The door between the scanner room and control room stays open. The control room has its own door to the hall—used by techs when they are not escorting patients, and by visitors to the scanner.

Separate doors for inpatients and outpatients help insulate these two patient "streams"—in view of inpatients' extra tubing, postures of discomfort, and attendants. Separate entrances are a consideration for patient comfort (and economics, via customer satisfaction).

At each scanner, outpatients are checked in by a receptionist and directed to a waiting area. They are eventually met by techs and escorted to the scanner room. Inpatients are wheeled by Transport staff. If stable, these patients are "parked" in the hallway (at Body CT, in an alcove)[39] and left until techs are ready to "get them on the table." Techs can see the hallway/alcove on a video monitor in the control room.

Patients constitute the material stream whose ingress and egress to the scanner room most closely corresponds to the scan itself. Other materials arrive in batches, variably timed: linen, syringes and needles, gauze and tape, bottles of contrast. There is outward flow as well: dirty linen, trash, "Biohazard" containers (for "sharps," etc.). The rooms are cleaned daily or more often, depending on need. Some flows are more infrastructural: air exchanges through special ventilation systems, water, electricity. Techs and nurses watch over contents of the cabinets, to ensure there are no delays for lack of materials.

Techs are the "host" occupants of the scanner/control rooms. Most non-patient visitors are radiology colleagues and staff, clinicians and students, and service personnel. These all enter through the control room door. Entry to the scanner room from the hall is largely reserved for patients. (In some hospitals, opening a scanner room door while a scan is in progress will stop the scanner.)

If the control room door opens and someone enters from the hall, techs notice quickly—the room is small. Familiar visitors get only a glance or a nod. The atmosphere in the control room is usually informal: greetings are subdued and matter-of-fact. "Hey—what's up?" a tech asks a nurse. Techs introduce themselves by first name and address residents they know, and some attendings, by first name as well.

Patients almost always enter the scanner room in company of a tech. The tech greets inpatients on the threshold or in the hallway, and steers the stretcher or wheelchair; outpatients are greeted in the waiting area. Indeed, most outpatients arriving at the Body Scanner have been greeted forty-five to ninety minutes earlier—when they started drinking contrast.

Sometimes patients in the waiting area are addressed collectively, with a name spoken as a question. "Mr. Stanford?" A man looks up from conversation. "Hi, I'm Zeke, one of the CT techs. You're here for a scan of your neck today?" The paperwork says what scan is to be performed (and usually why) and identifies the patient's doctor. Greeting a patient may involve announcement of the tech's name, or a handshake. On the way to the scanner room, the tech may ask if this is the patient's first scan and inquire about allergies. He asks cooperative, fit patients to lie on the table; he may enlist the help of a colleague in moving debilitated patients. The ambience is friendly, but motions are crisp and conversation economized.

Tech: How ya doin'?
Patient: Well, I wouldn't be here if I was good ... [*wan chuckle*]
Tech: Any diabetes?
Patient: No.

Tech: Any problems before with x-ray dye?
Patient: No.

Scanner and Props

The scanner room is fifteen by twenty feet: white, with acoustic tile ceilings and linoleum floor.[40] Its central feature is the gantry of the CT apparatus. This is the six-foot-tall box into whose bore the patient is conveyed. A horizontal table extends from one side, to support the patient during the scan.

The gantry, containing x-ray tube and detectors, is sometimes called "the doughnut." It can be tilted—defining a plane through which the patient's body will pass—the plane of "slicing." One can discern this plane when a thin laser beam is projected on the patient's body inside the gantry bore.

If tomographic slicing is a form of mechanical violence akin to the click of the photographic shutter—what marks it? The push of a button? Whirring at the gantry portal, at the plane of section? Likely the patient subject, listening, seeing moving lights, accords little importance to this plane. It is difficult to designate body parts on either side of this limen as proximal/distal, here/there, pre/post.[41] That which has crossed the plane has absorbed rays, cast shadows, will yield data;[42] it has briefly been the center of an orbit.

To a patient, the confining nature of the gantry bore may be of more con-sequence than the plane of sectioning—especially in regard to "posthumous shock." Some patients compare the gantry to a coffin. CT installations often have protocols to address claustrophobia (and MRI installations always do). Techs say many patients can be "talked through" these anxieties—though some need medication. In the Neuro CT control room, two techs discuss a claus-trophobic patient scheduled shortly. The clinicians reassured the patient the procedure would be brief. Tech: "They wanted to spiral this. I told them that was still ten minutes on the table." Doing the scan quicker may not solve the patient's problem.[43]

Within the gantry, rotating tube and detectors establish the origin and ter-minus of the scanner beam. What was once an "ice-pick" beam is now fan-shaped. X-ray tubes are expensive parts, and tube "lifetime" is important to techs and administrators concerned with materials management. Detectors translate a received x-ray beam to an electric signal, which is then forwarded to the computer. Detectors are of narrow profile to index a narrow column of beam; they are insulated from one another so that one detection event will not affect adjacent detectors; and they are arrayed in multiples to filter random "noise" from individual detectors.

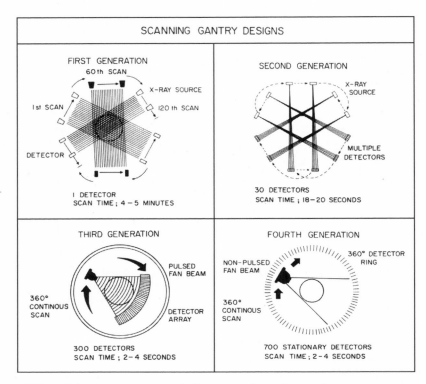

SCANNING GANTRY DESIGNS

FIRST GENERATION
60th SCAN

X-RAY SOURCE

1st SCAN

120th SCAN

DETECTOR

1 DETECTOR
SCAN TIME ; 4 – 5 MINUTES

SECOND GENERATION

X-RAY
SOURCE

MULTIPLE
DETECTORS

30 DETECTORS
SCAN TIME ; 18–20 SECONDS

THIRD GENERATION

PULSED
FAN BEAM

360°
CONTINOUS
SCAN

DETECTOR
ARRAY

300 DETECTORS
SCAN TIME ; 2– 4 SECONDS

FOURTH GENERATION

360° DETECTOR
RING

NON-PULSED
FAN BEAM

360°
CONTINOUS
SCAN

700 STATIONARY DETECTORS
SCAN TIME ; 2 – 4 SECONDS

19. Tube and detector arrangements in four generations of gantry design

The bed on which the patient/specimen lies is the "table." It is horizontal, at roughly the same height as a stretcher, perpendicular to the gantry. Its base remains fixed while a narrow tabletop portion slides into the gantry bore.[44] Movements are mechanical—involving gears and belts that techs control with buttons. The table is cantilevered, but there are limits to the weight it can accommodate, beyond which transport mechanisms may fail. Nor can the specimen exceed the diameter of the gantry bore.

Next to the gantry is the "MedRad injector": a mechanized syringe for timed injection of intravenous contrast. It is supported by a hinged arm that hangs from the ceiling.

Apart from the gantry, table, and injector, most accoutrements of the scanner room are along the walls. There are controls to activate a fire extinguisher system. One entire wall is occupied by the CT computer. There are color-coded ports for oxygen, suction, and electricity. Sinks are segregated clean from dirty—a classificatory scheme that extends throughout hospital spaces.

Likewise, on the phone on the scanner room wall, handwritten letters say "NO GLOVES"—to separate dirty hands from clean ears.

Walls hold cabinets for linen, basins, contrast, catheters, pads, phantoms, and more. There are carts with IV supplies—antiseptic, catheters, needles, syringes, gauze, tape, gloves. There are yellow boxes of emergency equipment. A few items are on rollers: a cloth laundry cart; a vital sign monitor.

There is a dedicated water supply to cool the scanner x-ray tube. Heating of the tube limits continuous exposures, the number of exposures in quick succession, sometimes even the number of studies performed in a day. It necessitates high air flow through the scanner room: if air conditioning fails (particularly in summer), the entire schedule may be canceled or reallocated.[45]

Between the scanner room and the control room is a leaded glass window. On the control room side is the Siemens console (Siemens is the maker of University Hospital's scanners), with knobs, switches, and toggles, keyboard, trackball, and mouse. Two large CRT monitors display scanner instructions and images. Between them, against the glass window, is the control box for the MedRad contrast injector (bearing cards of local support personnel). To the side are linoleum countertops, holding film folders and papers. Near each console sits the techs' CT log, a binder with pages bearing these column headings: Date, Name, Medical Record #, Tape, Exam #, Study, IV Contrast, Charges, Time In/Time Out, Comments. On a nearby shelf are manuals of CT Protocols, SOMATOM Operations Manuals, a SOMATOM Quality Control book, a Siemens Image Quality Guide, and a binder of technical reprints. Above the control console, surveillance monitors depict views down the halls outside, and behind the gantry.

Each control room has additional computers: one for the SMS Scheduling System, another for Clinical WorkStation and Internet. Both control rooms have fixed two-panel viewboxes. There is also a film printer—in Neuro CT, against the wall opposite the console; in Body CT, in a small room adjoining the control room (once home to a prior scanner's CPU)—along with film, contrast, soda (to mix oral contrast with), old logbooks, and magnetic media containing studies made by the previous scanner.

The wall phone closest to the SMS computer is surrounded by pages of phone numbers for clinical units and pager numbers for radiology staff. Other walls in both control rooms are papered with memoranda and reminders.[46]

The scanner rooms and their accoutrements have a high-tech, ergonomically rationalized feel. But among the proprietary machines, white surfaces, and printed labels, handwritten notes offer reminders that machines depend on their operators—who record signs of their adaptations—like the writing on the phone on the scanner room wall.

Positions, Careers

Instrumental practices of the scanner and control rooms yoke craftpersons and tools. Various careers mesh here: those of techs and nurses, physicians and residents, clerks and couriers, plumbers and service persons. Then there is the patient and his lesion. And the scanner. All these careers are brought to bear when a patient is strapped to a scanner table, when a Siemens serviceman unpacks his tools to fix a computer glitch, when a tech leans forward to push a button and say, "Stop breathing."

Most human agents whose careers intersect in CT wear ID badges specifying affiliations and titles. Some role differences are legible in forms of dress. Techs usually wear scrubs—though some wear casual streetclothes. Patients wear hospital gowns (if inpatients) or street clothes. Nurses often wear white coats; the Nursing Administrator wears a long white coat. Radiologists—including residents—wear business clothes: hard/polished shoes, pressed slacks or dress/skirt, dress shirts/ties or dress blouses. White coats are rare among radiologists, as are sport jackets. Interventional radiologists wear scrubs and sneakers; so do residents on call. These modes of attire mark roles—and some prospects for "role distance."[47]

Presently, in the Body CT control room, tech Sandra has returned from lunch to find on the table the same patient who was there forty-five minutes earlier. Melinda, her junior and shift partner, explains: "We had to bring her back and scan her prone." Seeing that Zeke, a third tech assisting during lunch, is having difficulty moving the patient back onto the stretcher, Sandra asks Melinda, "What are you doing?" and steps into the scanner room to help. Melinda, at the scanner controls, responds with a trace of irritation: "Eating bonbons."

The door to the film printer is open: Mark Hanlon, whose badge simply says "Radiology," wants to show Zeke something about the printer. He works for Sun Health, an independent firm, and comes often to maintain equipment. A tool briefcase with several tiers is open on the floor. The women techs laughingly grant Zeke his "boy talk" with Mark. Nonetheless, getting the printer fixed is a serious concern.

All told, there are nine or ten CT techs: seven full-timers, and two part-timers who work twelve-hour shifts on weekends. The techs correct me: "Tech" is short for "Technologist"—*not* "Technician." All University Hospital's CT techs are certified through the American Registry of Radiologic Technologists—with an "R" designation—meaning general Radiology. Louis also "has his CT"—but "doesn't wear it." CT has had an independent registry for the last year or two.

The techs came to their current positions by different routes. Some have a B.S. degree in Radiation Technology; some were trained in the University's own program. Most worked in other modalities before coming to CT. CT is claimed by some to be better than tech jobs in the "Main Department"—a job they can "stick with"—which one expresses as a "saying: 'Old radiology techs never die; they just go to CT.'" They feel they have good relations with the radiologists here, in part because one of the bone radiologists has a special interest in tech education, and his wife is on the faculty in Radiation Technology.

One tech, Zeke, reflects on his career. He began in Diagnostic, then went onto Nuclear Medicine. "Even went into the ultrasound room a few times." Then CT came along: "I got lucky, in a way." The machines were "kind of intimidating . . . but I got on-the-job training"—largely from the manufacturer. This was around 1980, when scanners were fewer and farther between. Zeke wound up "driving around a mobile scanner" for several years to bring CT service to smaller hospitals. (During this discussion, a passing clerk remarks for the microphone that she wants a raise.)

Conversation turns to divisions of labor between techs and radiologists. I ask if radiologists can work the scanner. One neuroradiologist can do quite a few things—including "put a disk in and retrieve his own study." But most would be lost at a scanner console.

Generally, radiologists are not around as much as they used to be. "We get a piece of paper"—a "protocol," boilerplate instructions for each scan. Techs remember when they "used to get much more detailed protocols, all orders written out, number of cc's of contrast per second." Protocols have become more rote—partly because they are more exhaustively specified in the CT Procedures book.

Two techs speculate that radiologists might be at risk of "phasing themselves out" as directors of scans.[48] More clinical specialists are requesting specific protocols, "like radiologists used to"—"chest with and without [contrast], by 5 mm." This, the techs acknowledge, is sometimes inappropriate—"bad instructions"—but the radiologists often respond: "Do what they say." Even for reading, radiologists are sometimes superfluous. "Ortho reads all bone films." Techs suggest there might be a checkbox someday on the requisition—"consult with radiologist"—making this an optional element of a radiographic study.

Techs are glad to talk about evolution of scanners. They speak of successive generations. Their Siemens SOMATOM Plus arrived in July 1996. Louis recalls the "Technicare" scanner from 1988–90. Zeke recalls a GE 8800, a "nine-second scanner" (nine seconds/slice) with a "two-minute reconstruction [time]." Then came the 9800, which made two-second exposures and "instant" reconstructions. The current Siemens scans at less than one second per slice.

A Radiology administrator reflects on growth of CT technology, and its persistence in the face of MRI, which some thought would replace it in the early 1980s:

> CT and MR were the first two technologies where it wasn't "pick it up throw it out the window and plop in a new one in its place" . . . It became a software development system. Yes, there were computer enhancements as computer technology advanced . . . And so we had some forklift upgrades, if you will, of computer systems . . . But the gantry and the tables and the basic scanning technology has remained constant in CT since its original inception . . . Except that the tubes have gotten more high-powered . . . And longer-lived, and can fire a lot faster, which is what's been responsible—and frankly detectors have gotten a lot better—which has made the whole physics of doing CT a lot better over the last ten years. So as you have replaced CT scanners through their normal evolution, you've been able to get into technologies which have been improved, but you haven't had to replace it.

Zeke recites virtues of the present Siemens scanner. In addition to scanning quickly, it supports heavier patients: 440 lbs. Girth is the limiting factor now. The field of view has improved. There is a longer scan length (nearly the whole body). And the tilt of the gantry can be changed remotely, from the control room—"finally." This, as well as more sophisticated techniques of "reformatting" data, makes the system much more patient-friendly. No longer must patients dangle their heads in uncomfortable positions for imaging facial bones; no longer is it "hard to get the last disk shot" for lumbar spines.

Now here Zeke is, squinting before trackball, monitor, and keyboard. And the patient, on the other side of the leaded-glass window, strapped down with four-inch velcro straps.

The standard position for CT patients is supine and immobile. In early days scans took hours. Body parts subject to movement—like the chest—were hard to image. Even though breathing and peristalsis are lesser factors now, techs still must issue instructions sometimes: "Hold your breath, don't breathe . . . now you can breathe." In University Hospital's state-of-the-art scanners, these instructions are pre-recorded, playable at the touch of a button.

Patient immobility on the scanner table contrasts sharply with the dynamism of the scanner. "Can't tape 'em tight enough," says a tech about agitated patients. For kids, "sedation is about the only answer."

To remain immobile is to assume, mimetically, a *deathly* posture. This contributes to the complex irony of CT's substitution for postmortem autopsy in procurement of diagnostic clues. Imitation of the cadaver seems to be requisite for peering into opaque living bodies.[49]

Sedation involves nurses. Nurses' work is "patient-centered." While techs assist patients on and off the table and start intravenous lines, they concentrate on "running the scanner." Nurses are guardians of patient safety and ensurers of "continuity of care"—from the wards, through imaging suites, and back.[50]

Adult outpatients usually do not need nursing attention. Hospitalized patients may, especially if sick or frail. Critically ill patients, from Intensive or Special Care Units, usually arrive with nurse attendants from their units. This is also the case for patients from the Emergency Department. But there are plenty of patients for Radiology's nurses to help with.

Nurses administer medications and participate in decisions about what to use. They are notified when IV contrast is to be administered to any patient with an allergic history. They are involved in all invasive procedures—like biopsies. And they respond rapidly to emergencies.

Nursing shifts are usually ten hours. One nurse covers both CT and Nuclear Medicine. I remark to the Supervisor that I am impressed so many nurses are assigned to Radiology; I do not remember them in institutions where I trained. She affirms that University Hospital is a "leader" in committing nursing expertise to radiology—part of a "state-of-the-art" department.[51] Radiology Department, she clarifies: Radiology nurses are not employees of the hospital's Department of Nursing. "Her" nurses are "special people." Eight of ten have been ICU nurses in the past. Turnover is minimal. She cites their contributions to administrative and academic activities. One recently helped codify hospital procedures for conscious sedation. She herself has contributed to a manual on Ultrasound certification and published a textbook chapter on abscess drainage.

Patient Contact

A patient lies on the Body CT table in velcro straps, arms over head, awaiting a biopsy. She asks one tech (a man) to ask the "other tech" (a woman) to "scratch my belly button . . . I need someone with fingernails." Though a defining feature of CT is a certain abstraction from the patient's body, there is

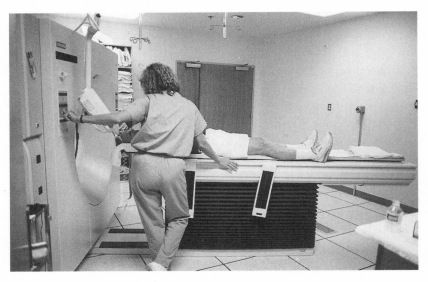

20. Tech positioning patient, Neuro CT

plenty of touching left in techs' interactions.[52] Assisting patients onto the table, strapping them in position, starting IVs, and assisting them back to stretcher or wheelchair are hands-on activities.

An inpatient enters the Body CT scanner room from the hallway alcove—an elderly woman, with eyes closed. She is losing her hair and has a nasogastric tube taped to the bridge of her nose. She is accompanied by two family members, both women. The tech invites them to help her move the patient onto the table. She has gloves on; they do not. When the stretcher is beside the scanner table, one of them begins counting, anticipating a synchronized lift. The tech interrupts: "No, no, no, we're not going to do it like that . . ." She picks up a two-by-six-foot plastic flex board leaning against the wall and places it over the space between the beds. "We're going to roll you up, Ms. Clark." She enlists the family members in pulling edges of the sheet to roll the patient onto her side, away from the scanner. She pushes the board under the patient's back. "Now we're going to roll you back." Pulling on the sheet, she single-handedly pulls the patient across the board onto the scanner table. Ms. Clark grimaces briefly, then rests; her kin exit. The tech pulls the sheet lengthwise, toward the scanner bore, to place Ms. Clark's head in a cushion; she fastens velcro straps across her chest; she plugs an intravenous line into the "med lock" in Ms. Clark's arm and checks the contrast syringe. As she steps back into the control room, she removes each glove with a snap.

Tech / patient contacts are polite, pragmatic, sometimes intimate, sometimes vexed. Techs have all been vomited upon, some recently. A few have been stuck with needles. Their encounters with patients are long enough for techs to develop likes and dislikes, sympathies, modes of abstraction. Some patients become familiar—for instance because of a series of scans over several weeks. "Not the lady with the helmet! Sally Nye? Oh no, we had to send her home [unscanned] last week."

Louis shows me the scheduling book and explains the code for marking patients' progress through the scanning process. A line is drawn through the box "when the patient is here." The box is blocked in black when the patient has been scanned. "I'm going to mark this 10:15 a no-show." Inpatients are scanned at the beginning of the day, at 7:30/8:00. The rest of the scheduled patients are outpatients. Most days, one tech remains at the controls while the other maintains the schedule, escorts patients from the waiting area, starts IVs. Zeke shows me how he "sends for" inpatients. He can "generate an order for Transport" simply by typing it into the SMS system (chapter 4, "Curating")—he does not need the phone. Though this summons a brief lament for "lost human contact," he appreciates the new system's efficiency.

Nurses have a different range of patient contacts. They are the only ones permitted to "access" special intravenous ports: "Techs don't touch central lines." Nurses often open the charts that accompany inpatients. One nurse reviews a chart attentively, to determine if a prior reaction cited by the patient was "really an allergic reaction." Rarely do techs open a patient chart; radiologists do, but usually only the laboratory section.

Verbal contact of staff with patients is respectful but attenuated. I rarely hear techs or nurses ask a patient why he is undergoing a study. Zeke asks a trauma patient undergoing a neck CT if his "neck hurts much"—because he walked into the scanner room with no collar on. Nor do techs solicit questions, though they answer questions patients ask. When a patient inquires "how they read films down here" if he carries them off to ENT Clinic, Lucinda explains that "they read 'em off the screen."

Performing a Study

When a CT report says a "study was performed," a tech has done the performing. "Performed" is spoken more often about CT than about plain x-ray procedures. This is partly because CT is an intervention of greater complexity. "Performance" is also a term that echoes in literatures of scanner development and salesmanship. But performance captures something important about tech roles in CT—aspects of improvisation that arise in use of the machine, with different patients, different circumstances, and dynamics of critique and assistance in teamwork.

A tech has positioned a patient and returned to the control console. She jumps up as if she has remembered something. "Got to clamp her Foley [catheter]." On another occasion, between patients, three techs are busy in the scanner room. Zeke is putting out sheets and blankets; Serena is lowering the stretcher; Marilee has just emptied a suction canister and is setting off to retrieve a supply of new suction tube tips.

At the console itself, techs' performances of scans are highly scripted—by "protocols." These are not the protocols of diplomatic etiquette or military SOP. Nor are they the protocols of laboratory bench research, or the experimental drug trial—though this last usage is closer.[53] These protocols are written recipes, for types of scan to be performed in particular circumstances. Protocols are described in a manual in each scanner room.[54]

The Protocol manual in Body CT lists five Chest CT protocols (Routine Chest, Staging Chest, High Resolution Chest, Pulmonary Embolism, Aortic Trauma) and eight Abdomen/Pelvis CT protocols (Routine Abdomen, Biphasic Liver, Pancreatic, Renal, Adrenal, Routine Abdomen/Pelvis, Rule Out Renal Colic, and Pelvimetry). Neuro CT protocols include six Head and Neck protocols (Temporal Bones, Sinuses, Tumors, Neck Abscess, Orbits, and Facial Fracture), among others.[55] Most protocols resemble, in format, this example:

Routine Abdomen

Purpose:	The Routine abdomen protocol is used as a general survey scan, when the clinical history is vague, when detailed thin sections are not called for such as in pancreatitis, or with the general statement "rule out abdominal mass."
Scan Extent:	Dome of diaphragm to iliac crests.
Scan Type:	Spiral
Table Speed:	8mm/sec (pitch of 1); May increase pitch if necessary to cover area of interest.
Slice Thickness:	8mm
Reconstruction:	8mm
Respiration:	Studies are performed with suspended respiration.
GI Contrast:	800 ml of oral contrast is given 60 min before the scan with an additional 300 ml given immediately before scanning.

iv Contrast:	150 ml at 3.0 ml/sec. Begin scanning 60 seconds after start of bolus.
Photography:	Print soft tissue windows, and lung windows of bases.
Modifications:	1. Evaluate Adrenal mass:
	Delayed images (1 hour) are recommended to further characterize any adrenal masses identified.
	2. Bowel vs Abscess:
	Delayed scans following additional oral contrast may be necessary if initial oral contrast administration is suboptimal.

To perform each scan, the techs "execute" the appropriate protocol: they program the scanner and the injector to carry out timed steps, by pushing buttons in a defined order. Most buttons are labeled; some are lighted. I ask how machine lore is passed around, how functions of buttons are learned, how techs arrive at best ways to accomplish certain tasks.[56] The question is deflected to the keyboard. "Basically, all the keys you're taught . . . The reformations, the special reconstructions, that's later . . ." Louis believes protocols have become more complex with progressive generations of scanners. "The recipe's gotten bigger than the cake."

One example of changing cake/recipe proportions involves the relation between scanning and reconstructing. Historically, these have been separate operations for the computer: raw data is gathered, then density distributions are calculated. Each slice was reconstructed before the next slice was scanned. Then "dynamic scanning" was developed, for blood-flow questions—and scans needed to be completed while contrast was still in blood vessels. In dynamic scans, reconstruction is delayed until all scan data is acquired. There are limits to the number of slices that can be acquired in quick succession without overheating the tube. Delayed reconstruction also presents post-scan choices for techs.

Tech:	Reconstructions have changed, you're right. Instead of just filming bone windows and adjusting the windows . . . we actually have to go in and review it and reconstruct it on a bone algorithm.
Investigator:	Hmm.
Tech:	Which—we only used to do that for like temporal bones if they were looking for a real bony detail—or bone work, we'd do a bone algorithm.

Investigator:	Mm-hmm.
Tech:	But now we're kind of getting in that trend where we're doing a different algorithm to give more detail versus just window-ing it.
Investigator:	Mm-hmm.
Tech:	It's true with Peds cases too, we also—instead of just filming lung windows, we go back and do ultrahigh algorithm on the lung. Like a hi-res chest to look at the lung markings for Peds cases.

[*segment omitted*]

Tech:	And Applications pretty much tells us what looks the best and what other institutions are using.
Investigator:	Mm-hmm.
Tech:	But we go in and play with it sometimes, and sometimes we change it.
Investigator:	And see what looks better.
Tech:	Right. Get the radiologist involved. That's another trend. I'm not trying to say it's good or bad, but I see in the old days when I first came here, the radiologist tended to spend more time in the actual scan room.

So techs may adjust the recipes that govern production of images from scan data. They are committed to producing finer images—yet feel burdened by, occasionally skeptical about, the increased sophistication of these image processing tasks.

I ask techs about transitions from one model scanner to another. One of them likens the experience to learning another language. For example, each CT company assigns a different name to its initial index view, with numbered slices superimposed on a longitudinal x-ray of the target area. What GE calls a "scout view" and Technicare called a "delta view," Siemens calls a "topogram."

Investigator:	Has the console changed? The actual thing you sit in front of?
Tech:	Oh yeah, yeah it has. It's gone to more of a computer room Windows base, whereas the old one used to be just hit a but-ton and type yes or no and just kind of follow—it prompted you. This one's more—open the window—it's multitask too, which is a lot different than—
Investigator:	Right.

Tech: The other one you kind of focus. This one'll do so many
                 things. It'll scan, it'll review, you can do other reconstructions
                 in the background, you can film in the background. It can be
                 doing all kinds of stuff while you're scanning.

[*segment omitted*]

Tech: It's not the pie job where you can scan and talk on the phone
                 at the same time.

Techs execute protocols, and may "play" with algorithms involved in reconstructions—but they do not choose protocols. Nor does proper execution of
a protocol guarantee adequate images. Supplemental requests by radiologists
are common. When faster scanners were finishing scans before IV contrast had
made it through the kidneys into the bladder, radiologists requested bladder
recuts. Now "we have to do bladder recuts on everybody." Repeated enough,
the ad hoc becomes protocol.

Serena the tech escorts an outpatient, Mr. Stover, to the Neuro scanner. "I'm
going to let you go ahead and lay on down."[57] When he is on the table, she says,
"Now I do have to start an IV and give you the x-ray dye like last time . . . all
right?" She comes into the control room and comments, out of his earshot:
"He didn't bring his creatinine with him. I called the lab—last one they had
was last July—it was point nine—and I paged the doc." On this occasion, Serena is working without a partner. With a partner, she might focus on scanner-
table activities, and her partner on the console.

It is mostly outpatients for whom techs need to start IVs; inpatients arrive
with lines already in place. "Let's see your arm here . . . Make a fist for me there . . ."
Snapping of gloves. "OK, Mr. Stover, there's going to be a stick right here now"
(fair warning—as well as device of hospital scientificity: accurate prediction
of pain infliction). The IV catheter is inserted into a forearm vein. "Open your
hand now . . ." The tech gently presses skin above the catheter and queries, as
fluid infuses: "That feels OK? All right." Gloves off.

The tech attaches coiled tubing from the contrast syringe to the IV catheter
and issues standard instructions. "All you have to do now is lie very still for
us—and we'll be through here—this is—probably take five or six minutes . . .
I'm going to take a couple of pictures first getting you lined up—and then we'll
turn the x-ray dial on, OK?" The tech washes her hands before she returns to
the console. The patient hears quiet keystroking and air conditioning.

To get the patient "lined up," the tech first "shoots a scout view" (Siemens's
"topogram") of the region of interest—in this case, neck. Then, overlaid on the

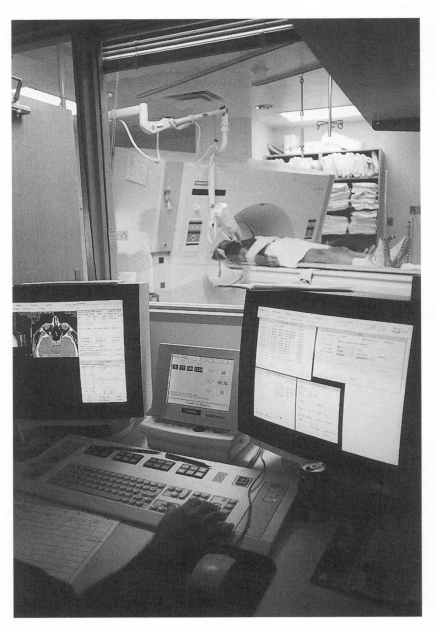

21. Scan in progress, Neuro CT

scout, the tech drags corners of a parallelogram to mark first and last cuts of the scan. She does this by spinning a trackball and clicking buttons. Angles of this figure will correspond later to the tilt of the gantry.

The tech comments as she sets scan parameters: "No dentures and no fill-ings—will enable me to shoot straight through . . . [not] have to angle around it." She selects a second scan region: a new parallelogram on the scout view. "Make sure I've got all my techniques right . . . Need a little bit more power through the shoulders." Then: "It doesn't like that . . . That was too much heat for it—wouldn't allow me to do it. Could have a two-minute delay between this one and this one."

Once the scanner is programmed—its likes and dislikes accommodated—the tech gets up from the control console and steps into the scanner room to advise the patient: "Contrast is going to start going in now." The tech returns to the control console. Several beeps. "That's the contrast goin' in."

Now the tech speaks through a microphone and a speaker on the scanner. "Mr. Stover—I'm going to ask you in a minute to not swallow during the exam . . . know that's kind of hard." The scanner whirs. "OK, Mr. Stover, don't swallow. Try not to swallow." Then: "OK, you can relax for a second." "OK, now once again, try not to swallow." Thin beep at the console. "OK, you can relax . . . You can swallow a little bit now . . . you did a good job."

The other tech, Zeke, has returned. He enters the scanner room and ad-vises Mr. Stover to relax for now. Serena, at the control console, explains: "He's going to load the contrast for the next patient. We know how much contrast we're going to need, regardless of the protocol."

Serena heads for the door, saying "I want to check these"—that is, ask the radiologists to review images on the monitor. She returns shortly and says to Zeke: "Yeah, I think he's done." Zeke enters the scanner room and speaks to Mr. Stover: "Going to pull you out of here . . . you doin' all right? . . . get that IV out . . . you don't need this IV for any other tests today . . . You have an ap-pointment in ENT clinic? . . . Going to throw a piece of tape on there . . . It's going to take us about ten . . . minutes to print up all these images . . . down to the waiting room . . . Drink lots of fluids today." When Mr. Stover has left, and before another patient is escorted back, the techs confer on logistics of upcom-ing studies. "A head, then a neck." One of the clerks steps in with news from the waiting area "Serena, Mr. Boorstin just started drinking [contrast]."

Serena is concerned about a new, unfamiliar error message on the moni-tor: "Cannot write to optical disk." She summons Zeke. Looking over her shoulder, he feigns irritation. "Better be good: dragging me over here." He has not seen this message before either; it takes them five minutes to sort it out.

Contrast Agents

"Contrast agents" have been used since the advent of x-rays. Contrast designates substances that are "radio-opaque"; they block x-rays. This is the primary locus of their agency: distinct from passive flesh, contrast stands out.

In CT, there is gastrointestinal (GI) and intravenous (IV) contrast. GI contrast flows in the gut, between mouth and rectum. IV contrast is distributed wherever blood goes—"opacifying" vessels and blood-rich organs. Prominence of vascular tissues is referred to by radiologists as "enhancement."

Some patients receive no contrast. Patients undergoing a head CT for concussion, for instance: fresh blood in the brain is discernible without contrast. For other scans (for example, abdomen/pelvis), both kinds of contrast are standard.

The techs are annoyed that an upcoming scan is delayed because the patient only began drinking contrast at 12:15. "The contrast was delivered to the nurse on the floor at 10:45," notes one. Later, they explain other contrast timing issues: They will begin scanning this abdomen about seventy seconds after the IV contrast bolus starts. The normal IV rate of 2 cc/second will be cut back because of this patient's fragile veins.

Logistics of "contrast agent administration" call on coordination of other agencies too. The Scheduling Office inquires about kidney function and allergies. Radiologists select protocols which specify contrast amount and timing. A patient drinks GI contrast before the study. A tech administers IV contrast at prescribed moments and rates. Contrast effects are distributed across images by distributed practices upon and within bodies. Intermingled flows of patients, images, and contrast make "enhancement" an institutional logistical project. As a fellow says about contrast in liver scanning, "timing is everything."

But there is an economics to flows of intracorporeal agents. IV contrast is of more than one kind, and the best kind is expensive. This has presented interesting problems, at University Hospital and elsewhere. IV contrast can stress kidneys—the reason for vigilance about a patient's latest creatinine. Some patients are allergic to contrast. And some patients are simply made uncomfortable by IV contrast—unto vomiting.

In the late 1980s, a new kind of IV contrast was introduced: "nonionic," or "low-osmolar"—versus the standard-issue "ionic" stuff. The new contrast was celebrated as safer and better-tolerated. University Hospital began using it—but it was expensive. Soon, for cost reasons, criteria for "selective use" were framed by the American College of Radiology and adopted across the country—including University Hospital since 1995. This occasioned controversy, as patients, physicians, ethicists, and policymakers worried about fiscal triage and iatrogenic suffering.[58] Several techs are not shy about voicing opinions—that adverse reactions are more frequent with ionic contrast, that they would prefer not to have to use it at all. Most say the main issue is nausea. One of them has begun starting more IVs in the hands, since hand IVs require use of nonionic contrast as a matter of policy.

Tech:	It's hard, we try to go by the guidelines, but I can probably tell you there are times when just a gut feeling that I really like this person and I don't want to make them throw up all over the table and—
Investigator:	Right.
Tech:	And they kind of hint that they know that can happen and they don't want to—I mean you have to almost make a personal judgment in some cases that you just feel like, I'm doing the right thing. Or—and there's other ways around that too, which I mean you can . . .
Investigator:	Do you ever feel constrained or look back on the constraints that were on you to give ionic contrast and say, I wish I hadn't had to give it, because this happened?
Tech:	Right, and yeah, there is. And then you have a reaction with hives, or the patient—
Investigator:	Mm-hmm.
Tech:	You know, we've had a few reactions. Or they throw up all over the place and the scan's ruined . . . And the next time they all—after a couple reactions it makes you a lot more hesitant to load it again.

This dilemma faced by techs is a daytime dilemma: at night and on weekends, ionic contrast is not used, because staffing is thinner. Nor is it used in the ACC scanner, where there is no radiology nurse and only one radiologist. Whenever ionic contrast is used, policy dictates that a radiologist be in the "scanner area."

Supervisors are aware that discretionary choices made by techs stretch the guidelines governing nonionic contrast use. Although the contrast budget of the hospital is large—$800,000 yearly, one tech says—there are solutions on the horizon. One administrator offers a longer view:

Administrator:	Competition has just hit the contrast media market . . . And what's happened is that there are now organ-specific agents that they [vendors] can make their money on . . . And so the generalized nonionic agents are—and they're coming out of patent and they're going into the generic mode now.
Investigator:	They're getting cheaper.
Administrator:	They're going to get very cheap. What's going to happen, probably in the next twenty-four months, as new contracts like the one we're negotiating come about and these gigantic discounts—I mean discounts larger than the price you're paying.
Investigator:	Hmm.
Administrator:	It's phenomenal now. Prices, again an irrelevant number. It's what you pay that becomes relevant.
Investigator:	Mm-hmm.
Administrator:	That I think we'll see a swing of the technology back to all nonionic contrast media.
Investigator:	Mm-hmm.
Administrator:	And where we'll be paying the buck and we'll start fussing and focusing will be on liver agents and you know . . . Feridex is a good example, which we use in MR as a liver focusing agent.
Investigator:	Mm-hmm.
Administrator:	CT is going in exactly the same direction.

[*segment omitted*]

Administrator:	And, so what's going to happen is these companies are going to say, "OK, nonionic contrast media is going to become, if nothing else, a way in the door. We're going to practically give you your nonionic contrast media to get . . ."
Investigator:	"But you've got to listen to my sales pitch for Feridex."
Administrator:	"And you've got to get the organ-specific agents on the menu."

Price is irrelevant: it's what you pay. Contrast enhancement, indeed. CT contrast not only highlights the vessel and the lesion but summons a whole palette of contrast agencies that promise to reshape the larger diagnostic economy of the hospital.

Materializing Images

"We print a lot of images," remarks a tech. Advances in scanner and printer speed have supported the desire, by radiologists, for finer cuts and more refined sequences of images. She is finishing a neck scan. "Have to do these necks in two different algorithms—soft tissue and bone . . . 180 images, something like that."

The number of images being printed in CT is watched by administration:

Administrator: Some of our body studies are going 300 images in CT now.
Investigator: That's phenomenal.
Administrator: And culturally, to me that feels like "We can do more so we will."
Investigator: Mm-hmm.
Administrator: Instead of looking at—continuing to look at the efficacy of the decision making.
Investigator: Of the finer cuts . . .
Administrator: And what they do is they do—they scan in fours and reconstruct in twos [millimeters] or some ungodly thing that just runs slice numbers off the scale. And Sally [the chief tech] says that now the printer is working behind. They're printing so many images now that the printer is running behind.

The techs confirm the lagging printer problem. Film printers, like scanners, have undergone improvements. The Siemens printer in Body CT is "a six-year-old camera"—with a "ninety-second process." "The printer downstairs [Neuro CT] is a forty-five-second." The Body CT printer is now the "weak link." "'Cause a lot of times you're scanning the third patient, and the first patient's films are . . . getting out."

Techs know about materials flows: the Neuro CT printer uses 175 sheets of film per day; film costs six dollars per sheet. They are amused by questions about recollections of the darkroom. They no longer engage wet chemistry, a longtime feature of film processing in and out of radiology, except to change bottles of fixer and developer on the processor. They load film in a fully illuminated room, using a "self-contained system."

I express interest that some patients carry off their own films. Notwithstanding the patient's de facto exclusion from the reading room, he may participate quite early in custodianship of images. It turns out that this applies to specific classes of patients:

Tech: Just about every ENT patient gets their films. Occasionally some of the Neurosurgery people want their films.

Investigator: Do they get uh—do you only print one copy of the study, or do they get a copy—

Tech: Just one, yeah.

Investigator: So they don't get read by—

Tech: What'll happen is the radiologist will read it off the, ah, monitor. We'll just link it in there to the reading room—

[*segment omitted*]

Investigator: Before you had the MagicView, what would happen?

Tech: They would read 'em, and the patient would have to wait.

Beyond the filming, each scan requires digital "archiving" on optical disk. Zeke holds up one of these disks to show its shine, and comments, "One time we had some that were actually transparent." It resembles an ordinary CD. The department keeps digital copies of scans for about a year, after which time disks are overwritten. Two weeks of data fit on one side of a disk—so about a month per disk. Just the images are retained—the "raw data" only rarely, because it takes three times more memory.

Practices which give substance to images are fundamental to the intellectual economy of diagnosis. They have become routinized, with automated filming, digital archiving. They are most noticeable when the automated rhythms fail to keep up with other strands of workflow: the computer won't archive; the printer is down.

Radiologists at the Scanner / CT Biopsy

Despite radiologists' diminished contributions to scanning practices in recent years, they still visit the control room occasionally. Some exchange pleasantries; some check the schedule. Sometimes radiologists come to discuss image quality. Simple requests, like "Please print images 56–72 for the Teaching File," tend to be handled by phone.

There are emergencies—contrast reactions, primarily—that engage radiologists in person. The most serious emergencies—"codes"—become busy and chaotic events, with many participants representing jurisdictions beyond radiology, and much crosstalk of inquiry and command. The yellow emergency boxes are opened on such occasions. Radiology nurses remain involved until the patient is transferred to a proper care setting (usually an ICU). Such events are fortunately rare.

Radiologists become most intensely involved in scanner work itself during CT-guided biopsies. Of several "invasive" procedures in diagnostic CT, this one has special relevance to the status of the tissual lesion.

CT biopsies are typically performed by the CT resident or fellow du jour. Biopsies require extra attention and time: the resident must disengage from reading room workflow; the nurse is involved; the attending or fellow must supervise. There is extra communication between the scanner room and the reading room beforehand—"the patient is here—needs to be consented"; "the patient is drinking contrast." Once the patient is "on the table," the resident is summoned for the procedure.

CT biopsies are tidy instrumental events: hit a target, avoid collateral damage. Yet they are procedures that presume some diagnostic uncertainty. Perhaps a prior scan, or other imaging procedure, has generated questions that a biopsy is intended to settle. Independent of the patient's interest, a biopsy result can vindicate a diagnostician's claim, legitimate a guess, overturn expectation. Biopsies exchange with curiosities and intrigues of a wider diagnostic economy.

From the patient's perspective, what matters most is how the biopsy result may translate to treatment or prognosis. Patients submit to invasions of a needle out of hope something helpful can be learned. Their benefit/risk deliberations are formalized in a consent process, which involves review of potential complications: bleeding, infection, lung collapse, death (the list varies, along with its accents).

Acknowledgment of possible complications is significant. Such an acknowledgment marks a particular existential position assumed by caregivers and patient—however fleetingly. They have nodded toward, signed off on, risk of dire consequence. This is one more face of death in the role of the CT specimen, in its supine immobility.

Two pairs of size 8 gloves, the resident requested. Latex-free, please. The nurse (who likes this resident—says she would not "wait on" just anyone) can only find size 7.5—which the resident accepts. (The techs have kidded him about wanting Latex-free gloves, saying he's "high-maintenance.") His tie is tucked in his shirt, and he stands beside the scanner table, with a sterile kit opened on a tray.

Mrs. Porter, deeply jaundiced, on her back, strapped at chest and pelvis, says she is allergic to tape and betadine. The resident reassures as he prepares her abdomen. He uses betadine anyway (it is in the kit), saying he will "take it off with alcohol right away."

Many CT biopsies require placement of skin markers, prior to making a scout image (or preliminary scan): a locus for insertion of the needle. Mrs. Porter already has a good external landmark—her umbilicus—that obviates previewing. The resident places sterile towels and proceeds to anesthetize the skin. "Little stick: one, two, three, four, five. Take a rest for a second."

Focused on the square of skin in a sterile field, the resident makes a nick with a scalpel. He picks up the anesthetic syringe, inspects it from two directions to assess its verticality, and then pushes it in, saying, "I'm going to numb deep right now." He injects anesthetic at several levels. Occasionally there is a faint hiss or wince from Mrs. Porter. The resident says apologetically, "The more I numb it now, the better." He steps back, then returns to lengthen his scalpel nick slightly.

Sterile hands in the air, the resident walks toward the control room doorway, and says to the techs, "We need to call Path—beeper number 5024." Then back to the table: the resident places the biopsy needle in the skin nick and instructs: "Hold your breath for just a second." He checks the verticality of the needle, then pushes it down. "Breathe shallow." He checks the depth of the needle (whose length he measured in advance). "We're going to go ahead and take an image right there." The resident pushes a button to move the table into the scanner, then steps back to the control room.

The tech speaks through his microphone: "Mrs. Porter, hold your breath." The scanner hisses. "You can breathe." These instructions are repeated three times, generating three images—on two of which the needle is visible. The resident and both techs inspect the images. Tech Louis: "I would go farther in."

The resident goes back into the scanner room to reposition the needle. In the control room, the resident from Cytopathology has arrived with tubes and slides. While the CT resident is still handling the needle, the attending enters the control room. "I presume we are close, because the pathologist is here—or the cytologist."

The attending inquires about the entry site. The resident joins in a discussion of needle position. The region is scanned again. "She is a *breather*," comments the attending. The needle tip is still not quite in the plane of the lesion, a deep collection of peripancreatic tissue. The attending wants to check the skin entry location, suggesting the needle is "five millimeters off," longitudinally—but the tech reminds him that "the interval" is one millimeter, not five. The attending warns Dr. Hastings, from Pathology, that this may take awhile.

Tech to patient, over the microphone: "Hold your breath, don't breathe . . . You can breathe." The images show the needle tip at the lesion. The resident goes back to the patient, advances the needle again, and aspirates. The attending murmurs, "If I were him, I would take it out." The CT resident does so, as if he heard. He exits the scanner room briskly, hands the syringe and needle to the pathology resident, and returns to the table, to hold gauze to Mrs. Porter's abdomen.

The pathology resident carefully pushes the syringe to extrude material from the biopsy needle onto a slide. "Twenty gauge [needle]?" asks the attending. "Yeah," responds the pathologist. He smears the slide, and then another.

Louis the tech speaks up: "BP [blood pressure] is going down a little bit— that says 94 over 39." The pathologist has not heard this, and queries "another pass" with the needle. The resident appears undecided. He cannot see the blood pressure monitor, but the nurse can. "81 over 43 now, Jim." The CT resident inquires of the patient, "Are you all right, Mrs. Porter?" Another pass is deferred; the pathology resident carries slides off to Cytopathology.

Over ten minutes, Mrs. Porter's blood pressure normalizes, and tensions dissipate. She remains on the table until it is clear that the biopsy material obtained is adequate. It is—though no diagnosis is yet disclosed to her. Before she leaves, to rest a few hours in Recovery, the area is scanned again to verify there is no internal bleeding.

The process in the scanner room takes an hour. The process is longer if one considers prebiopsy staging, readerly practices of the cytopathologists, recovery of the patient. It has entailed considerable tensions. On the table, the worry and discomfort of Mrs. Porter herself. The anxiety of the resident, regarding the technical quality of his work (under the attending's watchful eye), the risk he presents to the patient, the irreducible role of luck in sampling the target lesion. In the control room, the worry of the attending, who has entrusted delicate violence to someone less experienced—and the worry of the pathology resident, wondering if he will receive enough tissue to make a diagnosis. For the whole biopsy enterprise (in general—not necessarily in this case), there are background questions of a technical, also political, nature: is CT the best tool to have guided this approach to tissue? Should this have gone to Surgery?

There is a bit of shop-floor bravado in the ethos of image-guided biopsy: nearly any part of the body "can be biopsied with impunity with a #22 needle." Such confidence, most conspicuous among residents, is qualified by stories, and even jokes, of untoward outcomes. A mock dictation: "The colon was as-pirated, and then injected into the mass . . . Percutaneous nephrostomy tube to drain the pyonephrosis is suggested."[59] CT biopsy as a "minimally invasive" diagnostic option in the hospital is informed by these anecdotal forms of craft knowledge.

CT biopsy makes a connection between radiographic image and tissue. During the procedure, visible CT specimen (on screen) and about-to-be-procured tissue specimen are a mere ten feet apart—ten feet which are traversed by the radiologist several times. The image:flesh correspondence is bridged in the radiologist's bodily practice, shuttling with gloved hands in the air, process-ing advice ("I would go farther in"). His embodied translations help reaffirm an indexical relation between the dense needle-line on the monitor and the corporeal lesion itself.

The tiny bit of tissue removed from the patient is not cadaverous, of course: Mrs. Porter remains alive and hopeful. But it is handled as cadaver tissue would be: sliced and fixed on a slide. Biopsy knowledge pivots on death. The tissue procured by this 20-gauge needle is no less sacrificial (on the altar of micros-copy) than the cadaver of a turn-of-century CPC.

CT biopsy—and other CT-guided interventions like abscess drainage—is in some ways a return to some originary contexts of classical tomography. Among the key contexts reactivated here, in this atmosphere tense with bra-vado and dire consequence, hierarchy and institutional competitions, are the military contexts of foreign body localization that solicited the invention of early tomographic apparatuses. Early tomography was concerned with target-ing and extraction. CT biopsy is not so much a supplement to CT imaging as it is a return to one of its constitutive missions.

Reprise

Effects of machines: they co-opt gestures, imprint histories. They organize the attentions and routines of workers.

In standard histories of tomography, users of machines and their craft-practices are largely "out of the picture." In these histories, there is a tyranny of the geometrical/mechanical, and issues of intellectual priority, over messier contingencies of practice, like pushing a particular lever too slow. When it comes to CT, the complexity of the mathematics and engineering involved combine to make the machine superhuman—or, as one historian referred to an early MRI machine, "indomitable."[60]

I have focused on basic "gestural" characteristics of these machines in order to emphasize how they imitate, even co-opt, certain forms of handling of the specimen. But I have also tried to return to visibility some of the elements of handwork, tacit knowledge, and judgment that CT scanners' operators, and indeed specimen-patients, find ways to exercise. As in the epigraph: Jupiter the servant, circling a tree which Poe calls elsewhere "a magnificent specimen," predicates his universal claims on the regard of minute particulars.

To return to the CT scanner: the orbital gestures which produce a cut are violent like a camera pushbutton: they deliver "a posthumous shock." I emphasize this partly to reframe some of the pretensions of a term often spoken about CT and kin imaging technologies: "noninvasive." The harmlessness of ten minutes on a scanner table is undeniable. But the cuts inscribed on, through, in the aftermath of the specimen-patient are associated with phenomena that, together, conjure up (if one grants a bit of magical thinking) the missing cadaver.[61]

To review these phenomena: the scanner is tuned and calibrated on a "phantom." The patient is asked to lie still, often to stop breathing, within a space whose dimensions recall a coffin's. The set of image slices are referenced to a "scout view" as skeletal as any plain x-ray. Elsewhere they may be referenced to actual cadaver slices. And the diagnoses they enable can speak dire prospects for the patient.

Thus part of CT's capacity for "posthumous shock." But this formulation of Benjamin's, about photography, also relates to the infinite reproducibility of the image of a lived event. This suggests further aspects of the violence of CT's cutting. CT delivers images to a system of readers and processors from which the patient is, for the most part, alienated. The circulation of these images may indeed return something valuable to the patient: a shadowy intimation of personal fate. But diagnostic economies are not solely connected with the fates of patients. During the circulation of images in the hospital, abstracted from agencies of their production (tech and patient), images are exchanged for professional credit, research articles, teaching lessons—a whole range of prizes that constitute, vis-à-vis the diagnosis itself, exchange and surplus values. The violence of the scanner cut is in part its enlistment of the patient's shadow-fate in processes of reification and commodification.

3   DIAGNOSING

It is well said of a certain German book that *"er lasst sich nicht lesen"*—
it does not permit itself to be read. There are some secrets which do not
permit themselves to be told.

• • • **EDGAR A. POE,** "The Man of the Crowd"

What is diagnosis? Etymologies tell us: knowing open, knowing apart. This
makes knowing—or rather, *gnosis:* special, intuitive, elite knowing—something
situated and gestural. Not simply mental, but embodied practice. Poe: "As the
strong man exults in his physical ability, so glories the analyst in that moral
activity which *disentangles.*"[1]

    Carlo Ginzburg has described a way of knowing modeled on hunters' skills,
"venatic logic": reading animal signs, "excrement, tracks, hairs, feathers"—
perhaps "the oldest act in the intellectual history of the human race." He has
shown how this hunters' knowing, past-oriented and reconstructive, formally
resembles future-oriented kinds of divinatory interpretation in texts from
ancient Mesopotamia (third millennium BCE). "Both presuppose the minute
investigation of even trifling matters, to discover the traces of events that could

not be directly experienced by the observer." Both involve "analyses, comparisons, classifications." And both these forms of speculative knowledge are like those of medical semiotics, diagnosis and prognosis. This explains "how a diagnosis of cranial trauma reached on the basis of bilateral squint could turn up in a Mesopotamian treatise on divination."[2]

Ginzburg's genealogy of divinatory knowing, an "evidential paradigm," connects diagnostic enterprises of the late twentieth century to primal scenes of the reenacted hunt and the beginnings of writing. This is an important claim. But it also summons work of historical specification. Diagnosis is shaped in particular situations by conjunctions of discourse and habitus, semiotics and thought style, media and technologies, social structures and economic exchange.[3] Diagnosis is insistently particular, and historical.

This chapter addresses historical dimensions of contemporary diagnostic enterprises, especially those of pathoanatomy and kindred disciplines of morphological correlation that have shaped medical imaging. Following on the previous chapters' treatments of reading, writing, and cutting, it underscores some crucial role relations and ritual contexts of CT diagnosis. And it develops a pivotal conception of this book: diagnostic intrigue. Intrigue is a crucial feature of the thought styles of the CT Suite—and yet its status and effects owe much to historical ballast which the CT Suite, in its rhetoric and ambition, has been eager to transcend: crucially, the cadaver. My fieldwork and writing have been haunted by intrigue just as the CT Suite is haunted by the dead.

Detection and the Crowd

Walter Benjamin called Poe's tale "The Man of the Crowd" (1840) "something like the x-ray picture of a detective story."[4] This expression captivated me.

"X-ray picture" figured so well the skeletal austerity of Poe's story—"the mere armature . . . the pursuer, the crowd, and an unknown man." With admirable brevity the phrase articulated representation, medical diagnosis, decipherment. I knew Benjamin as cultural critic was interested in storytelling, in its rhythms and conditions. And I knew he was concerned about photography's influence on art and experience—in particular how (in another jewel-like phrase) the click of the camera shutter "gave the moment a posthumous shock."[5]

Yet this x-ray picture remark appeared not in an essay on storytelling or technology but in one about Baudelaire's Paris—about new aesthetic sensibilities that evolved in crowded European cities in the nineteenth century. Baudelaire, Poe's translator and "*semblable*," had celebrated "The Man of the Crowd" as portrayal of a social type, the *flâneur*—stroller and watcher, reader of faces and social ranks—a person who not only went shopping in the new arcades of commodity fetishism but went "botanizing on the asphalt."[6] The flâneur embodied a set of responses to new conditions of the crowd that could be somewhat disquieting: close proximity of bustling strangers, favoring looking over speaking and listening.[7]

It was in the social conditions of the crowd, wrote Benjamin, that the detective story found its origins. Masses threatened "obliteration of the individual's traces"—or they offered concealment: "asylum that shields an asocial person from his persecutors."[8] The crowd afforded opportunities for exercise of culprits,' as well as citizen-detectives,' faculties—not to mention techniques of the police. But the police play no role in this tale of Poe's, which catalogs forms of activity that serve the detective and the "type and genius of deep crime." Visuality and visibility figure prominently: casual watching of a crowd, gestures, clothing, more careful inspection, classifying, occupations, typing . . . then close inspection, tight surveillance, abrupt reactions, quick glances . . . intense pursuit. (Pursuit, *suivre*, Suite.)

Recalling Benjamin's x-ray figure: what is reduced to shadow in Poe's tale is, in part, authentic social relation. City life has become flat social text. The narrator is solitary and suspicious—of what remains hidden in the crowd he reads. In this the narrator exemplifies a type. He is no Dupin: but as type, perhaps he serves as a better exemplar of that bold claim by Borges: that Poe did not merely invent the detective genre; he invented the detective *reader*—"a reader who reads with incredulity, with suspicion, with a special kind of suspicion."[9]

Historians of the detective story confirm Benjamin's claim that the rise of the genre is bound up with the complexion of urban culture—but often it is crime, police forces, and criminology that they stress. What Benjamin emphasizes, as detection's condition and content, are aesthetics of alienation and shock.

Shock. This joins Benjamin's urban sociology with his critiques of technology. Both the city, with its jostling crowds, and the age's new devices of "mechanical reproduction"—phone, camera—interrupted flows of experience and inculcated quick gestures—small violences, reflex motions, jolts to lyricism. The click that makes a photographic negative (*cliché*) stops life in its tracks to re-present it to itself.

Poe's "The Man of the Crowd" indicates some sociological affinities of diagnosis. But it is, again, not quite a detective story. Benjamin suggests it is a more austere kind of case, an x-ray picture: flâneur-detective and unknown man, silhouetted against the crowd. This case is doubly austere because its relation of pursuer and pursued eschews standard props of the full-blown detective tale: corpse, scene of the crime. The detective story, which seems elsewhere, in Poe's own hands, to insist so much on the centrality of the crypt or the corpse, is shown in x-ray view, surprisingly (shockingly?), to have in its own skeleton none of these "posthumous" elements.

If detection does not need the cadaver, what does it need? Poe's own double and cryptic formula for "the type and genius of deep crime":[10] refusal to be read; refusal to be alone. I return to this formula at the end of this chapter, as it seems important to the ethos of culprit-pursuit in the Suites of diagnostic CT.

But first, to reflect further on felicity conditions of detection and move these reflections into the hospital, let us consider one of the grand theaters of diagnostic intrigue of the last century, the Clinical-Pathological Conference, or CPC.

Theater of Diagnostic Intrigue

CPC: at University Hospital this ritual unfolds at noon on alternate Tuesdays. Medical students, residents, attendings gather in the Old Clinic Auditorium. One of the residents stands at a podium and presents a case—a narrative of a single patient's illness and hospital evaluation. Her narrative is rigorously circumstantial: it omits a diagnosis. It follows a conventional format: complaint, history of present illness, past history, physical examination, laboratory data, hospital course. (In its classic form, the hospital course culminated in death; lately, this is not necessarily so.)

Following the resident's presentation, a radiologist briefly summarizes findings from relevant x-rays and other imaging studies. But he too avoids offering a conclusive diagnosis.

Then a discussant, a senior physician *not* privy to the diagnosis, offers a half-hour of learned commentary on diagnostic possibilities in the case. A provisional synthesis that makes sense of the available antemortem information. She hazards a diagnosis—an educated guess.

When the discussant is done, the audience members are polled for their judgments. Then a final phase begins. A definitive, final diagnosis—heretofore withheld—is delivered by the pathologist, from her findings at postmortem autopsy. This diagnosis is supported by exhibit of diseased tissue, or photographic slides thereof. A tissual lesion makes possible a retrospective re-diagnosis—corrective or corroborative, depending on what the discussant said.

In this pedagogical ritual of the CPC, there is movement—from confusing specificity of detail to the stable generality of a classified disease type. There is a sense of both fullness and closure: fullness of the individual (one patient, one illness); closure of juridical pronouncement.

But there is more: a "good" CPC performance, for a discussant, is one in which she generates a broad, interesting "differential diagnosis" (which disease entities could do this?), illuminates maxims of patient care (clinical "pearls"), and, finally, makes a daring guess about what disease this is—a guess which is subsequently verified by the pathologist, who demonstrates, if not a rare entity, at least a complex presentation of an ordinary entity. For other members of the audience who engage with the problem, however, a good CPC is arguably one in which the discussant diagnoses erroneously—while they themselves succeed, preferably so others notice, in diagnosing correctly.

The inculcation of diagnostic acumen—that precious asset of the clinician—requires its exhibition.

This explains why the CPC has been a ritual of guild consciousness and pedagogy since the mid-nineteenth century—transplanted from the Paris clinics of Pierre Louis to Boston in the early twentieth century. The CPC is a performative version of "case history"—like other case histories mobilized in everyday work rounds around the hospital—but one ballasted in a particular way, traditionally with a cadaver, and cast in the format of a detective story.[11] A narrative of circumstances and material traces, summoning exercise of interpretive acumen, eventually tested against the legible truth of the dead. As a teaching conference format, it has been preserved by departments of medicine. Though it remains a canonical genre—one is transcribed weekly in the *New England Journal of Medicine*—the CPC's prestige has eroded: it is not as well attended of late, indeed is no longer conducted in some institutions. Nor is its evidential ballast always a cadaver: tissue referred to by the pathologist is often a biopsy specimen, a few dead cells from a still-living patient.

But at University Hospital and wherever else it is conducted, the CPC, this form of case presentation which the hospital inherited from the nineteenth century, comprises both an epistemological structure and an economy of affective exchange. It reenacts analysis within a performative community. It exemplifies the hospital's stakes in theatrics of diagnostic intrigue.

Nature's Secrets

Doctors are detectives. This is a commonplace, a truism, even for medicine's laity, which does not attend CPCs. Many of the CPC's elements—unexplained death, puzzling circumstances, inspection of details, unto the microscopic—sustain a current vogue in forensic pathology dramas on TV. And radiologists are detectives too, in popular and professional imaginings. Says *The Radiolo-*

A Medical Mystery

A 20-year-old college student, newly arrived in the United States, was evaluated for severe abdominal pain. What is the diagnosis?

MICHAEL S. MAHONEY, M.D.
MARC KAHN, M.D.
Tulane University School of Medicine
New Orleans, LA 70112-2699

Editor's note: We invite our readers to offer their opinions by e-mail (mystery@nejm.org) or fax (617-739-9864). We will not be able to acknowledge the receipt of responses. We will publish the diagnosis in the Correspondence section of the October 29, 1998, issue. All replies must be received by September 24, 1998.

22. "A Medical Mystery," *New England Journal of Medicine,* 1998

gist's First Reader: "For the sake of their sanity, diagnostic radiologists must develop a taste for puzzles."[12]

Radiology penetrates hidden secrets. Once opaque, now transparent: this formulation recurs in venues ranging from *National Geographic* to histories of radiology to histories of art.[13] A radiograph exposes, opens up a once-upon-a-time mystery. This is the most general way in which x-rays, including CT scans, relate with secrets, with diagnostic conundra.

Yet the radiograph is not the end of the secret—and there is detecting left for radiologists to do. The diagnostic x-ray reproduces, mimetically, something of the one-time impenetrability of the opaque specimen. It is a transform of Nature's secret.

The rhetorical device of radiograph-as-Nature's-secret is appealing, in popular and journalistic venues. It occurs throughout radiological and clinical professional literatures. Trade journals often carry a "What's Your Diagnosis?" x-ray quiz—wherein a single radiograph is printed for readerly evaluation, with the definitive answer withheld—deferred to a following page, or the next issue. An example of the image-cum-puzzle genre is this "mysterious" image from the *New England Journal of Medicine,* from its weekly feature "Images in Clinical Medicine."

Such an image-conundrum is of course both game and pedagogical device: challenge, test of acumen, prospect of frustration or triumph. And it is a theatrical contrivance, a gambit to ensure readerly return, a buildup to masterful revelation. The stand-alone image, like any "fact" of Nature, is an artifice—divested of what props it up, of the expectations it defies or satisfies, of the history of its procurement. Intrigues, puzzles, mysteries are always culturally embedded.

Once upon a time it was symptoms-unto-death that materialized Nature's Puzzle. Now a filmic slice can do the same. How this comes to be: when and how a filmic puzzle replaces narratives and dead tissue—or quotes them, or imitates them, or exchanges with them: these are central concerns of this book.

Nature's Puzzle never really stands alone.

Death and the Gold Standard

In the CPC, testimony of the radiologist is traditionally a feature of the first half, deliberately circumstantial, a contribution to, but subordinated to, the conjectural reasonings of a clinician. This befits the status of radiological evidence as antemortem evidence, and of radiologists as consultants to clinicians in real-time diagnosis.

But this is odd. Radiological testimony trades on the legibility of fixed appearances—and in this respect it resembles the testimony of the pathologist. Yet resemblance between radiology and pathology has not gained radiologists the privilege of the last word in the theater of the CPC. Here, information drawn from the dead trumps that drawn from life.

This engages the paradox to which I referred in the introduction: the Clinic of explanatory lesions was founded on the cadaver and often still wants to privilege the cadaver—even as the cadaver seems to be on its way out. What does the conventional wisdom mean, when it says cadaver evidence is the "gold standard" in diagnosis?[14]

We inherit the gold standard from classical economy. Massive holdings in a central treasury, depot of value—ballasting the circulation of parchment, the volatility of print-monetary exchange. The moral weight of gold, the truth-value of gold, was much debated through the nineteenth and much of the twentieth centuries. The gold standard materialized vexed relations between Text and the Real.[15] But for even these relations to hold, some of the real had to be buried in a vault.

Are relations between the cadaver, as one-time touchstone of diagnostic signification, and gold, as one-time touchstone of monetary exchange, more than analogical? There is a detective tale of Poe's—"The Gold-Bug" (1843)—that addresses this question with marvelous acuity. Its key props are a gold scarab-beetle and the piece of parchment with which it is captured. A series of exchanges, misreadings and rereadings, pivots on the iconography of death. It emerges that on the parchment is a secret text, and decoding of the text is eventually consummated by the digging up of buried treasure. Decryption connects new species to ancient specie, collection to piracy. But the treasure hole also yields skeletons: and so decryption is also about the crypt. Excavation of meaning and value both take place under the sign of, in complex figurative relation to, death—and a hoard of stolen gold.[16]

One allegorical redux of this tale: gold and dead bodies lay together, in the middle of the nineteenth century, as the end and justification for chains of signification, for exchanges among different orders of inference and evidence. This fits the status of the cadaver in the CPC: not just another datum, another text: the proof-text.

How can radiographic evidence replace, or serve as, a gold standard? Must it quote the cadaver? Where is death in the diagnostic economy of the radiograph? We have already noted the status of the "posthumous" in Benjamin's critique of photographic shock. Historians of radiology note that early receptions of x-rays were marked conspicuously, in professional and popular sensibilities, by concerns about death. X-ray images looked spooky. Moreover, x-rays, it soon became clear, could themselves be dangerous.

But it is not just that skeletal shadows conjured up for viewers the genre of the *danse macabre*—or that early radiographers lost limbs and died prematurely. Writing of any sort—the graph, the trace—whether inscribed via x-rays or some inked stylus, testifies to the absence of a writer. This imbrication with loss is something x-rays share with all writing (and this relates to the shock Benjamin attributes to photography): they are vestiges, fossils, of experience.[17]

Is quotation of death sufficient for radiographs to serve as gold standards in diagnostic efforts? And what about CT scans? CT has been referred to as "a living autopsy." These remarks are excerpted from my interview with a senior radiologist:[18]

> . . . I can remember when I first came here . . . twenty-four years ago, I would religiously go to the Autopsy Conference which is held—up on the eleventh floor . . . And how often at the autopsy bench—Autopsy Conference—we would be surprised at findings, and in many cases. This was my experience when I was a resident and also a medical student at Hopkins as well, where the pathologist was held in great reverence. . . . Even at that point . . . , the old, now retired, but wonderful Professor of Pathology . . . head of the Autopsy Service . . . was bemoaning the decreased autopsy rates . . . Radiology even back then was frequently blamed for the demise of the autopsy, or at least decreased frequency . . . But at any rate, suffice it to say that I was very impressed with how initially—my first ten, fifteen years, including medical school and residency—how the pathologist was the guy you went to with all the answers . . .
>
> . . . But then in the early eighties this strange technology of CT came on board and it all changed . . . our whole way of thinking about pathology changed. And my attendance at the Autopsy Service stopped . . . Very rarely did I find anything at the autopsy table that I didn't know was going on. As far as structural changes— . . . So, it was during this time that the gold standard I think suddenly started to shift—you were asking about gold standards . . . from the autopsy table to the CT table. And nowadays, I have no idea what the autopsy rate is now, here, at this hospital or . . . whether we still have—I know we still have a radiologic autopsy conference—Morgue Conference, as it's called . . . I haven't been to it in fifteen years, come June. In part, I was doing other things too, but I found that there was very little that I really needed to know—
>
> . . . And I go to the CPCs periodically that are still limping along . . . [put on by the] Department of Medicine, and I participate in them . . . As the presid-

ing radiologist. And you know, they used to be the high point of educational experiences, and I'm sure you—the famous clinician gets up there and talks and he's shot down by the famous pathologist and so forth . . . And it was a game, but also a wonderful educational experience—and the radiologist was helpful in the early years, but not always as helpful as we would like. Now, gee, it's rare that we don't see the lesion and know what's already—the diagnosis, it's already on the [film] jacket—and we have to struggle to kind of keep from telling our clinical colleague in hints or body language what the diagnosis is . . . when we go over the films and images with him beforehand . . . So the whole dynamic has changed. The pathologist still, by tradition, is respected and shows the histology, and there are a lot of things that are seen with histology that are not recognized with traditional gross imaging, but there's very little, I must say, that are surprises. Things that are surprises on CT—in CPCs now are more . . . sort of functional things or infectious micro things that are kind of interesting. So, I would be pretty haughty and arrogant to say that radiology now is the gold standard . . . But I think it's getting there.

. . . CT has done more to revolutionize . . . because of its availability and its resolution and its appeal to non-radiology physicians . . . Although I'm sure you'll get a lot of arguments from pathologists about this, but from organ change, mass, structure, atrophy, cyst sort of stuff, sort of gross pathology issues . . . CT has . . . made a big difference.

CT, gold standards, the waning authority of the pathologist, indeed of the cadaver: "the whole dynamic has changed." The pedagogical game of the CPC, highly conserved ritual for over a century, has over the last couple of decades begun to unravel—in part because a more exhaustive portfolio of images, especially CT images, threatens to give away the surprise. The revolution to which computed tomography has contributed so substantially is not merely one of body imaging but also one of the ritual contours of diagnostic proof and pedagogy.

The Radiological Case

The cadaver's authority aside: the CPC has lost ground in the teaching hospital because other conferences now offer it competition. Some of these are equally interdisciplinary; most involve radiology; some are exclusively for radiologists and their trainees. These last are the place to gain a sense of the specifically radiological case presentation.[19] Here, legibility of lesions is still at stake, but within a different economy of attention and effect.

Compared with the CPC, a radiology teaching conference devotes very different proportions of time to clinical, radiological, and pathological considerations. The focus is on image review. Clinical circumstances are highly abbreviated, and pathological correlation can be limited to what is written on a folder. Radiology conferences are rarely organized around a single case, exhaustively discussed. They address multiple cases: *seriation* is crucial to radiological thought and teaching. And because radiological evaluation is (unlike postmortem autopsy) eminently repeatable, within each case there may be seriation of constituent "studies." Studies constituting a case may be presented in sequence from simple to complex, or according to the temporal unfolding of an actual "workup."

A case presentation in a teaching conference is conventionally made by a resident, summoned to the podium after being given opportunity to review the instant images. The presiding attending will often frame the case with an "indication"—that is, the reason for which a study, or series of studies, was performed. This might be a patient complaint, a clinical situation, or prior imaging. This indication may have been inscribed on the study requisition—and may be the only mention of clinical concerns. "This study was performed on an MVA victim who had prolonged loss of consciousness."

The resident lays a film upon the projection table and begins to speak. Tone and rhythm signal formal presentational mode. "Well. [*deep breath*] The study is a noncontrast head CT. [*pause—to expand field of view on display, readjust contrast, and move film from one corner to the corner diagonally opposite*] This says the patient is forty-two years old. This CT shows transaxial soft-tissue images from the base of the brain up through . . ."

Opening phrases of a radiological case presentation are orienting devices. The auditor/viewer is oriented to the genre of the study (CT), its "technique" of production (noncontrast). She is oriented to a key epidemiologic criterion, the age of the patient. And, with respect to a grid of slices, she is oriented spatially—to the plane of slicing and the portion of a bodily volume displayed. Crucial coordinates of radiological knowing: this technical modality meets a patient of this age in this representation of a volume.

Then the presenter addresses findings (see chapter 1, "Reading and Writing") in the film. Often the order has to do with obviousness. One can speak to conspicuous finding first; or one can "run through" a more systematic review, articulating normal findings, before "keying in on" an abnormality. System and thoroughness are important to radiological expert practice. To neglect thorough review of a film in the excitement of discovery of a particular finding is a rookie's error. Residents learning to present cases are held to particularly rigorous expectations. Yet there are also the virtues of brevity. Expertise speaks succinctly to key findings.

Findings are described in terms that render them comparable to other findings in radiological memory. An initial description—"ring-enhancing lesions at the gray-white junction"—strikes a balance between specificity and generality. The description is neither so particularizing as to preclude resemblance to *any* other finding nor so general as to allow resemblance to *all* other findings. A good description locates a finding in an intermediate taxonomy of appearances: it does not specify a final diagnosis. The description stabilizes the finding in a semantic field enough to ask: what disease entities can produce findings that look like this?

This is the question in "differential diagnosis." This term is not specific to radiology; it is common throughout clinical domains and indeed figures prominently in a CPC. Although the term properly references a process of selection and choice, it has in everyday usage become conflated with one of its intermediate products: an inclusive list of diagnostic possibilities.[20] The standard reply to the question "What's your differential?" is a list of disease entities that fit the description at hand.

In teaching conferences, the formality of case presentation may be relaxed so the attending can assist the resident in developing the differential. Building the list sometimes requires contributions from others in the room. Sometimes diseases are eliminated as soon as they are mentioned. Sometimes elimination does not begin until an inclusive list has been developed. Either way, a process of selection unfolds. A measure of suspense is created, and curiosities are engaged and intensified. Eventually the inclusive list of possibilities becomes a hierarchy of probabilities.

If the differential-expanding phase is founded on efforts to reconstruct a textbook list (in which pedagogical interventions are mostly remedial), in the narrowing phase, the attending often contributes particularizing facts, useful maxims, and observations from experience—supplements to textbook knowledge.

At some point, the field of diagnostic possibilities is narrowed to one, or the attending reveals the answer: the heretofore-withheld diagnosis. If the case came from the teaching files, the diagnosis was inscribed on the folder. Usually this is "path-proven"—though visual proof is not often displayed.

The final diagnosis may be offered in a perfunctory manner: amoebic abscess: next case. But it may also recast exchanges which preceded it. For the resident—or her peers—it is advantageous to have selected the correct diagnosis before it is confirmed—or at least to have answered many questions correctly en route. Failure is common, however. "Neuroblastoma?" asks the resident hopefully. "Uh, no," replies the attending, glancing at the jacket. "Hemophilic pseudotumor from a recent bleed." "That was my second choice," mumbles the resident—sarcastic, but not bitter. For the attending, a successfully handled teaching case is matched to the resident's experience and skill and affords opportunities to challenge, discuss, and instruct along the way. On both fronts, there are intellectual and affective stakes: suspense through withholdings, pleasures of discovery, lost knowledges recovered under pressure, acumen displayed. The final diagnosis is a key prop in this performative economy of thought and affect.

Unlike the clinical case presentation, the radiological case presentation is not a narrative of events and circumstances. It offers reconnaissances of images, followed by reconnaissances of diagnostic prospects. Each final conjectural effort aspires to a best fit with received classificatory criteria.

The very tidiness of the radiological case presentation mitigates against sustained intrigue. Cases unfold quickly—and repeatedly—over an hour. A teaching conference comprises a dozen or more cases. A radiological case presentation in such a conference—findings, differential diagnosis, final diagnosis—comprises but one turn in a rapidly spinning wheel of looking and testifying, describing and comparing, remembering and judging.

Poetics of Detection

The foundational works of the detective genre remain exemplary guides to its construction. Poe's tales are canny and often explicit about their form, the aesthetic effects they seek, and the cultural conditions that give them currency.

The detective story is prototypically a *sectioned* tale: divided, twice-told—one pass for clue gathering and conjecture, and a second pass of authoritative retelling, review of evidence. As a reader follows a detective's steps through puzzling particulars of a case, she shares the detective's intention to solve them. But the detective's solution usually comes first, and is mystifying, chastening— "I stared at the speaker in mute astonishment"[21]—so that the reader is eager for explanation and, in the retelling that ensues, can admire the acumen that the explanation reveals. In this sense the reader finds himself in the role of apprentice to a master.

As the reader follows steps of the detective, so the detective follows steps of a criminal—again, a kind of apprentice. Tasks of detection become mimetic overlay to, or recapitulation of, the primal scene of the crime: it is not simply that "the analyst throws himself into the spirit of his opponent,"[22] but often that the detective must reproduce, bodily, key gestures of the culprit agent. The butler, with the candlestick, in the foyer.

Speaking of a detective in the singular oversimplifies the sociology of detection—Poe's and others.' The cognitive skills at stake are appreciable "only in their effects"[23]—and so there is the appreciative narrator, also something of an apprentice, who engages the detective in dialog, assists in and reflects on analytic tasks, and recounts the tale in its proper phases. For this narrator-apprentice, partly soliciting and partly modeling responses of the reader, the originary mystery (whodunit—and how?) becomes encased in a secondary mystery, that of mastery (how did he—the detective—do that?). This social triad—criminal, detective, apprentice—is a spare, but crucial, social nexus. In "The Gold-Bug," the detective's apprentice role is even further developed, into two distinct roles, narrator-friend and live-in servant, such that the analytic corps itself has a triadic structure, its three members distinguished conspicuously by rank and habitus. Their synergies and differences are crucial to the momentum of the tale: detection bridges, is constituted through, hierarchy, divisions of labor, and misunderstandings.[24]

Cognitive work of detection, as comprehended by assistants, represented by a narrator, and reflected upon by readers, is entwined with suites of affect: curiosity, vexation, astonishment, to name a few—and with collegial competition, wherein conditions of one's pride constitute another's chagrin. Within the economy of intrigue, these affective states are not gratuitous by-products but are managed as important and positive phenomena—part of the very fabric of detection. The division between the empiricist circumstantiality in the first part of the tale, and the masterful reconstruction in its second part, shapes this affective economy.

To emphasize affect and aesthesis in this account of detection is not to diminish the role of logic in casework. The case serves epistemological purposes and relies on cognitive skills. But, again following Poe, an adjustment to standard accounts of logic may be helpful. Some of the best discussions of detective logic come from work in semiotics by Umberto Eco and Thomas Sebeok, who consider Poe's tales, among others, as early and canny precursors to theories of the pragmatist philosopher Charles Peirce. They cite Peirce's attentions to a third modality of logic, a supplement to the familiar dyad of induction and deduction: "abduction."[25]

Abduction is a name for hypothesis, a kind of conjectural thinking, applied to a puzzling situation. It is formally distinguishable from deduction, demonstration of a known rule, and induction, which infers rules. The easiest way to appreciate the distinction is to follow Peirce in considering how each modality makes different use of the same three logical elements but assigns them structurally different roles. The three elements are (using Peirce's terminology):

Rule   (general truth, law of nature)
Case   (presumed or hypothetical situation)
Result (empirical, observed datum)

Deductive reasoning makes use of these three elements in precisely this order: that is, it *begins* with the rule, and proceeds through a case, to arrive at an observed result which demonstrates the truth of the rule. So, to adapt an example of Peirce's:

Rule   All the beans in this bag are white.
Case   These beans are from this bag.
Result These beans [therefore, as observed] are white.

Inductive reasoning, by contrast, begins with a given case and proceeds by examing its entailed results, to allow the inference that some universal rule is operative. For example:

Case   These beans are from this bag.
Result These beans are white.
Rule   All the beans in this bag [therefore] are white.

The induced rule becomes more certain with multiplication of cases.[26]

Abductive reasoning begins with observed results and associates them with some general rule/s, to generate a plausible hypothetical case. In terms of beans and bags:

Result These beans are white.
Rule   All the beans in this bag are white.
Case   These beans are [therefore, arguably] from this bag.

This is the most risky, the most crafty, the most conjectural of the three sorts of conclusions. One seeks a case which offers the best situational "fit" between observed facts and general rules. The abduced case is not so much truthful as plausible.

In practice, the conjectural leap of abduction—what Peirce also variously called "hypothesis," "guess," "presumption"—is immediately resubmitted to inductive or deductive processes for corroboration or refutation. Whichever logical process is emphasized in a given setting depends on the combination of facts, rules, or cases ready to hand, and dispositions of a logician.

Three features of abduction warrant special note. First is its provisional and crafty nature: abduction is inspired and clever guessing, producing a case which fits, and hopefully explains, some preexisting fact.[27] Second, selectivity is key—including selection among an infinitude of possible rules (general concepts or doctrines) which might be brought to bear in explaining a particular fact. Third, abduction is insistently performative. This is not to say that the logical operation itself cannot be carried out in silence—the master detective often surprises, with the solution of a case arrived at privately—but rather that abduction cannot easily be appreciated or taught without exemplification, or without involvement of the body of the detective. Thus the detective's explanation about how he solved the case prototypically involves mimetic reconstruction of the gestures of the criminal at the scene of the crime. And in the bean/bag example: the abduced case arises from the comprehension—grasping together—of found beans in the one hand, and a received rule in the other.

Historicizing an Ethos

Borges's claim about Poe—that he created the detective *reader*—is, while fair tribute, a historical oversimplification. There is Ginzburg's analysis of a universal human conjectural paradigm. There is Peirce's account of logical operations— also universalizing. We know tales of curiosity, riddling, and deception from Greek myth, European folklore, early modernity.[28] If we suppose nonetheless that a particular kind of "suspicious reader" *did* emerge in the middle of the nineteenth century, what other historical influences were there—besides Poe?

It will still be hard to beat Poe at this game. He wove so many of these very influences into the fabric of his tales. This is a part of Poe's own cryptographic transmission: his tales are like little windup machines of cultural reflection. They offer glimpses into, clues about, and commentary on the cultural conditions of the intrigue he weaves.[29]

There is the rise of comparative anatomy. A text of Georges Cuvier's helped solve "The Murders in the Rue Morgue" (1842). Poe knew the "Introductory Lecture on Human and Comparative Physiology" (1826)—and understood the revolutionary status of Cuvier's work bridging fields of natural history and medicine.

Part of what was important about comparative anatomy was its embrace of dissection: the scalpel revealed forms of anatomic difference that mere superficies could not. Dissection (not so much new as newly invested) became a privileged means of relating exterior morphology to interior organs and mechanisms—thus Cuvier's famous boast: to be able to reconstruct the whole animal from a single bone, indeed, from a single facet of a single bone.[30] In turn, the interrelationship of organic systems, as revealed through dissection, provided impetus for a new way of thinking about life—as codependent processes taking shape amid environmental constraints—especially the constraint of death. Opened-up animal bodies were central to the genesis of "biology" as such.[31] Moreover: dissective technique in the nineteenth century marked continuities between comparative anatomy and pathoanatomy—that ur-technique of medicine's emergent detective ritual, the CPC.

Cuvier's work was invoked in Poe's preface to a shell-collector's guide—a philosophical meditation introducing a book to "aid in the collection of a cabinet." This preface articulated a *relational* approach to morphology: "as the Greek *conchylion* . . . embraces in its signification both the animal and shell." It also pointed to other domains of Cuvier's expertise: paleontology and geology. Fossils, Poe noted, bear the imprint of geologic time—both as "immense beds composed of the spoils of these animals" and as individual crystalline "medals of creation." And collectively, the cumulative reworking of "calcareous matter" by the "conchyliferous mollusca" constitutes a formidable, earth-moving force: "they alter the physiognomy of the superficial structure of the globe."[32]

Poe's fossilized shell here plays multiple specimen roles: comparative anatomic specimen, calcareous clue to long-decayed flesh; local geological specimen, clue to stratigraphic time; and part of the crusty face of the organism of the earth. A list of clue-following disciplines: comparative anatomy, stratigraphic geology, paleontology, phrenology. An array of conjectural scientific inquiries marking a shift from prior concerns of natural history.

The public face of this shift was the Museum of Natural History. This was no mere *Wunderkammer,* or cabinet of curiosities.[33] Museums like the Peale Museum in Philadelphia appropriated and installed specimens and artifacts in the role of *clues* for public decipherment—regarding earth's prehistory, a vertebra's relation to a gait, man's relation to animals, and so on. These museums were crucial institutional contexts not only for display of specimens, classified and labeled, but also for the inculcation of a complex conjectural sensibility, a new ethos of detection, for popular as well as scientific clienteles—a schooling in relations between visible features and hidden realities. These relations took two overlapping forms. First, there were projects in what Tony Bennett (after Thomas Huxley[34]) calls "backtelling"—narrative and epistemological reference to the "newly fashioned deep-times of geology, archaeology, and paleontology"[35]—engagement with an unknown that was specifically past, even prehistoric. Second, there were projects of constructing unknown wholes from

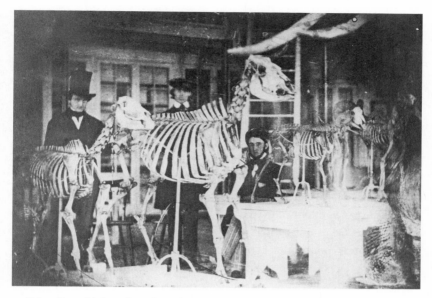

23. Edgar Poe with Joseph Leidy and assistant in Academy of Natural Sciences, Philadelphia, ca. 1842

known material parts, like Cuvier's anatomic reconstructions.[36] Poe eagerly engaged, and folded into his fiction and criticism, the new scientific fields developed from what had heretofore been "natural history"; a daguerreotype (figure 23) shows him in a museal space amid comparative anatomic specimens.[37]

Cultural historians have remarked on other museological dimensions of this nineteenth-century conjectural ethos. Egyptian archaeological artifacts were prized for evocation of a buried past. Hieroglyphics stimulated public enthusiasm for cryptography (exploited by Poe). And there was craniometric anthropology. Craniometricians were "backtellers" in paleo-oriented registers, but they also engaged more presentist comparative anatomical discourses, like phrenology and racial classification.[38] Phrenology became a cultural force that extended these museal preoccupations to the streets, through handbooks and parlor games, in Poe's day and beyond.

With or without "deep-time," conjectural reconstruction has continued to operate as a powerful feature of museological projects and their cultural spin-offs since the late nineteenth century. Reconstruction, of missing morphemes and hidden causes, was not so much invented in the nineteenth century as it was given an institutional home then, and new scientific roles, with a widening popular audience. Small wonder that this was the era when detective literature blossomed—and the era and ethos within which morbid diagnosis, based on new pathoanatomic methods, bloomed, in Paris, then elsewhere in Europe and America. The institution of the hospital was a key nineteenth-century locus, beyond the museum and the urban settings of lay detection, where conjectural reconstruction, abductive reasonings, assumed special importance.

Constraints on Curiosity

At the CT viewbox or conference podium, diagnostic intrigue involves more than standing outside of, and triumphing over, opaque Nature. Trickster lesions, wagers, errors: sometimes the diagnostic process is a game, with contestants vying for credibility and credit.[39]

How does the game in CT compare to other diagnostic arenas? Some radiologists express concern that CT reading is relatively uninteresting. One senior radiologist offers these concerns about the historical situation of CT and radiology, and his own "aesthetic" interests.

> I enjoy the diagnostic process, the intrigue, the curiosity, the satisfaction of search or not . . . Anyway, I tell my residents and they look at me askance— "fuddy duddy old fart"—I say "CT makes it too damn easy for you. It makes it too damn easy. You can see things sitting right there. You got a 3mm slice and got this darn mass there you know . . ."
>
> . . . Because the modern medical student and house officer is in a hurry and they want the answer and the slice is right there, there's the answer right there. Whereas the beauty of radiology, to me anyway—maybe because again I come from a different tradition—has been the curiosity and the question—the curiosity has been piqued by the subtle findings. I tell the residents, "You know, I like to be able to explain everything on this plain film." And if I see a problem and I can't figure out why it is, I'll feel like I've been frustrated. I like to be able to look at this and understand everything that's going on. If I have to use CT to help me, fine. But most of the time, I can sort of get a sense of what's going on without that . . . And it's the fun of that. Which is really exciting to me. It's the fun of it. Whereas CT, it's so cut and dried, it's so obvious, and maybe to a clinician it sounds great—the person never took a look at the plain film anyway . . . And was ignoring findings that he should have been seeing for years . . . And now CT comes along, and any Tom, Dick and Harry can look at a CT and find the abnormality . . . CT doesn't really give me—fifty slices of the body doesn't give me a real sense of what that body is . . . A chest x-ray or abdominal plain film does . . . in terms of the appeal to me.[40]

This radiologist, in the year of the Röntgen centennial, finds my questions about aesthetics a provocation to nostalgia. His testimony forefronts psychology, private affect—pleasure and pride of the diagnostician. But these reflections are also social, embedded in concern with training, the propagation of his craft.

The desire to understand, to get the answer, to see the whole picture, is associated with a process the radiologist calls "fun." CT poses a threat to fun, by making things too clear, too quickly—by turning search and inference into mechanical revelation and rote response, and by substituting quantity of information for the image of a unique totality.

This radiologist's concerns correspond with what Max Weber, as a sociologist of specialization and of modern institutional rationality, called a trajectory of disenchantment. In his famous essay "Science as a Vocation," Weber suggested, early in the twentieth century, that modern science is increasingly populated by those he called, quoting Nietzsche, "Specialists without spirit, sensualists without heart."[41]

Is curiosity a necessary casualty of modernity, mechanization, speed? Of laying out the lesion as a slice? It is more complicated than this. As suggested at the outset of this chapter, detection has thrived on alienation and shock since the nineteenth century. Some kinds of intrigue continue to bloom in institutions, at viewboxes, under conditions of divided attentions, interrupted gestures, repeated readings—and the general exclusion of patients. It is not clear that these cultural conditions can be assimilated to a grand narrative of disenchantment.

Detection in the CT Suite

How is intrigue present to the postmodern CT Suite? And how not? How does CT interpretation contribute to intrigues in the wider hospital? Agents, props, and effects of diagnostic cases are unevenly distributed in rooms, hallways, and documents.

Some situations in which CT scans are interpreted are scripted rehearsals: teaching conferences, employing rhetorics of detection sketched earlier in this chapter. Other situations are more improvisational, ad hoc—as when radiologists and clinicians come together, at various junctures in the diagnostic workday, to compare notes at the reading room viewbox. In the waning days of film, the reading room viewbox is still the privileged locus of radiological interpretation.

But not all imaging cases foster intrigue. Some images are straightforwardly matched with an image in a table or a textbook. How is this direct matching, sometimes disparagingly called "pattern recognition," related to the Peircean account of "abduction" laid out above?

If, in abduction, the gap between the generality of rules and the particularity of findings is to be filled by conjecture—by a plausible, but indefinite "case"—this gap is narrowed considerably when general rules are available that are especially particularized, or when findings are especially typical. The fit of the case becomes more automatic, its status more certain. In such a situation, in which the case is thoroughly determined by the pincer-grip of rules on findings, it is said that findings "are diagnostic." In such a circumstance, abduction is thinned, unto pattern recognition. Or findings which summon abductive conjectures from a novice can become snap classifications for a master—by virtue of experience with similar findings, and a more robust repertoire of rules.[42]

When CT diagnoses seem straightforward—"the slice is right there, there's the answer right there"—they dispel intrigue. Radiologists may also be alienated from broader diagnostic reasonings of clinical teams outside the reading room. When they remain cloistered procurers of findings, their contributions can be perfunctory, circumstantial. Radiologists can be oblivious to intrigues far from the viewbox.

But sometimes an extramural puzzle is delivered back to the reading room—as supplement to radiological fact-finding . . .

One day a requisition from the ER requests a chest CT to "rule out aortic dissection." The scan is performed, and, in the reading room, the thoracic aorta is pronounced normal. "Incidental note" is made of a "homogeneously calcified spleen."[43] The official CT report does not comment further on this finding.

The next day, a Medicine resident, struck by this report, enters the reading room. She asks the CT fellow: "What's the differential of a dense spleen?" The fellow recalls the case. "Probably a mixed hemoglobinopathy—something like sickle-thal . . . like a slow infarction. Also you can see something like that with old Thorotrast."[44] From a radiologic perspective: the finding is stark, the list of alternate diagnoses tidy, reportage matter-of-fact.

But the resident's eyebrows arch; she is arrested. "Oh my gosh! He's a sickler!" A sudden insight. Evidently this forty-seven-year-old man, admitted yesterday with chest pain, has no documented history of hemoglobinopathy (sickle syndromes are usually diagnosed in childhood). The patient left the CCU today, having "ruled out for MI," and this resident's team has taken over his care. "Has he had other pain crises?" asks the fellow, surprised that someone could suffer death of a spleen and not notice. "I don't know," responds the resident. "He *is* black, from Barbados—I've only just met him—thought

I'd start with this weird CT finding." She leaves quickly, elated with the sudden fit of this explanation for chest pain—and eager for further laboratory proof. Later she confesses having exploited the situation by making a "gentleman's bet" with the cardiology fellow. Though she properly deserved no credit for "making the diagnosis," she found a way to appear, however briefly, prescient; later she says she found the role of purveyor of this diagnosis—to her interns, her attending, and to the patient—eminently satisfying.

So this shard of intrigue (what to call it without the ingredient of suspense?) properly belongs to the hospital wards, and only incidentally to the reading room. The CT fellow had no reason to suspect that his finding would in one stroke both spotlight and explain a lifetime of intermittent pain. He sees this discovery as a "bonus." He appreciates some of the affective spoils of intrigue—surprise, satisfaction. Outside the reading room, the calcified spleen is a veritable prize, exchangeable by its purveyor for cultural values of various kinds. Credit, acumen—and tribute to a patient's stoic endurance.

Diagnostic intrigue is, indeed, no stranger to the reading room. This is a place where apprentices sit with masters and receive guests from other disciplines. Intrigue develops, as Poe's tales suggest, across hierarchies and divisions of labor. One can observe a case take shape in the CT reading room through a series of exchanges among clinicians and radiologists—over an hour or a few days. Often the first radiological engagement is not with the images but with a cryptic inscription on a requisition: what question is this scan intended to answer? An hour afterward films are reviewed by the resident in discussion with the fellow, then a bit later with the attending, and yet again in dictation of a formal report. Over the course of the day there can be a series of clinicians parading through the reading room to look at the same set of films, all posing different questions, thickening the plot, and soliciting reinterpretations which are, sometimes, modulations of the official reading.

Visiting clinicians shift the spectatorial sensibilities of radiologists. On the one hand, visitors constitute interruptions of an assembly line. On the other hand, they generate speculations, plot thickenings, and supplemental conundra which make new claims on the attentions of radiologists and may contribute, little by little or sometimes suddenly, to the development of full-blown mysteries. They make the staging of curiosity a more complex intrication than simply standing outside Nature's Puzzle. A mystery in the CT reading room is hospital sociality's puzzle. The reading room is where, to return to Poe's cryptic formula for "the type and genius of deep crime," that which "refuses to be read" also "refuses to be alone."

Rogues' Galleries

A key ingredient of the detective tale has been elided in the foregoing account. This is the contribution of the police. Though police work is not represented in "The Man of the Crowd" or in "The Gold-Bug," it figures in all three of Poe's Dupin tales. Police methods are often backdrop and foil to the wits of the citizen-detective. Dupin criticizes them: "What is all this boring, and probing, and sounding, and scrutinizing with the microscope, and dividing of the surface of the building into registered square inches—what is it all but an exaggeration of the application of the one principle or set of principles of search?"[45] But police methods became too effective to dismiss so summarily. The metrological scrutiny disparaged by Dupin reached an apogee of sorts in the late nineteenth century, a generation after Poe's work. European police administrators were at pains to deal with a problem presented by urban crowds: identifying recidivists—those who had, in the urban divisions of labor, chosen a career of crime. These culprits were not like Kidd the pirate of "The Gold-Bug," or the exceptional culprits of the Dupin tales (ourang-outang, foreign naval officer, the Minister D——). These were professional criminals, in whom police confronted, as Allan Sekula has outlined, "mastery of disguises, false identities, multiple biographies, and alibis."[46] The chief architects of police methods were, on respective sides of the English Channel, Alphonse Bertillon and Francis Galton.

Bertillon, director of the Identification Bureau of the Paris Prefecture of Police, developed an elaborate system for measurement of criminal bodies—height, limbs, other attributes hard for criminals to dissemble. From his records he identified his first recidivist in 1883 and over the ensuing decade processed 120,000 prisoners, sorted according to anthropometric "signatures." Bertillon's anthropometric archives were linked to photographs of criminals, especially of ears (thus a lateral view accompanied frontal view of each face). He identified recidivants "by yoking anthropometrics, the optical precision of the camera, a refined physiognomic vocabulary, and statistics."[47]

In England, Francis Galton's contemporaneous use of photography for policing involved photographic composites representing criminal types. This technique was not about differentiating individuals but rather about superimposing them, combining multiple portraits into one, to create a kind of "pictorial average" of typical features of some class. Galton sought, like Bertillon, to combine statistics and photography in novel ways—but in order to accomplish a kind of typological classification and visual profiling, one that ultimately proved more influential and useful for a racializing anthropology than it did

Pl. 59 (a) Identité individuelle avec Dissemblance physionomique.

N.os 1 et 2. — Même individu : la 1.e fois à 17 ans et la 2.e à 24 ans.

N.os 3 et 4. — Même individu à 14 mois d'intervalle dont 10 passés en prison.

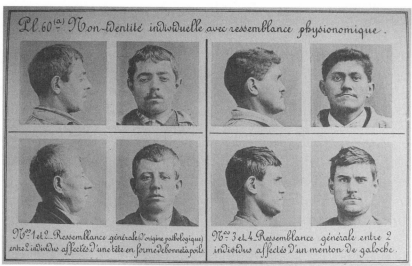

Pl. 60 (a) Non-identité individuelle avec ressemblance physionomique.

N.os 1 et 2. — Ressemblance générale (d'origine pathologique) entre 2 individus affectés d'une tête en forme de bonnet à poils.

N.os 3 et 4. — Ressemblance générale entre 2 individus affectés d'un menton de galoche.

24A AND 24B. Alphonse Bertillon, Resemblance versus identity, 1893

for criminology. Despite their differences, both Bertillon's and Galton's photographic projects were comparative and classificatory, trading on resemblance among images and their modes of juxtaposition.

CT interpreters, as well as other radiologists—and pathologists too—likewise trade in morphology, in visible appearance, in resemblance and juxtaposition. They face a problem analogous to that of police confronting criminal recidivism. They must track, identify, and map habits and associations of a set of culprits—diseases. Diseases are repeat offenders: they may take characteristic shapes or leave telltale traces. A good inspector with a good archive can correlate one lesion with others like it—and say yes, alas, this disease has struck again. But judging similitudes can be difficult. And a disease can be tricky: it can disguise itself, change its features, mimic another, change its modus operandi. The diagnostician strives to see through such tricks, to identify the culprit, to complete a dossier on it and file it in the archive again—regardless of whether its particular crime can be remedied (that is, regardless of prospects for treatment).

One of the confusing aspects of this project is that it mixes terms of practice. "Identification" of disease is a classificatory act: it means something like identifying a member of a species. When we refer to a disease "entity" we do not usually mean a hunk of contiguous biomass—an individual, or a biological specimen. A disease "entity" is a category, defined by family resemblance among various instantiations, a statistical and typological production. In this sense a disease entity partakes less of Bertillon's individuating ears than of Galton's composite types.

It should be no surprise that the very kinds of photographic galleries of culprits that served Bertillon in identifying individual criminals, and Galton in profiling types, turn out to be quite useful to radiologists. Two linked galleries (figures 26a and 26b) come from the Department of Radiologic Pathology at the Armed Forces Institute of Pathology (an institution whose importance to the radiological enterprise, and relation to the natural history museum, will be discussed in chapter 6, "Exposition"). They are examples of classificatory tools designed for use at the "point of service."

The first of these galleries, "Range of Disease," shows different appearances of the "same" disease entity—thymoma. This is like a dossier of Bertillon's, cataloging the disguises and *dissemblances* among portraits of a particular identified criminal. The other gallery, called "Differential Diagnosis," shows different entities that may share similar appearances with thymoma and visit the same site—here, the mediastinum (central chest). Each image in this gallery is chosen for its typical features: it is like a Galtonian composite-type. But

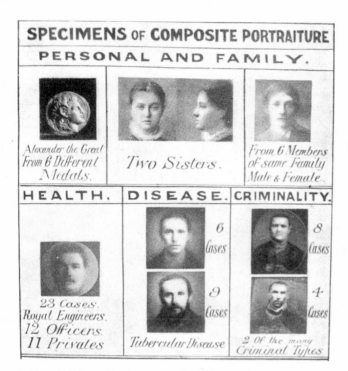

25. Francis Galton, Photo composites: criminal and other types, 1907

the whole array can work a bit like the police lineup—testing accurate recognition of the real culprit in the company of other look-alike suspects.

So radiographic casework in the hospital today resembles, somewhat uncannily, photographic casework in the prefecture of police a hundred years ago. What, again, are the grounds for, and terms of, this analogy? The pursuit of culprits; difficulties posed by comparing appearances; the salience and fragility of "recognition." And these are merely the optical, morphological dimensions of this analogy. Add correlations with other forms of evidence, the narrative reconstruction of the crime, of the illness . . . these are dimensions of the case in its larger, thicker senses, of which the imaging case may prove a modest—or a definitive—part.

Reprise

Diagnosis involves conjectural knowing—inference of significance embedded in minute details. This kind of knowing had a particular public efflorescence

AFIP SUBJECT REVIEWS

RANGE OF DISEASE

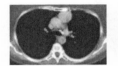

Homogeneous

Heterogeneous

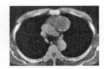

Calcifications

Cystic

Invasive

AFIP SUBJECT REVIEWS

DIFFERENTIAL DIAGNOSIS

Germ Cell Tumor

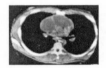

Hodgkin Lymphoma

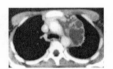

Lymphangioma

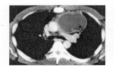

Teratoma

Thoracic Goiter

Thymic Carcinoma

Thymic Carcinoid Tumor

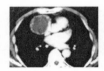

Thymic Cyst

Thymolipoma

26A AND 26B. AFIP tables: rogues' galleries for diagnostic identification

in the nineteenth century with the emergence of a detective, decryptive ethos—deriving some of its interest in the hidden and the not-yet-readable from the conditions of the crowd, and some from reconstructive projects of the natural history museum. Both these were concerned with work of collection, comparison, and classification—and morphological resemblance—the crowd through police culprit galleries, the museum through taxonomic tables and field guides.

Poe reminds us that decryption is related to death—the crypt, buried pasts, dead specimens. But he also reminds us how cathecting of a site or a specimen with curiosity or puzzlement is not strictly natural. "Nature's Puzzle" is something of a Romantic notion. Detection has social aspects, comprising hierarchies, divisions of labor, structures of apprenticeship. And detection has semiotic aspects, including risky operations of abductive inference and the recurrent prospect of *misreading,* which complicate signs' relation with the real, and tasks of expert judgment. And detection has performative aspects, as in the twice-told genre of the detective tale, with its odd dependencies on admiring narrators and mimetic relations between pursuer and pursued, which are important to its accomplishments and its aesthetic effects. Perhaps this contagious performativity is part of what Poe meant, in "The Man of the Crowd," by referring to a culprit's "refusal to be alone" as part of the "type and genius of deep crime."

Disciplines of comparative anatomy (at the inception of modern biology) and pathological anatomy (at the inception of modern medicine) were linked through practices of dissective cutting. The CPC assigns dissection a crucial role in medicine's theater of detection—structuring the diagnostic project around the gold-standard evidence of the cadaver.

But CT scanning: it brings a radically different kind of cutting to diagnostic projects. Its thorough antemortem slicing assists radiology in displacing the cadaver from a once-crucial role in diagnosis. CT also amplifies the "posthumous" shock of photographic reproduction. CT contributes to the brevity of the radiological case, which eschews extended chronologies of symptoms, signs, events: in the same conference timeframe that Medicine investigates one puzzling case, Radiology handles a dozen. CT can flatten logics of abduction to pattern recognition—and preempt puzzlements once enjoyed by readers of plain x-ray films. CT is, if not a disenchanting technology, then a technology whose relation to the putative continuity and organicity of "narrative," and to the ethos of diagnostic intrigue, is problematic.

To understand the place of CT (and its kin sectional imaging technologies) in the broader economy of hospital diagnosis—to understand what forms diagnostic intrigue can assume among radiologists, and under what conditions—it is necessary to address a fuller suite of CT practices and some of the historical precursors on which they trade. I engage diagnostic intrigue again in chapter 5, "Testifying and Teaching." But first I take up the archival practices that organize images, bodies, texts, in radiology, under the sign of Curating.

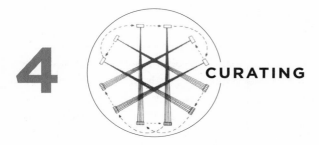

4 CURATING

"The description of the digits," said I, as I made an end of the reading, "is in exact accordance with this drawing. I see that no animal but an Ourang-Outang, of the species here mentioned, could have impressed the indentations as you have traced them. This tuft of tawny hair, too, is identical in character with that of the beast of Cuvier."
• • • **EDGAR A. POE,** "The Murders in the Rue Morgue"

Tuft and nailprints at the throat of the corpse: comparative anatomy meets pathological anatomy. Hair and prints would be tracker/hunter topoi—stuff of ancient interpretive practices—except that these hunters are bourgeois citygoers equipped with a text and a "*facsimile* drawing." They are not at the throat here—they're removed from the actual nailprints. Poe's canny drama connects a gentleman-scientist's lexical description with a graphic image: it subordinates *autopsis*, firsthand seeing—of the fatal wound, the actual beast—to library work.

Registering the dead, and traces of culprits, are curatorial concerns. Corpses, criminals, deviants were foundational cohorts in the history of modern archival projects.[1]

Fast forward to the CT Suite, Department of Radiology, late-twentieth-century hospital. Intensely archival domains—concerned with storage, organization, and display of images and documents. As ethnographic present, radiologic practice is shot through with archival past, with curatorial care. Archives in hospitals have metastasized well beyond their *locus classicus*, the library, and beyond their nineteenth-century formations—census, map, museum—to cite Anderson's prototypology.[2] In the CT Suite and throughout the hospital, traditional bureaucratic fixtures—rolodexes, file cabinets, bookshelves—are squeezed among computer workstations.[3]

Archives and Authority

Custodians of important texts and relics, of origins (*archē*) and their annotations, often derive from these a kind of authority. Curators also serve polities. In Western Europe until the Reformation, archival custodianships were functions of the Church and of courtly regimes—priestly and princely collections.

Archives assumed new forms in modernity, as European states comprehended far-flung dominions and organized testimonies, artifacts, and specimens for exhibition at the metropoles. Genres like the atlas, travelogue, and encyclopedia developed, along with, in universities and learned societies, new curatorial systems. These informed notions of scientific authority (related to, but often distinguished from, religious authority). In later modernity, archives produced new semiotics of power for nations and empires, and new, bureaucratized imaginings of community. As spaces, these were centralizing: they gathered findings from remote peripheries. As instrumentalities, they were hierarchizing: objects were brought from imputed chaos or primitivity into civilized, disciplined order. Centralizing, normative: archival structures, and their tools of assembling and classifying—demographic table, cartographic grid, glass case—materialized empiricist agendas of empire.[4]

Contemporary radiologic archives combine old and new techniques of collecting, indexing, and classifying. Lists, nameplates and barcodes, computer screens, file cabinets, shelves (for film and paper, magnetic tape and optical disks): *materia* curated in radiology are polymorphous, linked, undergoing exchange and translation. The shift to "paperless" and "filmless" systems, as important in radiology as in any bureaucratic discipline, complicates this further. Archives throughout medical imaging worlds are volatile and dispersed—a paradox, insofar as an ancient archival function is that of collation.

A woman on the elevator has a stack of film folders under her arm. She cheerfully confirms what her ID badge says: she is Sally the film courier, of the Department of Radiology. She is on her way from "Preop" to the "Main Department" to deliver these folders.

Nurses of Radiology say the courier's job is one of the most demanding in the department. This is partly about weight carried and miles walked—and being at the beck and call of many masters. The courier's distributed efforts have centers of gravity. But they differ day to day; and on some days there is such demand to move the collection around that she "just can't catch up."

Asked if she considers the Main Department to be the hub of the radiological archive, the courier replies no, that would be the Film File Room. It is next on her itinerary, so she escorts me there.

In the late 1990s at University Hospital, film is still the crucial medium of radiological diagnostic practice—though there exist "filmless" departments in other institutions, and digital archiving will soon make film almost obsolete here as well.

The Film File Room contains "old film" holdings of individual patients. At University Hospital this is in the basement. One cannot get here via the central elevators, but only from the staff elevators in the east hospital. The courier introduces me to the clerk at the file room window—and, across the linoleum countertop, submits her own request for studies. I save my ethnographic questions: clerk and courier are hurried. This film archive is not a museum, census, or map. It is more like a special-collections library. It loans out one-of-a-kind dossiers to authorized patrons, for specific periods, for review in institutional settings.

Patrons' interest in these holdings is distinctive: these dossiers document *culprit* entities. Images held in the film file room were made under circumstances of suspicion. Though culprit entities here are lesions, not patients, the folders are indexed to persons. Some are hospital inmates; most are at large. Such dossiers, registering individuals as bearers of morphologic traits, are in this respect successors to the anthropometric and photographic dossiers of nineteenth-century police.[5]

At this request window, however, culprit lesions notwithstanding, protocols of dossier exchange are not particularly inquisitorial—indeed, not so different from those of other bureaucratic institutions, where, over elbows on formica, requests are made by authorized agents to a clerk. File room curators do not voice diagnostic suspicion, nor do they refer to lives of patients in explaining regulations concerning film movement. The ambience at this countertop is more like the reference librarian's desk, or the DMV. When suspicion or curiosity appear, they are about logistics: "*How* many films by *when*?"

A request for past x-ray studies (older than a couple of days) will produce one of several responses. A folder may be "pulled" immediately from the stacks.

If it is not at hand, its location in a clinic or other node in the hospital may be specified. Some folders circulate more than others. This may be because of a flurry of appointments, a hospital admission, a disease crisis—or it may be because of institutional interests not bearing on patient care—a research project, a conference.

The clerk releases a folder in exchange for an ID number and destination. File clerks exercise caution releasing folders. They may not care about a patient's clinical circumstances, or ritual contexts for exhibition of a study (conference or rounds)—but they care to protect their collection. The proportion of the hospital's films circulated at any time is small: from the clerks' standpoint, all studies are supposed to be *here*, in the file room. Even with safeguards, films and folders disappear for days, sometimes indefinitely. Physicians are often culprits. "Physicians take films all the time"—off the viewbox, out of reading rooms; folders are signed out to persons who, on later inquiry, "don't know anything about them." From the clerk's perspective, his patrons, including doctors with IDs, are suspicious characters—irresponsible, unto thieving.

The curator is, after all, the one who best understands conservative handling of precious cultic objects. There is something, in this suspicion by the clerk of the doctor, of the librarian challenging the scholar, the monk challenging the priest. There is an aspect of Weberian *charisma* in this authority that derives from service to cultic objects—different from the authority of prophetic or priestly charisma, of revelation and elite insight, that is the purview of the doctor.[6] The clerk's legalism at the formica countertop, while a commonplace of bureaucratic administration, is not ipso facto bereft of enchantment.

Another response one may receive to a film request is that the films are in "deep" archives offsite. The hospital's full collection comprises 250,000 folders—so only those used within the last six months are kept in the basement. University Hospital's warehouse is a commercial storage operation known as "Starpoint." This facility, on the south edge of town, is not exclusively for medical recordkeeping but for storage of many sorts of commodities—storage being one of those services that can be generalized and outsourced.[7] With film, requirements are climate control, security, and timely accessibility. A courier from Starpoint delivers films several times daily on phone request; the hospital courier has never actually been there.

In this offsite warehousing, the storage function of the archive is separated from its display function—even more than in the basement file room. Starpoint has no visitors' gallery; it is not an "exhibitionary complex"[8] fostering civic pride, or instructing polities in the ways of Nature or of the Other. Though it has full taxonomic continuity with the circulating collection, and thus with an ethos of display, this facility and its collection are distinctly private.

A third possibility at the file room counter is that the desired film images may have been destroyed. Any study older than seven years is likely to have been "recycled."[9] There are rationales: cost of film and storage; recovery of silver salts; marginal benefits of older records to clinical decision making; statutes of limitation for malpractice suits. Some patients are surprised by this rapid turnover, thinking that their film records are a lifelong resource and that the hospital has an obligation to faithful custodianship. But films are not retained as compulsively as the paper documenting their interpretations. A decade-old cranial CT is, from a curator's viewpoint, obsolete.

One advantage film has over corruptible flesh—resistance to decay, information loss: "shelf life"—is subordinate to an economic of shelf space. The use value of filmic images is defined by the institution: no fixed relation holds between the lifespan of x-ray film and seasons of a person's life. Lifespans of films are better connected to the lives of diseases—polycystic kidney, colon cancer—seven years constituting a period sufficiently long to answer questions about "management."[10]

One file clerk remarks that he only makes a mental link between films and persons when a folder is marked "deceased." Yet even unto death, circulatory systems of bodies and of films are surprisingly contingent. Dying in the hospital results in updating of the Hospital Information System (SMS) with a "disposition code." But registration of this disposition on the film folder is not immediate. Whether a patient is living or deceased is not always apparent from inspection of the folder. Some folders circulate for conferences and research purposes irrespective of the life or death of the person they reference. Films are stored for seven years regardless.

This file room is not a final resting place, a catacombs. Films are neither skeletons nor headstones. Film archives are holding areas, transitional zones: loci of rigorous, but ultimately temporary, registrations of persons with silver-salted shadows. Mortality hovers here, but contingently.

The filmic shadows here include culprit entities—lesions, sites of clinical suspicion: cases. The life-or-death valences of some of these cases—and the irreplaceability, the singularity, of some films—accentuate the policial function of the film curators: they preside over a lockup. But their custodial meticulousness is disconnected from the culprit status of lesions. The suspicion the clerks feel is not directed toward diseases, about whose villainy they remain agnostic, but toward the clients who come to interrogate the prisoners, borrow the evidence.

Specimens

A radiology attending is showing slides in a GI Radiology Teaching Conference of what he says is lymphoma: "I think this [*on x-ray*] is a very soft lesion . . . this lesion was spongy, even the fixed specimen. Adenocarcinoma turns out to be hard . . ." Slide change. "So here's the gross specimen here . . . surface irregularity, clearly a mucosal process . . ." The photograph depicts something like a brown caterpillar with circumferential stripes lying on a green surface. Slide change: microscopy: dark gray on a bluish white background. "Path: there's villi . . . this would be beneath the muscularis."

Standard chain of associations: x-ray lesion, gross specimen, microscopic slice. This chain links archival modalities of radiology and pathology—"radpath." In chapter 6, "Exposition," I follow these associations to a special institution—the Armed Forces Institute of Pathology—whose public antechambers are the National Museum of Health and Medicine. There is a museological dimension to archival structures like this—even at University Hospital—where disparate media are yoked as exemplars of a taxonomic type.

The curator's term that best comprehends this association of different forms of materiality—flesh and film—is the one spoken twice by this radiologist showing the lymphoma: *specimen*. His narration of correlate slides—and my pilgrimage to the AFIP—exemplify how x-rays are part of a chain of specimenhood. But "specimen" usually references tissue, in hospital parlance—not x-ray images. Materials taken from bodies are specimens: urine, stool, sputum, lymph nodes—materials of lavatory and laboratory. Tissue has a special epistemological weight.

Yet x-ray film is heavier than one might imagine—dense, relative to what Poe once called "the specific gravity of the body."[11] To be sure, films are not so weighty as an adult body: folders for a few patients can be carried under the arm. Film folders for an afternoon clinic, though, or a day's output from a busy scanner might require a cart.

It is well enough to suggest that other institutions managing anthropometric and photographic archives (like police departments) share concerns with the hospital, but part of the utility of films in the hospital, as in the prefecture of police, is precisely that films are *not* bodies. Films are "of" bodies. Films can follow different trajectories, often faster, than the bodies they represent—without changing.[12]

How is x-ray film, as archival object, like a specimen? It was the museum of natural history that naturalized the specimen—in helping to found new discourses of taxonomy, and with them, modern biology. Its exhibition of the typical and representative rather than the singular and marvelous (of princely collections) involved new pedagogical agendas, and new aesthetics: instruction (of the public) summoned "visual interests" different from the veneration and wonder mobilized in previous generations of collection viewers.[13] One mark of this difference was the rise of the caption.[14] The labeled, exemplary specimen belonged to a modern museum ethos oriented toward public pedagogy and inculcation of pride in nationalist science.

Museums enable comparisons by structuring contiguities among specimens.[15] Much of this happens behind glass. This is in part due to persistent dangerousness of the specimen. A key museal technique of comparative anatomy, taxidermy, depended on the glass case to safeguard viewer as well as specimen.[16] Glass facilitated inspection while proscribing tactile engagement: its interposed surface influenced the standard "sensory regime" in the museum of natural history, and beyond.[17]

This natural history museum is an ancestor of pathology archives, and indeed of radiological archives. The stable, flat radiologic image on the viewbox, conjuring a pathologic deep-time, is a historical descendant of the taxidermic specimen fixed behind glass. X-ray film is more mobile, less fragile, lighter than the body to which it is indexed. Instead of a dead animal made to imitate life, the x-ray film holds shadows that imitate death. But x-rays still occupy, in the exhibitionary order of the hospital, the epistemological niche of the specimen. They represent an Other—which one can examine, puzzle over, and render legible and classifiable.

Images

Early modern image-curatorial institutions included churches and courts. By the eighteenth century, as image custodianship was secularized and assimilated to nationalist projects, images were exhibited before broader publics, often with pedagogical intent.[18] Galleries remained highbrow—testifying to the good taste of patrons or the state. They summoned efforts to discern: if not hermetic or sacred meanings, then, by the nineteenth century, what genre or school of "art" an image fit into. As image collections became emblems of national prestige, increasingly the images they held testified to the genius of artists and their curators.

Many image-custodial institutions value originality. Part of images' status as cult objects has derived from their uniqueness,[19] their materializing of talents (if not divine inspiration) of their creators,[20] their conveyance of special meanings.[21] The "aura" of the unique work of art has been predicated on masterful workmanship and authorial intention.[22] It has also been bound up with the attentions of its keepers—with ownership and its filiations—and with forms of special connoisseurship, expert appraisal.[23]

Neither auras of the cult images nor charismatic powers of their curators are stable. The original has been undermined, since the advent of lithography,[24] photography, and more recently digital reproduction, by the copy.[25] Mechanisms of reproduction have to some extent liberated images from artistic genius and given them autonomy. This can be described as a radicalization of substitutability, of exhibition value, at the expense of cult or use value.[26] But historians of photography note that images are not so easily decathected by mere reproducibility. Benjamin: "Cult value does not give way without resistance."[27]

In the teaching hospital, cultic contexts of image valuation are diagnosis and teaching. An x-ray image retains aura insofar as it references an original—a particular body/part. Body-film *registration*, as unique historical event, specifies and motivates the image:[28] it records circumstances of image "acquisition," establishes an indexical relation to a patient's dossier. The identity of the maker of the image, the tech, is mostly erased: if a residue of authorial intention clings to a CT scan, it is some combination of agencies of the lesion (chapter 1, "Reading and Writing") and the referring clinician. A relation of image to bodily event persists (through label if not likeness) until the film is recycled. If further aura accrues to an image through handling, it mostly comes from the radiologist, whose interpretation accompanies the image henceforth, as durably as the patient's name.

But ongoing relation of image to event—its originality—also requires the work of the curator. The curator's contributions to an image's aura are the obverse of its contributions to his charismatic authority. Custodial charisma stems from possession of and proximity to the cult object (not from revelation, as does the priestly charisma of interpreters).

As for possession: the hospital owns the film; the patient retains rights to her images through copying (at a price, if more than nine sheets). It is presumed patients will return to the same institution for future care. Hospitals maintain their own image archives, and films are not sent to outboard clinics except on request. These are economic aspects of CT images' cult value in the hospital.

Of course, diagnostic images can be liberated to serve other noncultic purposes, "exhibition values." Patients take copies home and post them on the refrigerator (prenatal ultrasounds) or hang them in a window. A researcher publishes images in a journal or a textbook—stripped of patient identifiers.

A vendor takes images to a trade show. These are quite different exhibitionary contexts—linked only by their abstraction from "proper" institutional cult values of diagnosis and teaching. Many of these are arguably reappropriations of images to new cultic uses.

A senior professor of Radiology whom I interview reaches for his wallet to retrieve a tattered copy of his own head CT. This is fondly proffered as a token of a bygone era (with left/right and black/white conventions reversed from today's CTs). Later I learn from a tech that this carefully tendered antique is not the original: the tech has already copied it on one or two occasions. This sentimental keepsake exemplifies supplemental functions, beyond diagnostic utilities, of radiographic films—as objets d'art, aides-mémoire. Nor are such personal/aesthetic values necessarily eroded with copying.

This summons the notion of the *fetish*. Though this term is popularly understood to stand for an object of intense personal valuation, I invoke it for its anthropological connotations, as object whose veneration has a *collective* aspect, yet which is close to, or partakes of, bodiliness. I also invoke the Marxist critique of commodities.[29] CT scans circulating in hospital diagnostic circuits are mostly bound by cultic orthodoxy—but those that escape the confines of the hospital, and especially those which make their ways to the popular media, can come to stand for more than they should in the estimation of consumers (see chapter 6, "Exposition").

I have been comparing medical image archives to galleries of paintings. But the threat of the copy raises one more issue in the museological legacy of images. This has been broached above in the context of rogues' galleries: the influence of photography.

In the latter nineteenth century, the spread of photography, internationally and across social classes, resulted in wide dissemination of optical codes of realism and aesthetics of verisimilitude (chapter 1, "Reading and Writing"). "Effects of the real" were, however, no longer connected with the original as by the lyrical brush strokes of the painter. Poses and other conventions—for example, of portraiture—shaped new reality effects, in cultured and class-specific ways.[30] The camera also gave form to alienation effects: principally, alienation from the time and scene of the image's capture, from the presence and durability of an original—an alienation commencing with, in Benjamin's formula, the mechanical violence of the camera pushbutton, the snap of the shutter, which "gave the moment a posthumous shock."[31]

There are pictorial conventions in CT image archives—though not those of perspectivism or bourgeois portraiture, or the identificatory poses of criminal mug shots. Radiologic imaging everywhere enforces specific poses: for example, the postero-anterior and lateral views of the standard chest x-ray. In CT, there is the "scout view" illustrating the plane of each cut along some axis of the body or head—and then the series of slices itself, displayed in some conventional order: axial slices, sagittal slices, coronal slices. But do these produce reality effects? In CT, these conventions do not contribute to effects of verisimilitude so much as to effects of regularity, method, mechanism. The series and grid specify the mechanical conditions of these images' production.

What of alienation? With CT, the mechanical conditions of production articulate a violence at least as shocking as the photographic shutter. This is the violence of the *slice*. Paradoxically, gestures that produce the tomographic slice are not so abrupt as the photographic camera pushbotton or shutter. But at the level of pictorial convention, what is laid out on the CT grid is potentially horrifying: a sliced-up body, *as if* cut on the macrotome, or by the guillotine.

The *as if* of these violence effects is crucial. CT is a "noninvasive" imaging modality. A cut-up body is not so much represented as it is *imitated* by the CT scan on the viewbox. What is cut by the scanner is a volume of data. The patient knows quite well that his body is unscathed: what is sliced on film is a simulacrum, a phantasm. How patients relate to their filmically conjured sliced bodies is not always clear, especially if they are excluded from viewing their images.

Like the photographic negative, the CT scan is copyable, without degradation. Simulacra are destined to be superseded by copies. It is not the body as such which is made obsolete by the scan. What the grid of slices renders obsolete is precisely the horrific phantasm it has conjured up: that victim of multiple guillotinage, the breadloafed cadaver (which the patient is not).[32]

The breadloafed cadaver is a mimetic referent of every CT study—but it is not merely a phantasm. Since 1994 the American public has known the "Visible Human Male"—an executed murderer, before he became a famous specimen— and, since 1996, his companion, the Visible Woman (sliced more finely).[33] The Visible Humans were national archival projects which both entered the public domain as collections of slices, from three different cutting modalities: the macrotome (shaving—as by the zamboni of the ice rink) for light photography; CT; and MRI. Their images can be accessed in various formats on the Internet—though the most powerful tools, including those pioneered by Gold Standard MultiMedia, are proprietary.[34] These images—CT images and their correlates—severed from personal health issues of their bodily referents, and through extreme publicity and commodification, effectively escape many institutional contexts of cult value in which CTs are normally embedded. These are anatomic, rather than diagnostic, images; students find them useful in

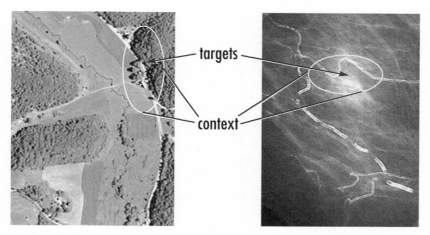

27. Satellite surveillance technology assists breast cancer detection

learning anatomy, but to the extent the general public prizes these images, it is for their exhibition value.

The Visible Humans, in materializing the phantasms of breadloafed bodies conjured up by CT's shocking images, may have returned us from the cult image to the museological specimen.

This is the case with x-rays: there is slippage and exchange between these roles, cultic image and specimen. X-ray films partake of some of the (art) museological attributes of images when they are indexed to an individual patient, with a certain "aura" related to their originality. X-rays partake of some of the museological attributes of specimenhood when they reference a lesion— something classified, as a material demonstration of a type.

What about the archival genre of the *map*—so often invoked in discussions of medical imaging? A map is both a special kind of image and a form of power. It is a tool of remote control—gathering distributed phenomena to a central site—an instrument of empire.[35] Maps have long facilitated projects of extraction, management, civilization. Maps are also military tools—and unsurprisingly there is significant historical entanglement of military mapping and surveillance agendas in projects of medical imaging. The origin of clinical ultrasound in military SONAR research is an oft-cited example. Another example is the use of particular kinds of neural-network computing, developed to find concealed military targets in satellite photographs, to detect patterns of unsuspected malignancy in mammograms.

One genealogy of body mapping is evident in the use of the term "atlas." The oldest usages (sixteenth century) designate volumes of large cartographic plates—for example, Mercator's *Atlas*, on whose frontispiece a scientist, in the place of the Greek god, holds the world in his hand.[36] "Atlas" only came to designate a volume of anatomic illustrations in the nineteenth century, trading on resemblance of folio formats and signaling apprehensions of "the body" as extensive territory. Anatomical atlases now encompass not only guides to normative bodily structure but also guides to deviant morphology—some exclusively through medical imaging.

Maps often comprehend multiple viewpoints. Unlike the Victorian portrait or one of Bertillon's profiles, a map is not necessarily founded on resemblance or on geometries of realism.[37] As instruments of spatialization, maps often distort—even as they address territory. Understanding relations of map to territory can require readerly expertise.

CT technology is indeed an agent of mapping. A scanner moves around in order to render a volume as flat images. Diagnostic CT makes body maps, which allow inference of causal relations and enable precise targeting.

In addition, CT scanners have themselves been mapped rather meticulously by health planners, across state and national, and international, jurisdictions, in documentation of their "diffusions" (chapter 2, "Cutting").

Filing and Tracking

The ambience in the File Room is calm, punctuated by phone calls. There are five staff members. One clerk is near the front, serving client requests; another is near the film stacks, refiling and retrieving films. There are women and men, black and white. There is discussion of someone's upcoming bridal shower; a TV is on, showing a soap opera.

The film shelving in the back is densely packed, with space for only one or two aisles amid five movable shelf units. The access aisle is shifted by moving the interior shelf units with a motorized crank. Each shelf unit is roughly thirty feet long, five shelves high; the three interior/mobile units have shelves on each side, so there are eight sides—roughly a thousand feet of film folders in all. The view down each aisle is striped with manila and colored, numbered labels. Between the front desks and the back shelves are additional workstations. Desk surfaces appear organized. Stacks of film folders do not suggest helter-skelter activity, or the range of preservation practices one might expect in, say, a museum basement.

Film clerks handle folders more than film. Annotations relevant to filing are written or pasted on the study jacket ("subfolder") or the patient's main folder. Tasks consist of marking new folders, sifting folders into slots marked by desti-

28. Clerk filing films, Basement File Room

nation and bar code, processing requests across the counter or over the phone. One clerk tells me they handle films at times—there is a slot for "loose" films, to be replaced in the appropriate folders—and that they also have to hang films on some viewboxes (rarely CT). Returning loose films to the assemblages they belong with is partly interesting, mostly "irritating." Film handling tends to be the responsibility of techs and doctors—those more directly connected with production or display.

Sorting is based in part on number matching. A numbered folder is "signed out" to a coded destination—until it is signed back in. Film folders not on back shelves are stacked on metal carts and racks. "Signing out" means exchange of films for a doctor's signature. This is not the eclipse of the radiographic signified, but rather a step toward new signification: it marks a film's release from custody, to the domain of reading and rereading.

One of the clerks is occupied with the Starpoint "purge list." This refers to films slated for removal to Starpoint after six months.

During their residence in the File Room, film folders are often signed out to reading rooms. Relations between the CT reading rooms and this archive are mediated by clerks, by Film Assembly upstairs, by couriers, and by phone calls and checklists. Frustrations about delays and absences are felt by clerks as well as by diagnosticians. In the File Room, the unfound and the potential-for-loss are constant companions—irreducible, despite ideals of total access and total control. Loss also produces discourses of intrigue: for the clerks, about thieving docs; and for doctor readers, discourses of speculation—perhaps papered over in formal reports with the phrase "no old films available for comparison."

One folder the courier seeks is not presently in the File Room, but in circulation upstairs. We are directed to Film Assembly, on the second floor, at the edges of the Main Department—a place more central, exposed, and subject to "real-time" clinical demands.

"Film Assembly" designates a place and a nexus of practices. It is a transfer hub and short-term depot for the circulating collection. There are folders for studies to be done, studies just done, studies bound somewhere soon. "Assembly" suggests a factory. Oddly, as factory, it has a decidedly more intimate and hectic ambience than the File Room—barely twenty feet square, with a service window and busy doorway.

At this moment, the hectic ambience is accentuated because "the label printer is down," and because there is an anesthesiologist in the middle of the room, impatiently pivoting, expressing dissonance between operating room time and Film Assembly time—as well as the subservience of film managers to those who work directly on patient flesh.

Countertops in Film Assembly are more crowded than those downstairs. Meshed with more demanding streams of circulation, the film piles are more volatile. Here there is a stack of folders just back from the Family Practice Center which "we have to match with the PPFs." There is a stack of films bound for outboard clinics: Mannsboro Orthopedic; West Fork Internal Medicine; Bonlee Regional; Parkville Radiology.

A list of inpatient workstations shows abbreviated codes for hospital services where there are drop sites for films—viewboxes associated with clinical and radiologic specialists. This and other schemata materialize some of the functions of a map, not routes or topologies, but distinct and separately accountable loci, places to which films travel and from which they return.

I meet Nellie Simons, Processing Assistant III, facing a stack of green-edged folders on a countertop. She chooses to sift this pile first instead of a second stack of flimsier manila folders which she calls "crappy"—"probably outpatient studies." Each folder in her stack is being "barcoded to the main File Room": registered for immediate transfer to that location, using a penlike wand—which she swipes across a bar on the folder, and then across a bar on a wall chart.

The film folder is like skin: a visible, tangible outer layer, upon which superficial signs are inscribed: color, number, sometimes a procedure date and a brief "reading"—enabling quick, summary grasping, more rapid channeling through the sociotechnical spaces of the hospital. Colors of subfolders correspond to particular modalities or divisions: Nuclear Medicine is red, Vascular manila, CT green, Neuro CT blue. Films inside each folder—requiring time and effort to sift through—constitute a depth of enclosed meaning, accessed only in special locations.

Aggregates of folders in Film Assembly are temporary: folders are slotted, accumulated over minutes to days, then moved to other destinations. Cabinet slots are contiguous, but the hospital locations they represent are not: there is no map of empire in this list of capital cities and client villages. My only access to the topology that links them—the sole materialization, as far as I can tell—is the film courier. Daily, she shuttles among these sites, and engages the practices of distribution at the ends of the shuttle.

The Courier

By 9:50 a.m., Sally the film courier has just been to Ortho and Ultrasound—delivering thick green/white computer printouts, including outpatient schedules for the next two days. She has been traveling nonstop since her day started, at 7:30, in the Ambulatory Care Center (ACC), a quarter-mile across the hospital campus. Now she has to place one film on the shuttle bound back to the ACC and has been unable to locate another film wanted there, an MRI. She stops in the hospital lobby to pick up a house phone, calls the ACC receptionist to alert them to the impending delivery, and its incompleteness. She then answers a page. "What's the MR [medical record] number? OK. And what's the patient's last name? OK, let me get your pager number." She scribbles furiously on a 3 x 5 pad (intermittently, all day). Nuclear Medicine needs a patient's chest x-ray "stat"; this is on the "Chest board" on second floor. On the way to bring that study, we pick up another film in Film Assembly to be delivered to Oncology Clinic, in an adjoining building. In Film Assembly we find that an entire box of films has been requested by a Surgical service. These will have to go on a cart. Meanwhile we are "off to Body CT," where every weekday (except Monday) the courier retrieves a large collection of folders and brings them back to Film Assembly to be placed in other jackets.

Upon return to Film Assembly, Sally consults a computer workstation. She moves one folder from a pile into a slot. She pages a Dr. Beck because the Neuro CT he wanted is not in its jacket. While waiting for him to call back, Sally looks up "other stuff" on another computer system, RTAS, the Radiology Transcription Archiving System. This system (pronounced *ár-taz*) stores transcribed x-ray reports and supports proofreading, signing, and cosigning. It can be searched by patient name, medical record number, radiologist, service, and more, depending on one's password privileges. At present Sally is retrieving dates and numbers for studies on her list.

The process only appears confusing, Sally assures me. The fundamental principle is simple. "If it moves, sign it out." This maxim has a particular meaning in Film Assembly, but it applies in all domains of Radiology. Films must leave a trail.

After Sally has spent several minutes in Film Assembly, there are some "pre-ops" (mostly chest x-rays) to be "scooped up" from PreCare, a patient screening station. The visit there involves taking down some of yesterday's preops. Then, at 10:30, the Starpoint delivery has arrived.

The courier's day proceeds like this. It is chaotic, but governed by a hierarchy of urgencies and proximities—some specifiable, others summoning inarticulate craft and judgment.[38] I ask Sally about finding her way. She likes to "kill two birds with one stone." I ask her how she knows which of two competing errands is most important. "You find out pretty soon who can't be kept waiting." I speculate about the range of affects she encounters—since I have observed, on the reading room end of conversations, mostly politeness. She acknowledges that impatience and indignation crop up.

Film Assembly informs us that another folder we seek is over in the ACC, where the patient had an appointment earlier today. The folder will be on the next shuttle. Films ride the shuttle bus "just like patients," says Sally—except that they "ride shotgun" in a special slot near the driver. When the bus from the ACC brings films, the courier is paged. There is also film traffic to and from the Family Practice Center, a half-mile down the hill, served by a different shuttle. Films also travel in a dumbwaiter, along the hospital's vertical axis.

The trajectory of the courier, if followed comprehensively and mapped, would clarify topological relationships of CT viewboxes with the rest of the radiological enterprise at University Hospital.[39] It would become clear that the CT Suite is not one contiguous array of rooms but a set of distributed and discontinuous sites. Sorting through what the courier is carrying would also clarify the fact that for CT interpretation, the relevant filmic archive is not merely old CTs. Old chest x-rays, cervical spine films, skull films, ultrasounds, and so forth are all relevant at times.

In the figure of the courier, her itinerary, and skills of organization and diplomacy, it would be possible to map many logistical and social relations among filmic repositories subserving diagnosis in University Hospital. The courier bridges hierarchies of prestige and urgency, and various loci of viewing and reviewing that structure much radiologic professional activity. Of course, there are limits to the insight the courier might want to bring to bear. Too much interest in nuances of a particular case could slow down the courier's service to needs elsewhere. The good courier is efficient, polite, disinterested. She might speculate on the conditions of her service—on whether a resident's tone seems to exceed the authority of his station—or whether a particular attending's abruptness seems a function of mood, or character, or conditions. But ultimately, like the mailman, she does not read the mail, nor does she judge those who do.

Filmless Images

CT images are reconstructions of data arrays: their origins are digital, and their careers of exhibition and storage can also be digital. Images on monitor screens, hard drives, tape, and optical disk are commonplace in CT and other sectional modalities—even in the 1990s. (At the time of my fieldwork at University Hospital, digital image handling was still partial, supplemental, for end-users; now it has become the norm, and film has all but disappeared.)

The digital image-handling system most relevant to CT at University Hospital in 1997 is called "MagicView." This is a product of Siemens, maker of the hospital's scanners.[40] The MagicView system connects the consoles of the CT scanners (in their several hospital locations) with a dedicated workstation in each reading room. An emergency chest CT done in the Neuro scanner can be transmitted to the Body CT reading room, across the hospital, for rapid review.

"MagicView" is a proprietary term—lettered on each monitor, and appearing on introductory screens at bootup. It is also the pet name for this system throughout the department—an everyday shop-floor term here, like the trade name of a pharmaceutical. The generic name for such a system—Picture Archiving and Communication System, or PACS—would be recognizable in other hospitals and to any radiologist, but "MagicView" might not. Most PACS though have equally catchy names.[41]

Archiving is only one of the key functions of digital image management. PACS functions extend from image "acquisition" (scanner) to display (reading room and beyond) to archiving (server, tape, or disk—always redundantly). This archive is not so much a common center as a set of nodes in a more tensed structure of image "distribution."

The magic of MagicView is about facilitation of image movement, with a saving of footsteps. It is also about warding off loss. "No more waiting for the 'sneaker-net' to manually bring (or lose) the patient's film," claims Siemens.[42] The trope of magic bespeaks an ideal of effortlessness—wave the wand, click the mouse—in making images do one's bidding. One Siemens trade journal advertisement shows a genie emerging from a lamp.[43]

MagicView is still new enough in 1996–97 to surprise its users, who continue to discover new tricks and facilities—and to make new mistakes. This is magic as reversal of constant conjunction, or ordinary causality.[44] Magicality of the MagicView is a "contagious" magic—a magic of contiguities,[45] of new spatial juxtapositions, of effectivity across distance. A button moves an image dossier between buildings in the blink of an eye. A dozen studies are contained in a small box, and then gone.

The MagicView system represents the front end of a revolution in radiological workflow. Within just a few years of my study, University Hospital has joined the ranks of filmless institutions.[46] Thus the status of MagicView in 1997—adjunct to filmic archives and courier labor—is curiously symptomatic of a transition that was, and elsewhere still is, incomplete. Films have not yet become a historical curiosity; they are still essential elements of radiological work. But all who handle the MagicView at this historical juncture understand the autonomy and mobility of the electronic image, the exactitude of digital duplication, the infinite possibilities in reformatting, the "multitasking" enabled with simultaneous distribution of images across institutional locations. PACS promise dynamism and mobility of images that would run Sally the courier ragged.[47]

There are still debates about virtues of film versus digitized images—though far more often in conventional radiology than in CT. These debates sometimes concern technical aspects of perception—resolution, dots per inch (of hardcopy), pixels or lines (of CRT display)—thresholds of detection. They sometimes concern psychological factors influencing readerly decisions. They are sometimes logistical, concerned with rapidity of image handling and interpretation, and elimination of image loss. They are always concerned with error—and money.

PACS do not transcend the modern (nineteenth-century) archive in all respects. Images are still kept in "folders." Image interpretations still require "signatures." And even in CT, there are still images that radiologists prefer to read on film.

Teaching Files

Some repositories of CT films serve other agendas than diagnosis. Films retained for education and research have different shelf lives than those serving patient care. For teaching, films are often retained longer than seven years;

for published research, image collections can achieve a kind of presumptive permanence.

One of the more enduring collections is the Medical Student Teaching File—celebrated by the department as "a unique conglomeration of over 5000 cases deemed interesting by students over the years." This contains CTs alongside many other kinds of radiographic images.

A second repository is the main CT Teaching File. This is an eight-drawer cabinet in the Neuroimaging reading room, filled with film folders, organized by organ system primarily and by disease category secondarily. Each folder contains a "case"—a series of related images (not necessarily an entire "study"), showing some CT anomaly, variant, or "classic" finding. These files are used by residents and attendings preparing for conferences, and by residents studying for exams. Each Body CT resident prepares two cases for them during her "rotation" on the service: "The write-ups should include clinical history, discussion of radiological findings, and pathological diagnosis."[48] Preparing a study for the files requires effort and diligence: obtaining copies of films; erasing patient-identifying information; inscribing on the folder teaching points about the case; finding the proper location in the cabinet. Archiving in this teaching file is reserved to a small subset of cases.

A resident asks the scanner tech to print an extra set of images:

Resident: [*on phone:*] Hi, this is Mary. Um, we've got another teaching file. It's teaching file day. [*laughing*] I know. Hopefully, last one. All right. It's on Larry Rausch and the images from 10–15, 65–69— he's on the monitor, 26–33 . . . Thanks a lot. Bye-bye. [*hanging up phone*]

Copying films constitutes extra work for the tech. From a certain perspective (dedication to patient throughput), this request (*again!?*) is comical. It turns out that techs know many of the "teaching" studies they reprint are destined, not in fact for resident teaching files, but for other repositories. The laughing deferences of this resident acknowledge a favor.

Other "teaching files" include the professional repositories assembled in the academic offices, and sometimes homes, of radiology attendings. These dispersed archives include discipline- and project-specific collections assembled over years by individual faculty members. Often these films become fodder for conference presentations, or evidence in published case series, book chapters, and the like—which eventually redounds to the credit of the department and the individual faculty members who "own" them.

Investigator:	. . . About teaching files—
CT attending:	Mm-hmm.
Investigator:	—those large collective files that are down in the reading room.
Attending:	Yes.
Investigator:	Seems, on the one hand, it's glorious to have—a central place, but they're also perhaps a little insecure . . . what I've been told by people is things leave those files, and are a little volatile.
Attending:	Yes.
Investigator:	Does that oblige you to maintain your own? Your own file?
Attending:	Well, let me tell you one thing. It's a matter of how—of your feeling of possession and how you feel towards having the films yourself. For example, I don't have a feeling to having the films myself because the way that I do it is I bring up the film myself, here in my office. I label the ones, then I make slides of the most interesting sections.
Investigator:	I see, so it's all—
Attending:	Then I have slides made out of them, and then I show them in conference and put them in the teaching file. So my slide collection is basically out of those files—each case I take one or two pictures of it and I keep for myself.
Investigator:	The slide.
Attending:	I keep it in a slide. But a lot of people feel that perhaps if you eventually go to publication or something they would like to have the original films because of the quality of the prints and other stuff comes better from the original film. So they feel that they have to have the films themselves.
Investigator:	The original, yeah.
Attending:	And they will treasure them and keep them themselves. I myself find them . . .
Investigator:	Heavy and—
Attending:	Heavy and a hassle to deal with. And I myself find that even keeping slides is a primitive way of doing it, because if we could have somebody who would digitize the images and save them in a zip disk or something, that would be the best way to do it.

Many Radiology faculty members retain images on 35mm slides—in carousels or binders. In the case of CT, each slide contains one or two, rarely four, slices.

But 35mm archives cannot always replace bulkier collections of "original" x-ray films. One of this attending's colleagues is known for his extensive home "hoard" of cTs and MRIs. His notoriety stems from the expense of film and tech time—and the comparative restraint exercised by other faculty in their collecting.

The attending's wishful reference, above, to a digital alternative signals how limited the reach is, at this historical moment, of the digital imaging revolution.

Some form of personal image collection seems requisite for the academic radiologist. His possessiveness is complex. Benjamin insisted on collecting as a form of practical memory and magical ambition—as well as a violence, a death: "It is the deepest enchantment of the collector to enclose the particular item within a magic circle, where, as a last shudder runs through it (the shudder of being acquired), it turns to stone." But also the collector aspires to connect his acquisition with "that place beyond the heavens which . . . shelters the unchangeable archetypes of things."[49] Classificatory schemas of the hospital, even those of teaching files, may be inadequate to the needs of a particular attending. But also at stake is the charismatic authority of the curator, deriving from care for, proximity to, "originals."

In any filmic teaching files, the central file or those maintained by individual attendings, films are collected as exemplary cases of particular diseases—often qualified as "classic" or typical, "beautiful" or "interesting" instances—or as deceptive, tricky, confounding. These are all potential forms of teaching value. The classification of these cases may be by organ or region, by diagnostic category, by imaging modality, or some combination.[50]

Neuroimaging attending (off duty, visiting): How come you get all the darn good cases and I never get any . . .

Neuroimaging resident: We've had the greatest cases today.

Neuroimaging fellow (on duty): I've made you about six teaching files.

Attending: Yeah, my days are the most boring days. It's like follow-up—follow-up stroke, and follow-up tumors. My days are horrible. My days are the worst days in the world.

Fellow: We've had a bunch of nice ones today.

Resident: Five good ones.

Attending: Yeah, I know!

The focus in teaching files on exemplarity of images—rather than on welfare of patients—entails a deregistration of images from primary diagnosis, and reregistration in a different symbolic economy—one of types and examples—and different ritual contexts. Teaching film folders are not color- or bar-coded. Usually images stored in them have patient identifiers eliminated.[51] Most patients whose images are in these files are unaware they have been deemed interesting and teachable.

Telling patients about the exemplarity of their pathology could be unkind—which underwrites this comic episode in the reading room, looking at scans of an alcoholic's brain:

Neuro CT attending:	You gonna fix him—her—him?
Clinician:	Negative.

[*group laughing*]

Clinician:	[*viewing film*] Ooh.
Attending:	[*quoting clinician, laughing:*] "Negative. Ooh!"
Student 1:	That's always a bad sign when they: "Oh my!"
Attending:	"Good teaching file!" It's always a bad sign. [*laughing*]
Student 1:	"Ooh" . . . you don't want to hear that. [*laughing*]
Attending:	"Is this your head?" [*laughing*]
Student 1:	[*laughing*]
Attending:	"Is this your head?" [*laughing*] "Ooh, great teaching file! Mmm, interesting." [*laughing*]
Student 2:	. . . should get something in exchange . . .
Attending:	[*laughing*] A free—right, right.
Student 2:	Free hospital stay, five bucks, lunch, a shave, a . . . [*inaudible*]
Student 1:	Free follow-up MRI. [*laughing*]
Fellow:	Right.
Student 1:	"Ah, still interesting." [*on follow-up MRI*].
Clinician:	So disrespectful . . .

In filmic archives serving research projects, identifiers tend to be retained long enough to facilitate correlation with other imaging studies and clinical information. Before the point of publication, identifiers are excised.

University Hospital teaching and research enterprises do have some digital image archives at their service. CT studies are archived digitally, and these can be accessed later. There are research projects ongoing, among Radiology, Computer Science, and clinical departments (notably Neurosurgery), developing novel modes of display. A subset of teaching file images (some without hard-copy equivalent) is available online, for review by residents and students.

Here is one case in an online teaching resource at University Hospital.

History:

> This is a 49 year old male with a history of a large pleural based soft tissue mass in the right hemithorax.

Images:

[*Four images: plain* AP *x-ray from mid-chest through abdomen, and three* CT *slices through right lung base*]

> Multiple masses are seen in the right hemithorax, some of them adjacent to the pleura. There is increased thickness of the pleura. The left lung is clear. A large pleural-based soft tissue mass is seen in the right hemithorax. Several enlarged right retrocrural lymph nodes are present.

Discussion:

> The mesothelioma is the most common primary tumor of the pleural space. This tumor may be benign or malignant. Malignant mesotheliomas are almost always associated with asbestos exposure. Whereas the benign form may be cured by surgical resection, the malignant tumor is aggressive, spreading by local extension, and usually encasing the lung and mediastinum . . . Surgery is not possible in this case, and chemotherapy and radiation have been unsuccessful. Prognosis is very poor with a median survival of one year. CT is helpful in establishing the extent of the tumor . . .
> [*truncated*]

Diagnosis: Malignant Mesothelioma

A conventional case: abbreviated history, followed by images. The images are presumed sufficient to make a diagnosis. Students and residents are intended to examine the images first, before reading the salient findings—and then to use the text to inform reexamination of the images.

A link on the front page of this teaching collection encourages submission of new cases. "Send case history, images, and brief discussion of entity to . . . Images can be accepted on film, as 35mm slides, or as 256 grayscale PICT or GIF."

To date, University Hospital teaching files have been local resources—troves of institutional experience, assembled with care, invested with pride. They are unique collections. But the rapid growth of digital image management, and of the Internet, make it likely that more teaching files will become available online. Will this erode the value of University Hospital's own teaching files?

Will the proliferation of online images from different institutions, and the discovery of overlap and commonalities among them, lead in general to diminished local investments in collecting? Or will local institutions and individuals, in service to trainees or to their own charismatic authority, continue to curate unique and original images? I return to these questions in chapter 6, "Exposition."

Curating Bodies

Radiology Departments manage bodies in order to make and manage images of bodies. Many of these management practices can be understood as curatorial, even if they do not commit bodies to long-term storage. Bodies-to-be-scanned must be registered, scheduled, disciplined. Radiology departments instruct patients how to prepare for their studies; they assemble patients in waiting areas before and sometimes after their scans; they orchestrate comings and goings. But for all this regimentation, the CT Suite's custodial relations with bodies are relatively brief, unfolding in minutes to hours—though planning and follow-up may unfold over days to years (for billing).

The administrative term which best summarizes the role of bodies in CT curatorial enterprises is "throughput." This Fordist/Taylorist term (see "Charting" section below) refers to movement of materials—not for their own sake but to maximize utilization of machines and staff time. To be sure, patient comfort and satisfaction are monitored; machinic efficiency is not an exclusive good. A tech reflects: "Kind of an assembly [line]—not in a bad way . . . It's become that, I've seen over the years too, because the scanner is so much faster now. We don't have time to go in there in the middle of the scan and go 'Are you OK?' . . . It's like get them on, scan them for ten minutes straight and get them off . . . And go on to the next one. It's really become more of that since the scanners are so much faster now . . . Not in a bad way, we still try to be as nice as we can." If one understands images and "readings" as the quintessential radiological products, it is clear that bodies are raw material, subject to managerial adjustments, just like film, chemicals, contrast, and x-ray tubes. An administrator's view of body throughput: "And so we now have two equal CT scanners in technology and capability, for bodies and for heads. And of course, it's just gotten the latest upgrade of software, which is ultrafast. Our CT studies are down [in duration]—we're doing sixteen and eighteen patients in an eight-hour day—in body studies, which are long studies . . . Heads, we're turning them over so fast, you know . . . you spend more time putting the patient on the table and off the table than you do doing the scan."

One also hears "patient flow" and "patient stream." Here an administrator describes aides facilitating flow:

> [We have] one [aide] in the Main Department—because of the disparate geography between the front hallway, where patients come through the department, the back hallway where the real staff flow goes on—we have to have someone out front to keep an eye on patients in case something happens to someone. We've had—in the time that I've been here, we've had a patient to crash out there waiting for Transportation to come . . . And we have a second person over in the ACC, not necessarily for the critical nature of patients, but because of patient flow. Patients flow from an outer area into an inter area—an inner area where there are just dressing cubicles . . . So patients have to be assigned to a dressing cubicle, and the aide knows which room which patients are in, and takes care of some duties back there, and then patients flow into the exam rooms where the staff are, and back out to the dressing spaces, so our aide over there is critical in helping patient flow work the way it's supposed to.

Patients flow to the CT scanners from inpatient wards and outpatient clinics. Some clinics are lavish in referrals to CT—like Neurosurgery and ENT; CT-intensive inpatient domains include the ICUs and the Emergency Department.

Patient streams shape institutional spaces. Here an administrator reflects on a space allocation decision that was made surprisingly late in the recent move of Neuro CT to the NeuroSciences building: the decision to install a Neuroimaging reading room there as well.

Administrator:	Very logical move, frankly. It just hadn't dawned on anybody, because to a large extent around here you plan a building almost not knowing who's going to be in the building . . . And from an ancillary support service perspective, it is virtually impossible to plan around that kind of information . . . When you're trying to decide stuff like how much capacity do I put in these buildings.
Investigator:	Yes.
Administrator:	You know: do I need one CT scanner, do I need two CT scanners, do I need three x-ray rooms, do I need four . . . It's virtually impossible until you actually open your doors, and you start seeing patients.
Investigator:	Clinics start to function and you get a sense of volumes.
Administrator:	And then what'll happen, as has been prone to happen in other buildings on our campus, is that the set of

assumptions and decisions that were made when you open a facility last for about six months . . . And then all of a sudden there's this big flip and things change dramatically.

[*segment omitted*]

Administrator: You're up to here and you're drowning and there's no way that you can do any more . . . So we decided when we realized what the volumes were going to be for Neuro CT and for Trauma, we decided we would move the Neuro reading room down there. So we quickly designated . . . we had the certificate of occupancy for the building—we turned a crew loose to tear out the space and create another reading room.

So the relation between spaces of the CT Suite and the flow of bodies—volumes posing a risk of drowning—is more dynamic, and resistant to planning, than the notion of archival management might suggest.

Consider a different relation of patient "volume" to institutional space. One patient is scheduled in both CT and Ultrasound for a biopsy on Monday. Ultimately he will undergo one procedure in one setting—but presently he has been "double-booked" because he is "so big." The referring clinic surmised the patient's size might pose a problem. Radiologists on duty that day will decide about optimal logistics of approach to this patient's body and lesion.

The insertion of the patient's body into radiological logistics begins in Scheduling. There a meeting is "booked" between a patient and a hospital "resource": in CT, a large fixed machine. Scheduling staff members answer the phone, enter information into the computer, print and mail appointments, and process other papers. There are three workstations in the office: all have a telephone, two computers (a RadCare workstation and an SMS workstation), letter files, and slots with green/white rolled printouts. Other items on the desktops include interdepartmental mail envelopes, floppy disks, an adding machine, a Selectric typewriter, lunch bags. Walls are busy with telephone/pager directory lists—as well as artwork by children of staff members. A large five-drawer filing cabinet stands against one wall. There is a microwave and a coffee machine.

In one corner of the Scheduling office there is a pneumatic tube column. Forms and schedules are placed in cylindrical containers and launched to destinations throughout the department—mostly on the second floor. This system has been around "for a long time"; a change to a computerized tube system is anticipated.

29. Radiology Scheduling

All but one of the Scheduling personnel I meet are women. It is difficult to speak uninterrupted because they must answer the phone, sometimes several times in succession. "Radiology Scheduling—can you hold please? Radiology Scheduling." The Processing Assistant III with whom I have been talking gets on the phone with an outpatient clinic. With phone between ear and shoulder, she types a few keystrokes to open an appointment screen. She keys in a medical record number. "Why are we doing this?" she asks. More typing opens a screen that says "Resource Next Available Time." A time is chosen from two alternatives offered, an appointment slip configured for this study: a mammogram. There is a field on the slip for "Patient Comment." This means miscellaneous information for, not from, the patient: "Do not wear powder." At one point the scheduler says, "I'll ask—hold on a minute." She asks colleagues in the room: "How much does a mammogram cost?" No one is sure—and there is no definitive reference to consult—but eventually someone suggests: "About seventy dollars, give or take."

Another caller requests an Abdominal CT. The scheduler says, "Hold on one second" and reaches for a green pad. Returning to the phone, she says:

> You need an abdominal CT? [*pause*] What's the Medical Record number? [*pause, writing on green sheet*] Has he ever had a CT before? Any allergic reaction to contrast? Diabetes or renal insufficiency? That's OK. Hold on one second. [*takes another call, transfers that to a colleague*] [*back to original caller*]

I'm sorry. What, a creatinine? OK—that'll work. [*puts pen down, reaches with right hand for keypad; keystrokes through three screens*] Clarence Nutter? And when are you lookin' at doing this? [*pause: typing with two hands*] Friday I've got a 2:45. Is that OK? [*pause*] What's the physician code? [*pause*] What clinic are you with? Pager? Just a CT of the abdomen? [*pause, typing*] Why are we doing this? [*pause, typing*] OK—I'll put it on there. Thank you, bye-bye.

Exigencies of screen and form, of standard procedure and habit, do not always match idiosyncrasies of referral and patient circumstance. In most cases, there is the particularizing question: "Why are we doing this?" This embeds the scheduler in a team ("we") and involves her, however briefly, in the genesis of the radiographic puzzle—as a proxy for the radiologist's curiosity, which may be expressed later in the reading room.

When I ask about the Scheduling office as a communication hub, several schedulers nod knowingly, and one indicates that it "can get pretty crazy—especially if it's clinic day." Next to a bin of Patient Processing Forms (PPFS) is a pile which has just been retrieved "from all the reading rooms" and brought back to the Scheduling Department, to be kept for three months. In explaining this, the Processing Assistant explains how CTS are scheduled. Rarely do patients call for their own appointments: calls are made by physicians, clinic secretaries, or, in the case of inpatients, by ward clerks. CT scans, like other radiology studies, require that a referring physician submit a short (3 x 8 inches) white requisition form. The PPF is produced when the Scheduling clerk enters information into the Radiology Information System (RadCare). PPFS and requisitions are then conveyed to the CT technologists, who log the request into their scanner's local schedule.

The process varies for emergency and inpatient studies. For CT scans requested by Emergency Department physicians, a requisition is faxed to the local X-ray Department in the Neurosciences Building (where the Emergency Department occupies the ground floor). A Neurosciences x-ray clerk enters the information into RadCare and generates the PPF. For inpatients, since fall of 1996, a new Order-Entry System has been in place. Ward clerks now have access to booking screens at inpatient workstations and can schedule scans directly. The Scheduling clerks here are not surprised there have been problems with this new decentralized system—and while they are pleased to have been relieved of a modest communication burden, they also note that the scheduling process was smoother when they controlled more of it.

The Procedures Book for Scheduling contains, among other items, phone numbers of community institutions—ambulance and taxi services, nursing homes, hospitals and HMOs, home health agencies. These institutions of custodianship, care, and transport mediate the engagement of the Scheduling enterprise with patients' bodies. A patient first meets this enterprise as a textual or a vocal representation (requisition or phone request). These indirect registrations launch her future bodily trajectories. Fleshy presence is anticipated in limited ways: "*How* big is the patient?" Or: "Clear liquid diet for at least six hours before the study."

"Resource Next Available" comprises bodies besides those of patients. There is radiological staffing to consider. The scheduling office takes schedules of techs and radiologists into account as aspects of the resources they allocate. In addition to telephone/pager directory lists on the walls, there are photographs of current residents. Elective (nonurgent) CT are not done on weekends, when only one scanner is operating and resident staff is thin. Scheduling of a CT-guided biopsy requires at least one phone contact with a radiologist.

A Radiology Nursing Supervisor reflects on nurses' relations with Scheduling:

Nursing supervisor:	If there's CT biopsies, [we know] which biopsies which day . . . There's always been a nurse involved in that. So that they can either—
Investigator:	With the biopsies.
Nurse sup:	Right, call the patient if we need to, be sure that Day-Op is set up, all those things are taken care of . . . And that's all information relay between the Scheduling office and the nurse. And then when the daily schedule comes, you know, everybody walks in [to the Scanner Control Room] to see what's on the schedule. They have so many add-on's though that . . .
Investigator:	That's striking. Almost—especially with inpatients.
Nurse sup:	Absolutely.
Investigator:	It looks like a quarter, or less, of the inpatients that they wind up doing at the end of the day are scheduled.
Nurse sup:	Right.
Investigator:	So does . . .
Nurse sup:	And that's because when you all [clinicians] make rounds upstairs, you find things, and the first thing . . . And Henry Chair believes in same day [service] . . . So

	if you want something, we're going to give it to you the same day . . .
Investigator:	That makes for, I would imagine, an enormous sense of non-control over volume.
Nurse sup:	I think the techs sense that; the nurses don't.

Everyone wants to see the schedule, even though its predictive value is limited. It cannot account for new demand for images on daily hospital rounds, a demand which the Department Chair is committed to satisfying, even though it places stresses on staff.

As with bodies of patients, the bodies of Radiology staff are only indirectly engaged with the Scheduling office—via phone, or forms. Schedulers come to know some staffpersons better than others. When I inquire if the influx of new residents in July imposes new burdens, one scheduler says that she actually has "the most fun with the new interns"—and likes to "pick at 'em," to "get them less stressed out."

Once patient bodies are launched into the flow of a departmental workday, they face a different set of archival structures. The work of Scheduling is mostly done: patients' names are on the schedule du jour in the scanner rooms and on viewboxes. The general form of the body archive on the day of the scan is the queue.

Inpatients enter a day's CT patient flow from the disciplined setting of the wards, where every engagement with the institution is logged on forms stamped with their plastic plates. Inpatient movement is orchestrated by Transport staff who work for the hospital (not Radiology), upon telephone request from the CT tech.

Outpatients arriving for scheduled Body CT scans are directed to the Imaging "front" desk on the second floor, down the corridor from the main Radiology Department. At this desk, in a cubicle adjoining a waiting area, a clerk greets each patient and completes a Body CT Booking Form (aka "green sheet"), which assembles standard information to supplement the sometimes meager cues on the referring clinician's requisition. This sheet helps, says a procedures book, "to avoid delay." Body CT patients are directed to a separate waiting room down the hall. There they spend the hour before their scans drinking contrast. According to the Protocols book: "Patient should drink continuously up to the time of the scan to ensure good opacification of the bowel without breaks in the contrast column." In Body CT, patient flow through the waiting room is orchestrated around intestinal contrast flow—roughly a liter per hour.

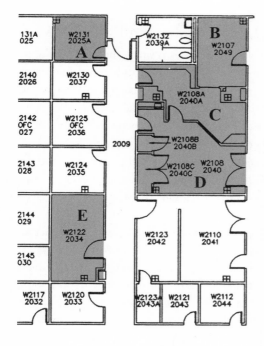

30. Body CT Suite:
A= waiting room;
B = computer room;
C = control room;
D = scanner room;
E = reading room

For Neuro CT, in the Neurosciences Building, one waiting room is shared with other radiology services. There, oral contrast is not given so often.

In both waiting rooms, patients awaiting their appointment with the scanner sit alongside patients who have just undergone scans and are awaiting dismissal, perhaps awaiting films to be carried to clinics. It takes time for techs to print images and check them with the radiologists. These waiting patient bodies constitute a small pool, an eddy, above and below the patient stream pulsing through the scanner room.

The waiting room for Body CT opens on the same hallway as do the reading room and control room. The waiting room is roughly ten feet square, with blue-gray carpet, a half-dozen padded chairs, and pastel prints depicting wetland scenes. There are several magazines on endtables: *Harper's, Home.*

Bodies of the CT Suite are, in their collective identity as a "flow," invested with more dynamism than specimen bodies of the natural history museum, fixed behind glass. But CT patients' eddying out in the waiting room, or the Transport queue, is also a slowing of pace, prefatory to their temporary submission, in the scanner room, to rigorous fixation. There patients assume poses and hold them, as if subjects for portraits. Patients are glad to avoid old-fashioned dissection, taxidermy, in management of their illness, if they can submit to temporary confinement instead.

The fixity of the patient's body in the scanner is surely the key pause in its flow through the CT Suite. Here the body of the patient fits best into the epistemological niche of the fixed specimen. (Though the whole body of the patient merely accompanies the specimen: the sick part—head, abdomen.)

Bodies emulate fixed, cadaverous specimens best during collection of biopsy material—when scanner and staff are enlisted in removing a tissue core or cells from some suspicious lesion. Tissue is conveyed promptly to the Pathology Lab—and rigorously archived there, as specimen. This process, considered in more detail below, speaks to the paradox of the apparently missing cadaver in the era of "noninvasive" imaging.

Curatorial practices of radiology engage body parts. A big girth, a high creatinine, an unpowdered breast, an empty stomach: these are forms of embodiment addressed by the curatorial interventions of the scheduler. The scheduler even registers the patient's prior embodied experience, through her question, "Why are we doing this?" Under the sign of resource availability, schedulers then match patient embodiments with staff embodiments. Techs, custodians of bodies in the waiting area and scanner rooms, orchestrate flow: patient flow, contrast flow. Once patients are dismissed from the waiting room or returned to inpatient wards, further curating of their embodiments is out of Radiology's hands.

Transcribing

Between the fleshiness of bodies and the flat text of x-ray reports, a crucial mediating practice is transcription.

CT interpretations are dictated telephonically, by radiologists at a viewbox, to digital audio files; the recordings are then retrievable from any phone in the hospital by dialing 6–6831 plus a medical record number (all part of the RTAS system, discussed above). The verbiage of dictations is also rapidly transcribed. Transcripts are disseminated digitally through the Hospital Information System and printed on paper for patient charts.

Radiology's Transcription area is on the second floor, down a hall from the Main Department. On the day I visit, in late afternoon, there is one transcriptionist present. At the moment, she is concerned because a report is delinquent and "there is no dictation on RTAS for it." She makes a note for the supervisor "to do further research."

To demonstrate the transcription process from start to finish, she logs onto RTAS with a two-digit "Typist ID." The oldest dictated, untranscribed report "on the system" will be transcribed next. The screen says "Session ID." The dictation is an attending's addendum about a CT, which provides a name and date but no order number. "I need that to bring the report up." The transcriptionist prefers not to consult a different computer at the moment. She proceeds to the next report, dictated by a resident. With a pedal she controls the audio: she plays it for me on speakers to demonstrate sound quality (mushy). The patient is Norris Farrington, #94887. The person dictating is a second-year resident, Githens, who speaks intelligibly but quickly. His voice and intonation are recognizable. The study is a CT of chest, abdomen, and pelvis. The transcriptionist types the report in half-sentence segments, with frequent pauses and rewinding. Each line of the dictation is numbered on the screen. After the report is typed, the transcriptionist runs a spell-check. The text report is then placed in a "queue" in the system for Dr. Githens to review and sign. Because this is a CT, which the transcriptionist assumes has been reviewed already with the attending, she places the report in the attending's queue at the same time, for "verification."

Not all transcription is done here in this room. Much is performed by staff working off-site, from home. The transcriptionists' supervisor—also the Nursing Supervisor—says: "I have three full-timers, I have three temps that work with no benefits and then I have . . . an outsource service that works for me— she has about four or five girls." One benefit of this variety is that the transcription service can accommodate fluctuations in volume of dictations, as well as staff vacations. "They [off-site transcriptionists] take this little box home . . . If they're local they dial in through the regular telephone. If they're long distance I have an Internet connection for them and they, they listen to the report, type it and then they send it off, it comes back to the mainframe and goes to the RTAS system." In-house temps are paid $10 per hour; "outsource" transcriptionists are paid 13 cents per line (65 characters)—closer to $15 per hour.

Transcription/nursing supervisor:	We do, in a month's time, about 45,000 reports.
Investigator:	Wow.
Trans supervisor:	That includes addendums that go to them and those kind of things. In, in a normal day, like yesterday—let's take yesterday—they dictated 680 minutes, which is equivalent to about 800 reports.

Trans supervisor:	See, I'm very lucky. Our um, RTAS, our RIS[52] system, generates everything. So, every transcriptionist, I can check at any time during the day.
Investigator:	I see. Right.
Trans supervisor:	This person today on 7/1 typed 91 reports for a total of 102 minutes.
Investigator:	Wow, OK.
Trans supervisor:	Ramone typed 101 for 91 minutes. So, there were 624 reports done 6/25, but you have to remember reports vary in size. You can have a chest x-ray which is two lines, to an MRI that's three pages long. So, that's why I pay people—I pay different . . . And I won't ever give Transcription up again . . .
Investigator:	Mm-hmm.
Trans supervisor:	I've learned too much and I like it too much and it's easy monitoring. I know what everybody does.

This archive of minutes of dictation and lines of typed report is an index of departmental productivity. But what are these pleasures of "easy monitoring"? Is there, even in this utterly bureaucratic function, a lingering curatorial charisma deriving from proximity to such crucial fruits of radiological labor?

The supervisor's aggregate data—200–300 minutes of dictation on a weekend day, 600–800 minutes on a weekday—can be sorted by radiologist as well as by transcriptionist. "Oh, we have some long-winded attendings," notes one transcription staffer: "They're great docs . . . But they can't—you know—Len Marcus, you know he's from New Jersey, man, he 'Ruhruhruhruh' and you're going, 'Gah'—and then you get some of the foreigners who you can't even understand." In addition to differences among individual doctors in style and verbosity, there are differences among cohorts of doctors. In general, minutes of dictation per report fall as residents advance in training.

The RTAS system also situates dictations, transcriptions, and signed reports on a timeline. Timely completion of reports is part of radiologic service quality. The supervisor follows this closely: "At one point, we were seven to ten days behind . . . When I checked it at ten this morning, we had about three hundred minutes on the system, so they were typing late yesterday's stuff. If we were down there now, they're probably typing this morning's stuff. And that's how Dr. Chair likes it." This monitored textual throughput follows the bodily throughput of waiting rooms and scanners. The queue works across both do-mains. The fact that oversight of transcription falls to the Nursing Supervisor further entangles textual and bodily flow. Beneficiaries of rapid textual report-ing are different, however, from those of patient throughput. Timely reports are appreciated by client clinicians. Rapid patient flow optimizes use of de-partment resources and staff. Sometimes these benefits flow to patients—by now, themselves, somewhere downstream.

Charting

CT reports are printed on paper and delivered to patients' charts. This occurs several days after a scan, in the hospital—sometimes weeks, in outside clinics. The paper record remains the final assemblage of patient data in many clinical settings—though this is changing. The chart is not so much for the patient (who rarely reads it) as for professionals.

The medical chart as we know it, a stack of paper concerning one patient's institutional experience, is an impressively recent innovation. In the 1950s, barely a fifth of American physicians maintained regular patient records. In hospitals, even in the early twentieth century, records tended to be maintained by department rather than patient. Organization of data for individual pa-tients in individual folders—the "unit system"—was only implemented after 1915, at the behest of professional organizations, litigants, researchers, and re-view boards.[53]

Mechanization, and the model of the factory, were important in shaping these hospital procedures.[54] The factory was organized (like the prison) by new technologies of monitoring and classification. Frederick Taylor described "scientific management," regulation of gestures of factory workers with the stopwatch, in the same year Röntgen described x-rays.[55] Hospitals emulated fac-tories: in application of cost accounting; in courting of middle-class clienteles; in inclusion of efficiency-conscious businessmen on administrative boards. Even hospital organization by divisions related to new accounting needs.[56]

Diagnostic work in hospitals was likewise influenced by bureaucratic mech-anisms. Around 1920, punch cards enabled quantitative case analyses that were previously difficult.[57] From 1900 to 1925, the length of the patient record in Amer-ican hospitals doubled (even as length of stay was decreasing). This was due to more specialists, professionalization of nursing, and new administrative units (with departmental forms).[58]

Graphical representation of clinical information became more promi-nent after the 1880s—beginning with temperature.[59] In the first quarter of the twentieth century there was a tenfold increase in graphs per patient-day. There was also an increase in preprinted forms systematizing collection of admission information. Radiology was thoroughly implicated in standardization of graphs and forms. From about 1917 onward, radiologists used standardized report forms. Typewriters became fixtures of radiology departments, conferring on radiographic interpretation an appearance of system and scientificity.[60]

Delivery of a printed x-ray report to a patient's chart is now standard American hospital practice. University Hospital's time frame for this delivery, several days, is typical.[61] It is also typical (at least among larger hospitals) that this same information is digitized in a computerized record. In the University Hospital system, one can retrieve the typescript of a CT report from a Clinical WorkStation terminal in a clinic across town, via dial-up, shortly after it has been electronically signed.

Such are the logistics of conveyance of the radiological interpretation to the chart, for perusal by clinicians involved in care of a patient. There are reciprocal relations of patient charts to radiological production: namely, conveyance of charts to radiology when x-rays are made or interpreted. In general, records are not sent from outlying clinics when patients are referred to University Hospital for x-rays. Within the hospital, whether the chart travels with the patient is subject to variation. Key portions are always available to CT radiologists at the viewbox, on the adjacent Clinical WorkStation.

The Literature

Interpretation of a CT scan involves texts beyond a chart: it involves various cumulations of radiological reportage—published cases, clinical trials, review articles: "the literature."

For many purposes, the literature pertinent to a particular scan is crystallized in a textbook—at the viewbox, on a shelf in office or home, or in the Radiology Library. At University Hospital, this library is down the hall from Body CT; it has a four-button entry lock. This library is described (for residency applicants) thus:

> A sizable budget is allocated to the library for purchase of new books, audiovisual materials and computer equipment yearly. Some of the more sophisticated and rapidly changing areas in Radiology such as magnetic resonance imaging, ultrasound and mammography are represented by the newer acquisitions. The library currently houses all of the commonly used Radiologic texts, plus some of the more classic texts (approximately 1000 total). To supplement

the array of Radiologic texts, the library contains a full complement of ACR teaching files and all of the current Radiologic journals. Video disc teaching files from the ACR [American College of Radiology] are also available with 2 monitors at the residents' disposal. The residents have access to the medical student teaching file, a unique conglomeration of over 5000 cases deemed interesting by students over the years. Books are available for overnight use and on a rotational basis. In addition, two MacIntosh computers and two PCs complete with a laser printer, FAX, modem, CD ROM, video discs, monitors, expanded hard drive, etc., are available to the residents on a 24 hour basis. Residents also have copying privileges and a copying machine solely for their use.

A "sizable budget" is important because radiology books are image-rich, large, and expensive. The presence of canonical references—"classic texts"—is assured. There is also a range of nonprint media, as well as devices for their "access." And perhaps the most important item: the reproducer of the canon (hand of the scriptorium's monk): the copying machine.

The library is fifteen by twenty feet, with shelves along walls and a table in the center. Texts are organized by organ system and modality. On the table, one of two texts lying out has "Atlas" in its title. Though the library is often unpopulated, at the moment there are two senior residents here, reviewing for upcoming Board exams. Journals are presently shelved.

As impressive as these walls of tomes are, some parts of the literature are not in departmental holdings. More can be found in the Health Sciences Library of the University. There is overlap among these library collections, but there are holdings specific to each. Compared with many other departmental libraries of the University, the Radiology Library is robust, nearer its scenes of service, and accessible twenty-four hours a day.

Accessibility of the library is not so crucial in 1996–97 as it was five years prior. Most of the key rooms of the CT Suite provide Internet access to the Web, to Medline, and to a growing assortment of electronic image resources. Collections of printed texts have begun to surrender pride of place to servers and networks—like the gold in the national vault made obsolete by e-commerce.

Radiologists consult texts and journals, but they also write and edit them. University professors all publish (or perish)—and academic radiology has as fevered a rhythm of scholarly production as any other department.

The Chairman was editor-in-chief of a definitive text on Body CT. For a recent revision, the Director of Body CT wrote a new chapter on Spiral Chest CT, a development of the mid-1990s. Other faculty contributors revised their chapters as well.[62] Most major textbooks require periodic revision, to reflect developments in the field and to compete with other books covering the same material. A textbook of canonical status would otherwise not stay definitive for long.

One radiologist reflects on one of his own books, which has drifted out of fashion but never been superseded. His fine microradiographic images extended the insight of contemporaries working at a coarser scale, and influenced successors. Even though the climate of interest has shifted, his book remains a source of pride.

Pace of textbook productivity affects unexpected aspects of authorship. One radiologist, renowned for the fact that he illustrates his own texts, suggests that saving time is one reason: "It's just line drawings. Actually, it helps. I like doing them because it really helps me to understand what's going on . . . and it saves me time. If I wait for Medical Illustration to do a drawing . . ."

The other important, and arguably primary, category of research publication is the journal literature. Journals are published biweekly to quarterly. Some serve more specialized readerships than others. At University Hospital, all residents and academic-track radiologists are involved with journal article production. Recent papers are posted on bulletin boards, discussed in conferences, celebrated in the department newsletter.

In preparing to interview one neuroradiologist, I consult his publication record. Medline lists eighty-plus papers he has authored or coauthored in the previous four years. A dozen are single case reports; many are descriptions of small case series (two to eight cases) of particular disease entities. The rarer the disease, the smaller the series—or the longer the list of collaborators, across more institutions. Some studies address one imaging modality; some compare findings across several modalities; some emphasize histopathological correlation; one involves postmortem imaging. A few are review articles (culling as much available literature as possible). Of these papers, this radiologist is listed as sole author on six and first author on twenty-four. Roughly two-thirds of the papers are coauthored by a University Hospital colleague. Some are coauthored by residents. This is a highly productive faculty member—and his productivity is linked with the productivity of the department overall.[63]

It is interesting that there are no large randomized trials in this particular slice of the radiology literature. Such trials are an important form of research in "evidence-based medicine," especially about evolving technologies. Large trials of many patients—a "population" rather than a case series—diffuses intersubject variation and provides the "power" to demonstrate small benefits.[64] There are large trials in radiology—and this researcher has collaborated on some—but they are not such staple elements of the research repertoire as they are, for instance, in testing new pharmaceuticals.

Large-scale research and statistical methods devolve from governmental interest in the welfare of citizenries.[65] One example of research involving x-rays and large populations concerns screening. More recently, influential refinements in statistical techniques have clustered around what is referred to as "decision analysis" research.[66] This is important for radiology, whose technologies are inserted into complex matrices of decision making, but not often proximately connected with what researchers call "outcomes"—in health research, traditionally, morbidity or mortality.[67]

Volume of publications is not the sole determinant of one's scholarly status. One University Hospital radiologist reflects on his editorial appointment with a prominent journal at a young age. "[The editor-in-chief] said, 'Well, you really seem to like it—your reviews of the literature, your reviews of manuscripts are always on time and very good.'" This radiologist's reflections touch on reactions of his colleagues to his appointment, on how long it takes to accumulate respect. It seems that election to the status of "peer" in peer review processes may involve impressionistic personal judgments as much as metrics of experience or excellence.

Peer review is professional scrutiny of research papers submitted for publication. Publication is the prize of successful review. A second level of scrutiny comes from the readers of the journal. Whether these readers believe or admire a particular paper is marked by further bibliographic citation. Citation by other researchers is a hallmark of scientific credibility and influence.[68]

Many other textual items exert powerful influences on CT practices. Within University Hospital, these include policies and procedures for nurses, techs, and physicians, and maintenance and safety records for scanners. Documents for training, examination, certification—of students, residents, nurses, and technologists . . . There is not space here to consider them all.

Reprise

Archives of the CT Suite do not constitute a room of their own: they are distributed throughout the hospital. Nor do archival centers merely serve containment, conservation. They are nodes and eddies in exchanges and flows of bodies, texts, and images through the hospital and beyond, supporting broad economies of scientific and commercial valuation. Curating comprises tensions between center and periphery, between fixed holdings and circulations.

The recent reorientation of radiology toward filmless systems—images stored and circulated digitally, only selectively transferred to film—exerts a constant pressure on workday routines and on the imaginary of "the Archive." Conversely: even given the volatility and speed of electronic information, and the hypervisibility of bodies' interiors, visible and palpable surfaces of objects and machines remain important. Craft practices of radiological curators still address folders, drawers, locks, keyboards, file slots, closets.

Radiological curators inhabit a range of roles. Techs, film managers, couriers, schedulers, transcriptionists are tasked with storing and distribution of objects of radiological interest. These curatorial roles are divided and hierarchically distributed, but everywhere there are powers that accrue to them, deriving in part from the value of the objects they are assigned to handle.

Among radiology's archival objects: bodies are registered and organized into streams for monitoring. Images are stored and distributed, on film and electronically, in folders. Texts summarize policies and protocols, convey interpretations to clients, and accumulate guild knowledge.

But bodies, images, and texts are not categories whose edges are prefigured. They exhange and overlap with one another within the CT Suite. The body in the waiting room is thoroughly texted: scheduled, identified, registered, and pre-subscribed to protocols for contrast. X-ray filmic images are captioned, gathered in labeled folders, assembled in labeled piles and slots according to a logic of institutional distributions—all via text. No map of these distributions is printable, but their topology is embodied in the efforts of the courier.

I have reflected on two kinds of museological object, specimen and cult image: these help situate CT films with respect to historical conventions of display. The notion of the specimen forefronts issues of fleshy embodiment, morphological comparison, and taxonomy; and it is faithful to CT images' frequent linkages with—substitutability for—biopsy materials. The notion of the cultic image forefronts issues of originality and elite interpretation—as well as potential for images to be appropriated to noncultic uses, exchange values—and commodified, fetishized.

The following chapter takes up some of the ways links are forged between bodies and images—corresponding to links between various diagnostic disciplines—in radiological "correlation." It also returns to the ethos of diagnostic intrigue as it takes shape within socialities of the CT viewbox. The chapter concerns pedagogical and testimonial practices in the reading rooms and in radiology conferences.

5   TESTIFYING AND TEACHING

The mental features discoursed of as the analytical, are, in themselves,
but little susceptible of analysis. We appreciate them only in their effects.
• • • **EDGAR A. POE,** "The Murders in the Rue Morgue"

At University Hospital, diagnosing is braided with training. Educating doctors is a task of the academic center. When production of diagnoses is yoked
to reproduction of expertise, certain diagnostic processes are expanded, and
expounded upon. Diagnostic pedagogy becomes conspicuously performative.
Radiologists teach to testify, and in testifying they teach. Trainees are instructed
how to bear witness, how to project credibility and authority;[1] and all radiologists, even outside academe, in all engagements with client disciplines, are engaged in ongoing, and reciprocal, instruction.[2]

Most radiological testimony, and teaching, unfolds in conferences and at
reading room viewboxes. These sites have common features. Both are relatively
insulated, from wards and the world. Both serve disciplinary as well as cross-
disciplinary colloquy. And in both settings, performances vary from the didactic
to the agonistic to the intriguing. But these settings differ too: the conference

room is removed from the shop floor, from the press of clinical urgency and demands of throughput. The conference schedule is more predictable, subject to forms of control that viewbox shop work is not. In this chapter I begin by considering distinctive testimonial and pedagogical routines in conferences. Then I consider correlative work at the reading room viewbox.

Conference Discipline/s

Radiology Conference Room: second floor, main corridor, near central elevators. Radiologists and their trainees come together here to view images and discuss diagnostic topics. Clinicians and their residents come here too, for conferences serving particular services and specialties. Other rooms throughout the hospital and medical school also host such conferences—involving radiologists in discussion, consultation, teaching—but none so intensely as this one.

In chapter 3, "Diagnosing," I suggested that radiological-clinical (rad-clin) conferences are heirs of sorts to the Clinical-Pathological Conference (CPC). Architectural corollary: the Radiology conference room is heir (of sorts) to the anatomy theater. These days, of course, anatomy theaters are not much used, for CPCs or any workaday events: they are museum exhibits, reserved for formal occasions.[3] As the anatomy theater has become more ornamental, the Radiology conference room has become more crucial. The conference room is as old as the Department of Radiology,[4] and its schedule is expanding.

Rad-clin conferences are not heirs to the CPC because they have replaced it.[5] They extend the CPC's legacy of lesion-centered diagnosis. They are theaters of intrigue and proof.[6] And though they tend to keep the cadaver in the closet, they are haunted by—obliged to quote, or emulate, or refer to—practices of pathological correlation.

Some Radiology conferences are more for teaching, some for diagnostic work. Most serve particular disciplines. Divisions of bodily region and technique split radiology and also connect particular radiologists with particular clinicians. This is one way in which CT scanning slices up hospital culture. Experts at conferences often include nonradiologists. Many conferences are thus cross-disciplinary, and correlative.

In Neuroimaging conference, neurologists describe visual problems and asymmetric pupils to radiologists reviewing images of the base of a brain. In GI Imaging conference, an endoscopist who has inspected a colon confers with radiologists about an abdominal CT. Special testimony about fixed appearances—testimony of a sort that, in the CPC, was brought by the pathologist on

31. Anatomy theater as museum

behalf of the revelatory scalpel—is now brought by the radiologist on behalf of the revelatory beam.

Rad-clin correlations are not unique to the teaching hospital. Even in non-teaching hospitals, clinicians and radiologists come together for case review—as in Morbidity and Mortality conference ("M&M").[7] But in the academic hospital, such gatherings are frequent and intensified and yoked to specific pedagogical agendas. Rad-clin conferences amplify hierarchies of training—attending, resident, student—as well as divisions and hierarchies of expertise: specialist / generalist. Even "working" conferences are pedagogical: radiologists teach clinicians (and vice versa), specialists teach generalists, veterans teach novices.

Conference Media

At University Hospital, the hallway by the Radiology conference room serves mostly professional traffic: despite the room's centrality, few passers-by are laypersons. There are two doors, front and back—usually shut, whether or not conferences are in session. No signs identify the room.

Inside, there is a viewbox centered in the front. It is fixed, not mechanical: one set of films must be taken down before the next can be mounted. There are three video monitors, at least one visible from every seat in the room, wired to a central display-top beside the podium. This display is a horizontal viewbox upon which a single film can be placed, and projected by a zoom video camera overhead. There are also slide projectors (on a cart, in the closet) and a large projection screen. There are blackboards on three sides.

The thirty-three seats in the Radiology Conference Room all face forward. The rearmost seats are over twenty-five feet from the viewbox. There is a phone on the rear wall, with a cord long enough for the headset to be carried into the hallway.

A generic room: no features so distinctive as the anatomy theater's radial gallery. Suspended ceiling, fluorescent lights, sheetrock, composite tile floors. Even the media accoutrements—jacks and hookups, cameras and monitors and screens—are in only a few respects specific to radiological agendas.[8] There is no fixed computer. And no drain, nor sink. At the moment though, overhead, a tile is missing—and alongside wiring boxes, a splice in a five-inch rubber pipe is visible: hospital fluids flow alongside the clean, dry agendas of this room.

At morning conferences, food is sometimes served: bagels and cream cheese at a morning Imaging/Pathology conference—where GI cases are discussed. And at noon conferences, residents and faculty often bring lunch.

The important materials entering and exiting this room, apart from presenters and participants, are cases: small stacks of film in folders. Some of these folders bear nosologic labels—craniopharyngioma, miliary tuberculosis—and others bear patients' names. A typical dossier of conference cases comprises these folders, sometimes cover sheets listing diagnostic considerations, sometimes yellow sticky tabs.[9] The case dossier imposes forms of handwork: case preparation is often recognized with thanks or applause—and many comments are directed toward the cases' materiality: quality of film copies; images one wishes one had; old formats from antiquated machines. Here the resident is fumbling to retrieve a film from a folder containing many sheets of CT images, while an attending, from eight feet away, in ambient light, remarks, "No, it's not that one either."

Apart from handling films, there are skills of multimedia presentation to master in conducting conferences. The film display top requires some switch, knob, and joystick facility to adjust brightness, magnification, contrast. Presenters take stock of where viewers are sitting: they may advise residents to come forward—"I don't want you to make a back-of-the-class mistake"—or offer a challenge, "You should be able to make this diagnosis even from the back." Disruptive movements of the display are often commented upon. "Did that slide just go into focus—or did I?" Radiologists are committed to finessing visual displays—and their meticulousness can border on vanity. Even bodily tremor becomes a pedagogical object: residents are taught, when pointing out findings on the display top with a pen, or pointing at the screen with a laser, to stabilize their hands casually on the film or podium.

Teaching Conferences

In radiology shoptalk at University Hospital, "teaching conference" means the noon conference: five days a week,[10] a routine event in otherwise disjunct schedules of faculty and residents. Attendings come intermittently: conferences are designed for residents. Attendance is expected (not demanded—out of consideration for clinical exigencies). There are fifteen to twenty participants in most teaching conferences. Lunches are welcomed: studying images and taking the midday meal are commingled practices for many academic radiologists, here and elsewhere.[11] Noon conferences are not exclusive: visiting professors, clinical residents, and students on Radiology rotations constitute a small extradepartmental faction at many sessions. There are teaching conferences outside the noon slot as well, including some with more rigorous, and exclusive, formats—like Hot-Seat Conference (discussed below).

Most noon conferences have a topical focus: a body region or category of pathology (the esophagus; abdominal trauma), an imaging modality (CT), or both (diffuse liver lesions—best evaluated by MR). The range and order of these topics are adjusted by educational coordinators, to ensure that residents are prepared for Board exams.

Each conference has one or two organizers responsible for presentation of cases and structuring of discussion. Faculty who are not presenting often attend sessions relevant to their fields of expertise.

The schedule of conferences is a lattice of disciplines—a kind of departmental census. A two-month sample is shown in table 1.[12]

Within this distribution of disciplines, attendings and fellows know they will be responsible for a number of teaching presentations every year. Most have "canned" talks they give—on topics of interest, topics they have reviewed in the past, active projects. Most talks are built around a case series—a carousel of slides, a set of folders from one's vertical files. Teaching conferences are a raison d'être for personal film files and for the central Teaching File.

"Talks" oversimplifies the variety of presentational genres in these noon conferences—and the briskness of discourse. A case series in a slide carousel

Table 1

Monday	Tuesday	Wednesday	Thursday	Friday
Neuro	NucMed	Corr Imag (Gyn)	Vasc	Res Mtg
MRI	Peds	GI (Esophagus)	Bone	Cardiac
US (Renal Tx)	Mammo	Vasc	Neuro	Chest
CT (Pancreas)	Peds	GU	Bone	Chest
Corr Imag	Neuro	Nuc Med	OB U/S	Res Mtg
Vasc	Cardiac	GU	Bone	Peds
Dr. [Guest]	Dr. [Guest]	Neuro	MRI Teleconf	Chest
CT	Peds	GI	Bone	Chest

can serve as the hub of a lecture, dialogue, or sharp argument. Residents are called upon to present cases they have not seen before and face attendings' questions. In some teaching conferences, particularly hot-seat versions, interaction becomes inquisitorial. One resident may question another—indeed a first-year may interrogate a third-year. Question-and-answer interaction can be cordial, but it is always asymmetric: the balance of power favors the case preparer.[13] In any case, teaching often assumes roles and voices besides the dully didactic. Salesman: "Could I sell you on . . . the bowel loops looking somewhat separated and possibly kinked?" Sideshow huckster: "You are about to be astonished and amazed at the breadth and depth of diffuse liver pathology revealed by MRI!" Primary schoolmarm: awarding gold stars for performance under interrogation—a half star here, a full star there, on a chart on the wall.

Most conference attendees come on time. At the beginning there is milling around, greetings, shoptalk. One resident conspicuously shuffles through page proofs of an article he has written with one of the faculty; another, in scrubs, hears a beeper and asks no one in particular: "Is that me?" The attending hosting has finished reviewing her stack of folders and is lounging and conversing in a chair toward the front.

Teaching conferences start with announcement of the topic—some rationale for the cluster of cases to be presented: "interesting cases from Body CT over the last two weeks." Case selection criteria may be omitted if there is a broad mix—or as the host presenter chooses. Keeping cases under wraps preserves surprise, displays pedagogical authority.

Once a conference is underway, particularly at the outset of each new case, attention is directed toward the presenter. Residents are aware they may be called to testify at any moment. Yet they also attend to other tasks: filling out forms, thumbing texts, eating.

Hot Seat

"Are you really allowed in here for this?!" A senior neuroradiologist expresses surprise that an outsider should attend Hot-Seat Conference. He is not convening this session—nor does he even stay for it—but he clearly feels protective. I reassure him that I have permission, and that I have been there before.

What is risky about a stranger auditing this event? Chiefly that the hot seat is so uncomfortable for its occupants. The hot seat is a role in a theater of inquest.[14] In it, gaps in one's knowledge are displayed, one's reasons are repeatedly questioned. The hot-seat role challenges and often embarrasses its subjects— in the process of teaching them rhetorics of testimony and inciting them to study harder. Taking the hot seat is a rite of passage—a series of rites, throughout residency.[15] One can understand departmental reluctance to expose this ritual to outsiders who might mistake it for gratuitous hazing.

"Norman: you ready?" A grimace: "Nope, but I'm here." There is no prescribed order; and if one is called on frequently, there is no point arguing unfairness. Grace under fire is one of the lessons. Enthusiasm would also be unfashionable: worse, might appear as brown-nosing. The standard posture in approach to the hot seat is dutiful submission.[16]

Norman accepts three sheets of film from the attending. He carries them to the front of the room, places them on the viewbox, and switches on panel lights. An abdominal x-ray with intestinal contrast, and two sheets of a Body CT. He stands before the films, silhouetted. Few details can be made out from the back half of the room. Norman takes a minute and a half to study the images, with the room hushed. This is his time, to inspect and to gather wits. This focuses Norman's performance anxiety, at the same time that it reemphasizes for the group the work of *autopsis*, seeing for oneself. There is no formal time limit to this personal survey, but there are conventions—and it can be cut short, at the attending's discretion.[17]

In this instance, the attending offers no history, no instruction. The omission of remarks commits Norman to more thorough, more systematic review.

Norman takes down the films, switches off the lights, and carries the films back to the lectern, to the display. He places the first of the films on the viewtop and touches a switch to display it on monitors around the room. He remains standing. Though radiological method favors sitting, the hot seat is, on this conference stage, more figurative than literal. The hot seat is a mythic topos—a place of risky testimony, of inquisitorial demand, which can, in fact, migrate to any chair in the room.

"What do you see?" asks the attending. Norman knows a presentation begins with description of findings and survey of negative (normal) findings. But attendings sometimes give specific cues, and waiting for these cues may spare the resident from being interrupted in an irrelevant line of testimony. Norman is showing the plain abdominal contrast study: "Small bowel followthrough showing nice contrast in terminal ileum . . . narrowing of terminal ileum and some filling defects . . . along mucosa you can see several defects . . . ulceration . . . Along here—a question of a fistulous tract between terminal ileum and sigmoid colon. (I don't know if you can appreciate that.) In addition . . . linear contrast . . . could be consistent with pseudopolyps."

The attending interrupts in a peremptory voice: "This is not pseudopolyps . . . This is Crohns, with effaced narrowed small bowel."

Norman resumes his description—despite the fact that the attending has contradicted his previous testimony, supplied a new and briefer description, and short circuited to a final diagnosis—" . . . air-filled terminal ileum, you can see how that wraps down . . . whoops, too dark [the display] . . . loss of normal *valvulae conniventes,* and normal mucosal [*pauses, searching for term*], mucosal, ah [*pauses—discomfiture: chuckles from others*], features."

The attending contradicts again: "That's just debris in front of the stricture."

For all the attending's brusqueness, though, it seems this resident is doing fairly well. Norman returns more confidently to "a nice demonstration of a fistulous connection." The attending seems pleased and asks him to "look at the CT." Norman places one CT film on the display and adjusts magnification to show just a few slices. He slides the film to display images sequentially: " . . . start to see on this image—that there is . . . transverse colon is coming across here . . . bowel wall a bit thickened . . . colon is not compressed yet . . . more evident further down [the serial images] . . . You kind of get this inflammatory mass in the right lower quadrant . . . further down . . . there is stranding . . . Contrast cannot get into this area because of all the inflammatory changes . . . It's tracking down that whole area . . . actually it's pretty neat—we can see that area of fistulous connection . . . It sort of looks like a tornado."

At this last there is a faint murmur from one side of the room, perhaps reaction to the informal tone or fanciful simile, as if it were somehow presumptuous, or inviting of further contradiction.

The resident is finished, and the last word is left to the attending. He summarizes succinctly: "A pretty classic case of Crohns, both from GI as well as CT findings." For all the tones of refutation, he has mostly ratified the resident's account.

When Norman has returned to his seat, and while the attending is scanning the room for the next victim, a resident sitting behind Norman leans forward. "Was that fun, Norman?"

This is a relatively simple hot-seat confrontation: decent description, brief interrogation, a few sharp contradictions.

Hot-seat Conferences are limited to two afternoons on the weekly schedule, but their performative conventions extend to other teaching conferences—and to the reading room viewbox. Some conventions require brief explication.

Expert Witnessing

Radiological testifying draws on ritual spaces from outside the hospital: saliently, the courtroom.

Courtroom and conference room are related by some resemblances. In both, cases are presented. Evidence is introduced, testimony given, judgments formed. Careful attention to historical particularities is linked to a body of precedent. Elite, mannered traditions are renewed.

Courtroom and conference room are also related historically. Radiology as a clinical specialty,[18] and prior to that as a "determinate science,"[19] owes its legitimation to the court.

X-ray photographs had entered courts within months of Röntgen's discovery.[20] There was initial skepticism about their admissibility: "like offering the photograph of a ghost."[21] In the inaugural U.S. case, jury and judge were persuaded of x-rays' veracity by radiographs of common objects and their own hands.[22] In most early cases, radiological evidence was limited to obvious demonstration, convincing to lay juries, of fractures and foreign bodies. Acceptance of radiographic images was largely piggybacked onto legal precedents for photographic images[23]—presuming accessibility to lay interpretation. By 1920, x-ray evidence was admissible in criminal and malpractice cases throughout the United States. Radiological professional societies arose around this time—though at first they largely addressed radiation safety.[24]

By 1930, lawyers were encountering increasingly complex images and entering increasingly subtle interpretations as evidence.[25] More lawyers and judges objected to providing juries with x-rays alone. They contested the "authentication" of images (chains of custody linking them to subjects' bodies) and effects of magnification and projection on verisimilitude. In 1935, a judge in Idaho observed that a particular x-ray image "would be utterly useless as evidence without being explained."[26] This designated the necessity of the expert interpreter. Courts became alienated from notions of self-evidentiality of x-rays and competence of lay interpreters—and by 1940, some argued that the expert interpreter alone could be sufficient, that images might be superfluous. A San Francisco radiologist argued in 1938 that "roentgenology is not a picture process, but a medical procedure." Therefore: "It would be much better if competent interpretation of the X-rays were presented to the jury rather than the misleading shadowgraphs themselves."[27]

The courts' endorsement of the indispensability of the radiological expert was a powerful ratification of radiological professionalism. In the mid-1930s, the U.S. Board of Radiology was constituted to certify radiological experts. The radiologist has ever since remained one of the scientific experts par excellence of the courtroom.[28]

Yet there is a certain irreducible instability in the role of the radiologist-cum-expert witness. Scientific neutrality encounters some of its phenomenological limit-conditions in the forensic theater of cross-examination and the summoning of competing accounts. The court structures a measure of defensiveness on the part of any witness. The witness is subjected to adversarial inquisition—into credentials, degrees of certainty, forms of ambiguity, alternative explanations. Expertise is reconstituted, sometimes deconstructed, in such rituals of inquisition.

Thus one precedent for the conference-room hot seat: the role of the radiologist in the court witness stand, sweating a bit as he offers expert interpretation. Radiological testimony has taken some of its prescribed rhetoric, its poise under pressure, its capacity to explain, from the courtroom.[29] The witness stand, a theatric form for making truth, is a locus where partisan agendas are balanced and x-ray evidence is put in dialogue with other, nonradiographic, forms.

Teaching to See and Say

Radiographic case presentations have recognizable broad contours and recurring elements (chapter 3, "Diagnosing"). But in practice, most teaching conference presentations do not unfold as uninterrupted monologues, from typical openings to proper consummations. Presentations are interrupted often—to amend errors, redirect inspection of images, translate observations into proper terminologies. If case presentation in the teaching setting opens with the tacit question, "What do you see?," it rapidly devolves into a process of amendment, cross-checking, and return: "*Now* what do you see?"

Looking harder to see more clearly, and choosing words to say more clearly, are in radiology twined practices. This is nowhere more apparent than in teaching exchanges about descriptions of findings.

Attending:	Mick, why don't you take this one? [*displays* MRI *images on monitor*]
Resident:	[*from seat:*] Mass in left cerebellar hemisphere . . . actually compressing the fourth ventricle . . .
Attending:	OK—there's a little bit of the fourth ventricle left. But actually, let me tell you just to tell you—while you're thinking . . . this patient came in with headache, and the CT was done, and it showed, you know, massive hydrocephalus . . . So you're right . . .
Resident:	This is enhanced.
Attending:	Yes.
Resident:	And, ah—there's heterogeneous enhancement of mass.
Attending:	Well, as far as enhancement goes—I would say that this is mostly *homo*geneous. I mean, it's not like you have cysts, or, y'know, dark areas . . . it's, you know, quasi-homogeneous.
Resident:	OK, so, fairly homogeneous enhancement of a solid mass . . . Yah, well, circumscribed borders . . . Low signal on T2.
Attending:	Somewhat isointense with gray matter, I would say.

This pedagogical exchange began with agreement about the location of a mass, then discussion of how the mass takes up contrast. It is not clear that this resident (third-year) fully subscribes to the attending's assertion of homogeneity; compromise is struck through equivocations—"quasi-," "fairly." Then there is the resident's claim about low signal intensity (darker gray) on T2 images. The attending implies this is insufficiently precise—by transposing the description into the language of tissue type—asserting that it looks like the gray matter of the brain.

Such a series of correctives is typical of teaching interactions aimed at finessing description of findings. The goal is not yet to name the culprit disease but to develop a verbal representation of a finding, a description. There is general movement toward terms of greater precision. It is common for changes or challenges of name to compel reinspection of the images.

Another case:

Attending:	Art?

[*resident accepts film sheet from attending*]

Attending: That's probably all you need.

[*Resident holds a single* CT *film with two scout views and a 3x3 grid of sections. He carries it to the viewbox; in a few seconds he brings it to the display top and zooms to one frame.*]

Resident: This is a . . . pleural mass.
Attending: That is a . . . loculated pleural effusion [*directly contradicting*].
Resident: [*promptly incorporating attending's correction:*] So there is . . . atelectatic lung and pleural fluid . . .
Attending: [*wanting more, faster:*] What can you say about this effusion from this CT picture?

[*three-second pause: resident looks distraught*]

Attending: This is exudative and it's organizing. [*tone is summary, peremptory:*] If this were ultrasound, what would you see? [*to other participants too; there are a few tentative suggestions—but he answers himself:*] Septae, multiple echogenic . . .
Attending: [*back to resident:*] Now, what I *really* want you to talk about—

[*pregnant pause: resident looks distraught again*]

Attending: What is this process right here? [*pulls pointer from his shirt pocket, telescopes it out, and points to the edge of the heart*]
Resident: Pericardial stuff.
Attending: Pericardial *stuff.* [*derisive tone*]

[*Resident does not offer a technically proper term; instead, he launches a differential diagnosis: things which might make the pericardium look this way.*]

This taut hot-seat exercise of inspection and describing, in conference, is not pressured by urgencies of the clinic (this case was consummated long ago) so much as by the urgency of inquisition. The attending prodded one seeing/saying into a reseeing and resaying, and then a third—with impressively few words.

First the resident saw a density and named it: "mass." The attending called the density "effusion" (fluid collection)—thereby indicating that the resident mistook liquid for solid. With one additional adjective—"loculated"—the attending pointed out that the resident also failed to note the liquid's heterogeneity. In view of these corrections, what the resident initially saw/said as one thing, he subsequently saw/said as two things: fluid and compressed lung. His shift of perception from oneness to twoness, and change of name, appeared, in this choreography, part of a single quick gesture.

The attending proceeded to show that even this revised seeing/saying was incomplete, on two registers.[30] Then he undid the focus of the resident by asking him to look elsewhere. Confusion made the right backdrop for the attending's quick-draw of a pointer. With one hand, without leaning from his center-front position, he redirected attention: over here, edge of the heart. The name given by the resident to what he saw at the tip of the pointer was so nonspecific that it was both error-free and ludicrous. With mere repetition, and shift of tone, the attending highlighted its unradiological generality.

The difficulty of the task before this resident had less to do with discovery of findings than with finessing their description. But modulations of description entail modulations of seeing.

Examples of this pedagogical situation, finessing description of a finding, would be easy to proliferate. These are lessons that seem to be about redirecting or recalibrating the gaze. But each such lesson presumes, and trades on, a Rosetta-table of equivalences between appearances and terminology: between what imaging is said to provide—in CT, shades of gray corresponding to tissue density—and the anatomic corpus, the body of nameable parts and their Greco-Latinate qualifiers.

Terms sought in these exchanges are not nosological. What the attending wants is not specification of disease entities but concise description of a topologic or morphologic situation. Where is the lesion? Liquid, solid, gas? Does it contain calcium? How big? Does it enhance? Terms sought here are meant to stabilize a conceptual entity which can then be submitted to more refined diagnostic consideration. Description of the instant lesion is meant to be comparable with descriptions of other lesions, to support judgments of similarity/difference. A first-order task, prefatory to classificatory and conjectural exercises to follow.[31]

The inquisition which confers such urgency on the seeing/saying of residents is one into acuity of perception, into familiarity with normal appearances, into facility with terms. The hurry—the need to think quickly on one's feet, to get up to speed—highlights the importance of logistics in radiologic work. Urgency pertains to patient needs but more immediately to rhythms of work on clinical services, to needs of busy colleagues. These urgencies are folded into the demeanor of the attending: he exhibits the impatience of the institutional culture radiology serves.

Pedagogical interventions can be gestural. When a resident has inspected images for too long without comment, the attending asks: "Any subtle lesion you want me to show better? Subtle, subtle, subtle." The resident looks away briefly, the attending taps the film to show the finding to the rest of the room—and the resident's ensuing discomfiture provokes chuckles. If no answer comes, of course, the indexing finger will be brought to bear for the resident too. In another conference, the attending happens to rest his hand on the film the resident is inspecting. "Oh—don't be fooled by my finger," says the attending. "You didn't put your finger on the pathology," comments a second attending, chuckling. Experiences of not-seeing, seeing-at-last, and having-to-be-shown are universal for radiologists-in-training.[32]

A finding clearly seen and described may still need a proper name—a test of memory. "What's this sign?" asks the attending, in conference—unanswered. "It's the 'delta sign'—for a thrombosed sagittal sinus." Some of the names sought are comfortingly simple: "a two-dot shunt" refers to a manufacturer's designation (high- versus low-flow devices); "target sign," in CT of bowel, re-fers to a bull's-eye appearance worrisome for colonic intussception. Some of the names are Greco-Latinate concatenations. Many are eponyms—"Springle's deformity": elevation of scapulae.

Naming can become arcane. Neurology residents resist a suggestion that they should remember the "open operculum sign" by saying, "No, that's Rad-speak." To this, the Neuroradiology fellow replies, "No, it was a Neuro attend-ing who taught me that." Names have boundary and membership entailments. In one Neuroimaging conference, a Radiology resident misnames a skull de-formity: "schizencephaly." After being corrected by the attending—"brachy-cephaly—," she retorts: "It's a coronal synostosis—whatever, whose 'cephaly' it is." Laughs all around: she has, notwithstanding her error, trivialized an entire set of morphological terms and reclaimed a measure of authority by using a sophisticated developmental term.

Occasionally attention is paid to etymology. Moments after the above ex-change, a senior radiologist seeks to translate classical arcana: "Turn that [skull x-ray] upside down. *Scaph* means boat hull. See this?" He outlines a curve on the viewbox and provokes nods of appreciation. On another occasion, the same radiologist engages discussion of forefoot deformities, and the definition of "varus" and "valgus," by quoting a false (albeit memorable) etymology:

> A long time ago an old Irish radiologist told me *varux* meant barrel. Imagine holding a barrel between the limbs—limbs draped around it. [*stands and em-braces the air*] It turns out this is not so . . . [*chuckles*]

In teaching conferences, supplying an arcane name is a bonus, particularly if it demonstrates specifically radiological knowledge. Missing a common name is a problem. There are, for all that, unnamed findings. A young, meticulous radiologist addresses a senior colleague: "Dr. Montrose, what is the exact name for the ___?" and proposes two alternatives, both highly erudite. The senior radiologist replies, "I have no particular name for it." Sophistry deflated. On another occasion, a senior radiologist comments on the appearance of the adrenal gland in hypotensive patients—"bright like a light bulb . . . This sign has not been reported." An unnamed sign exemplifies opportunity: there remain findings to be named—works of description to be pursued: the Age of Discovery in radiology is not quite over.

Differential Diagnosis

"Differential diagnosis" is primarily constructing a list—"what entities can look like this?"—and ranking it by relative probability. The process is reconstructive when such lists already exist in the radiologic literature.

Attending: There's a standard differential diagnosis for things in the suprasellar region—but I mean with *this* appearance you ought to be able to narrow it a little bit more . . .

Resident: Oh yeah. Anything that's white on T1 and T2 and doesn't enhance is gonna be proteinaceous or blood.

Attending: Yeah, basically . . . So what would be your main differential diagnosis?

Resident: . . . could be . . . epidermoid?

Attending: . . . How about the location of this thing for an epidermoid?

[*someone else answers*]

Attending: Usually not midline, OK. What would be a mass of high signal intensity in the midline?

Resident: A dermoid.

Attending: A dermoid. OK . . . do dermoids happen in the suprasellar cistern?

Resident: Uh, I'm sure they can.

Attending: OK. So a dermoid is in the differential diagnosis. What else?

Ingredient to differential diagnosis: description (suprasellar mass, nonenhancing, protein or blood); beginnings of a list; exclusion from the list of patently unlikely items; and the recurring question, "What else?"—often echoing long after a resident has exhausted her memory.

[*A resident in hot-seat conference has described a* CT *finding: dense liver. The attending asks for its differential.*]

Resident: TPN [Total Parenteral Nutrition].
Attending: Why?

[*lame reply*]

Attending: What else?

[*Attending leans back against empty/darkened viewbox, finger on mouth, hint of a smile.*]

Attending: Jim, help him out.
Resident 2: Thorotrast?
Attending: Yes. Everyone who has had [a rotation on] CT should be generating a differential diagnosis of six or seven items.
Resident 2: Amiodarone.
Attending: Good.
Resident 2: Multiple transfusions. [*attending nods*] Hemochromatosis. [*nod*] Intravasation of barium. [*nod*]
Attending: And then there's what this is . . .

"Differentials" are key tools of the radiologist, and residents spend hours committing them to memory. They are things to be "carried around." Sometimes residents use mnemonics or acronyms. There are, however, attendings who disparage such prostheses, preferring that residents support their recall in other ways. One approach is to consider each category of disease in a customary order—infectious, neoplastic, and so on—thinking methodically through possibilities. This device is accepted by attendings as a legitimate aide-memoire. No parlortrick, or pocket card, but resolute method.

The attending often consults others to fill out a differential. Group work fits well with the "what else?" mode of inquiry: it stages the fact of distributed ignorance while modeling an ideal of comprehensive knowledge. The full differential is something that all uphold as standard, even if, momentarily, all are inadequate to it.

Sometimes an attending feels the specificities of a case are too restrictive to allow generation of a sufficiently broad differential. Or there may be particular disease entities which the attending knows can look like the lesion on display and would like residents to associate with these appearances. She may say, "What if I told you . . ." to modify the instant case. "What if this was a kid?"

Counterfactual thought experiments amplify resonances between described finding and the archive of entities to which it might correspond.

When it comes to ruling out items in the differential, attendings intervene more directly. They may cite age of the patient, or location of the lesion. "Mets to spinal cord are extremely rare."

Though relative likelihood is crucial to thinning the differential, it is unusual to hear precise quantitative or statistical terms. Prioritization tends to be a semiquantitative exercise.

Attending: How common?
Resident 1:   Uncommon.
Attending: What do you think, Nora?
Resident 2:   Mycotic . . . Multiple developmental . . . uncommon . . .
Attending: Give me a number.
Resident 1:   Twenty percent.
Attending: Twenty percent is "uncommon"?!

Occasional demand for a number reminds residents of authoritative epidemiological and actuarial modes of representation—even if these are not retrievable in all, or even most, cases. The attending supplies this answer: 10 to 20 percent.

One recurrent expression in prioritizing exercises is decidedly aesthetic: an attending "likes" or "dislikes" a possibility. "Oncocytomas . . . in kidney? I don't like that." To a resident: "What do you like about them for abscesses?" means: explain what, about these images, supports the diagnosis of abscesses. Liking can also be folded into solicitations of the collective. The resident asks, "How about vasculitides?" and the attending turns to the rest of the room: "You like that?" He winces and turns back—"No, we don't like that"—and chuckles to hear himself mimicking an MC's solicitations of a quiz show audience.

In narrowing a differential, attendings ask honing-in questions. "What kind of vasculitis affects the internal carotid arteries? Think about it." In narrowing the differential, attendings also offer more emphatic propositions. Rules: "Medulloblastoma goes to bone." Morals: "A lesson for you first-years . . . about trauma CT: look for bowel wall edema. It requires no intervention, but does need close observation." Proscriptions: "Do not confuse omental hypertrophy with omental caking of malignancy!"

One salient rhetorical mode in narrowing the differential is interrogative. Another is didactic (attendings offering facts, maxims, rules). But there is a further rhetorical structure which often takes shape during this process of identifying the culprit disease entity. This is the mystery, the puzzle.

Intrigues in Conference

I return to the genres of detection, the economy of tracking and conjecture, introduced in chapter 3, "Diagnosing." During the discovery phase of a teaching case, when findings are made, the resident is examining evidence for the first time—like a detective at a fresh crime scene. During the presentation phase, describing these findings and generating a differential, she assumes the role of courtroom witness, supplying expert testimony (if haltingly, under pressure). What often takes shape amid these performative conventions is a whodunit.

There are differences between diagnostic conundra in radiology teaching conferences and in the prototypical genre of the CPC. In the CPC, there are more devices of narrative protraction—history of illness, hospital progress. These establish a longer chronological march, within which circumstances line up in causal chains: a plot. In radiology conference presentations, however, chronology is abbreviated, even crystallized—to a symptom or brief context of ("indication for") a study. Other devices besides chronological narrative contribute to intrigue.

Some of these devices are not obvious at first: withholding, or concealment—by attending, from resident—of information that could aid a diagnosis. Withholding key facts may summon a fuller differential or prevent too-quick solution of the puzzle. Indeed, withholding helps produce the puzzle.

A resident is given a chest x-ray to review—and her comments at the podium focus on hilar soft tissue densities. The attending gives her a CT on the same patient, and she focuses on slices through the hila. On his prompt, "Now look at the abdomen," she places a second CT panel on the display-top. "There's a fetus!" Pregnant women are not usually exposed to CT levels of radiation without reason. "It was a surprise to her too," notes the attending—and then he extracts several lessons, regarding justification for pre-x-ray pregnancy testing, and the utter inappropriateness of this study. Mere review of the chart would have disclosed the patient's *prior* diagnosis of sarcoid (explaining hilar densities)! This observation, that whoever ordered the CT failed to consult the record, resonates as consummating diagnosis—of professional neglect. By withholding the diagnosis of sarcoid, the attending has not only mimetically reproduced in the resident the nearsightedness of a neglectful clinician but has at the same time laid a backdrop for his own moral judgment.

Withholding information is not unique to the Radiology teaching conference. In the CPC, deferral of the testimony of the pathologist is an architectonic form of withholding. And testimony of the consultant radiologist is sometimes constrained. But apart from these withholdings, in the CPC it is unusual for important historical, physical, or laboratory evidence to be withheld: it would be unfair to the discussant-cum-detective. Strategic withholding of key information is, however, common in radiology teaching.

Yet one often hears the attending offer "a clue." Ostensibly this assists a resident toward a diagnosis—but it has the intermediate effect of contributing to the rhetorical construction of mystery. The structure of a mystery may be as simple as image and one such clue.

The attending places a scout film of a chest CT on display, long enough for the resident to make out an interstitial pneumonitis. "And this is the clue," says the attending—pointing to the shadow of a heart monitor wire. Pause for puzzlement. "This is amiodarone [lung toxicity]," says the attending. "The patient had an arrhythmia." In this case, neither the clue, nor the limited time for head scratching, was sufficient to help the resident reach the diagnosis. But the clue was enough to conjure up a fleeting mystery.

A hot-seat case is introduced with cryptic comment: "Timely—but only if you've been listening to the news." A resident haltingly finds and describes a dense liver on CT. A second resident supplies a differential diagnosis, six or seven entities (cited several pages above). Attending: "And then there's what this is." Ten-second pause. "Here is the clue. What's the patient's name?" The resident reads the corner of the film: Norris Beaufort. No response from any of the residents. The attending refers again, abstractly, to "current events." Minimal time for head scratching. Attending: Blanche Beaufort? Just convicted for killing her husband? This—is *arsenic*.

There is a faint collective "whew." The sentiment is not clear—horror? surprise? This is no mere simulacrum of a detective case, but a real one! The sudden criminalization of these images gives the case a sensational aspect. For a brief moment, courtroom and conference room are in even more direct communication than usual.

Citing the Literature

"The literature" is referenced in teaching conferences in various ways. It may be straightforwardly quoted. "This slide on the thrombosed sagittal sinus differential is from Ann Osmond's book." Sometimes it is a heuristic device: "What would the legend say, if you looked it up in a book?" Presumed is a conventional view of bioscientific publication: solid facts, laid down one upon another: edifice of positive knowledge.

Sometimes the literature historicizes radiological authority—as when the attending reviews "textbook anatomy" of the pancreas—"head, neck, body, tail"—and remarks that these divisions were "created" by imaging (not surgeons, as might be assumed). Or the literature's own periodicity, waves of fashion: "Five years ago, there was a run of academic papers on this [delayed contrast uptake in the liver]." When an attending showing diffuse liver abnormalities on MRI asserts, "Most of this material is not described in the literature yet," one understands these images constitute a cutting edge of knowledge accumulation.[33]

Authority of the literature may be forefronted ironically. In one conference on neurologic topics, an attending's interest in persistent skull foramina (infant-like holes in adult skulls) leads to another's speculation about vestigial orbits—"remnants of eyes in the back of the head, of some phylogenetically lower species . . . [there are] photoreceptors in the pineal gland." A resident interrupts: "You saw that on *X-Files!*" The attending retorts: "The Discovery Channel." The ensuing laughter does not suppress genuine interest. It is understood that the vestigial orbit theory was culled from some scientific text or another—and if one wanted, one might pursue it in the library.

Invocations of the literature serve to encourage trainees to "hit the books." Residents read at home (supposedly two hours nightly) and in the library between hospital duties. Attendings who chose the day's cases may know that a category of disease has been well represented. "So, colloid cysts: a good thing to read tonight."

Sometimes residents read during conferences, from textbooks, on the topic du jour. When the attending poses a question about the current differential— "Pauline, when do craniopharyngiomas occur in life?"—the resident responds with unusual confidence: "There's two peaks." Others laugh when they realize she is reading from the book in her lap. Later, when questioning proves unfruitful, the attending turns back to this resident: "Pauline, where else does the book say that medulloblastoma tends to go?" In this instance the attending knows the answer; on other occasions, invocations of the literature suggest limits of individually retainable knowledge: "Believe me, we had to look this up: it's a case of moia-moia."

In teaching conferences, the case at hand is continually tested against canons of accumulated knowledge. Is the current case typical, unusual, rare? Is a particular diagnosis likely or unlikely, given what the literature says about patients of this age, or gender, or geographic location? The very structure of the case series produces a sense of contribution to, and reliance upon, this body of knowledge.

Case Closure—and Hinges

The case is still open: images on monitors. The differential has been narrowed—with questions and clues from the attending and insights supplied by resident, peers, books. Along the way residents have learned rules of thumb, "pearls" of practical wisdom. Suspense has been created, curiosities engaged and intensified.

At some point, either the field of diagnostic possibilities is narrowed to one, or the attending reveals the answer. This final diagnosis is a prop in this theater of diagnostic intrigue—as in the CPC.

But is the final diagnosis so final? A favorite intervention of attendings: "I want your logic here." It may happen that the consummation of the case is something other than a diagnosis. The attending may be less interested in a resident's identification of the lesion than in inference of clinical context. Or, as in the case of the surprise fetus, there may be a moral tale hinging on the diagnosis. Cases end; but radiological teaching demonstrates many tactics for undermining finality of diagnosis.

And some cases have no final diagnosis. They raise questions of proof: how would one get to the bottom of this? Cases remaining unproven, ambiguous, or contested are especially common in working conferences.

Teaching cases often demonstrate "classic" appearances of diseases. "Nice demonstration of intussception." Classic cases trade on family resemblance, typicality of appearance. Their images serve as did Galton's composite photographs of a criminal type (chapter 3, "Diagnosing").

Teaching cases also include unusual lesions. An attending refers to MRI images of transfusional hemosiderosis as "exquisite"—both because of their technical refinement and the rarity of the disease. "You need a hundred pints of blood to get that film." Another attending showing developmental abnormalities of the spleen regrets *not* having a slide showing "splenic-gonadal fusion: if anyone hears about one, there is a reward out for this." In both these cases, rarity is figured as economic value—preciousness of an objet d'art; bounty on a fugitive.

Rarity of disease does not necessarily mean difficulty of diagnosis. An attending apologizes for an "outside scan" (another hospital's)—"the only case of this I've seen since I've been here": rarity constitutes a clue, and the resident makes the diagnosis in five seconds: "Echinococcus." But rare diseases that resemble common ones are tricky. An attending shows CT slices through the diaphragm, and the resident describes a "miliary" pattern of small densities. This summons standard diagnoses: miliary TB, sarcoid . . . But the differential runs dry, and the attending prompts: "a tannery worker—worked with cow carcasses." "Oh," says the resident, "it's brucellosis." Indeed, says the attending, "it's the only case of brucellosis that I've seen here on CT."

Rarity can be an attribute of imaging appearance, not just disease entity. One "unusual volvulus" (on CT) is deemed worthy of extra effort to demonstrate it "in 3-D reconstruction"—an image genre still uncommon in 1996. The uncommon finding summoned an uncommon mode of representation.

Another case: "Very unusual pathology. You won't see many of these in your career." Double implication: this is a privileged moment, a "trophy case"; and, this is a curiosity, more astonishing than enlightening, so anomalous that it may not illuminate principles of classification.

A sort of case favored in teaching conferences is one that anchors review of a class of disorders. A CT showing a left-sided inferior vena cava (IVC) introduces a discourse on IVC abnormalities "worth knowing," with illustrated discussion of embryology, classification, and more. In such cases, the imaging study is presented briefly, the drama of lesion detection and hot-seat challenge subordinated to didactic agendas. Nearly any image can summon such a review—of development, pathophysiology, epidemiology, and so forth. The hinge between imaging cases and didactic agendas is extremely plastic.

Didactic presentations in teaching conferences are, in 1997, mostly slide shows. And radiologic slide shows are, according to robust tradition, double-barreled: two carousels. "I pair everything because everything should be paired," says one attending, introducing a talk prepared for the upcoming RSNA conference. The projectors are duplicates: identical bulbs and lenses. This is not merely a function of radiological compulsiveness about display. This is methodical consistency: ensuring that differences between projected slides devolve from the slides themselves, not some technical aberrancy.

The standard rationale for side-by-side images is that it facilitates comparison.[34] One can show both views of a chest x-ray at the same time—or, with a CT, two slices of the same lesion or organ side by side. Or noncontrast view on one side, contrast on the other.

Possible comparisons proliferate rapidly. One can compare views from different modalities—CT and ultrasound—referencing "the same lesion." Or a comparison can involve different patients, with the claim "This is the same diagnosis." Multiplying cases this way conjures up, as conceptual bridge and metastructure, the generality of the disease class—irrespective of dissimilarities between images.

Another genre of side-by-side display: similar images, different diagnoses. Here what is conjured is not a nosological class but a visual type: *lesions-that-look-like-this*. Even if similarity is not attributed to trickiness of a culprit entity—"a very tough lesion"—a visual type associated with different diagnoses is manifestly duplicitous. At the same time, this is commonplace: precisely such a notion of visual type anchors the starting point of every differential diagnosis.

The dual projectors can show, alongside an image or in place of images, any genre of textual or diagrammatic representation found in the literature. This series of genre pairings is from a conference on esophageal disease:

Left	Right
Text: lye stricture: epidemiology & stages	Image: endoscopy
Text: table: caustics vs. acids	Image: endoscopy
Text: history: dysphagia	Image: XR: BA swallow
Text: nasogastric intubation esophagitis ...	Image: [?]
Text: pill esophagitis ...	Image: endoscopy
Image: XR: BA swallow	Image: endoscopy
Text: Candida esophagitis ...	Image: endoscopy
Image: XR	Image: XR
Image: XR	Image: endoscopy
Text: history: leukemia, odynophagia	Image: XR: BA swallow
Text: CMV esophagitis ...	Image: endoscopy
Text: history: AIDS, odynophagia	Image: XR: BA swallow
Text: infectious esophagitis ...	Figure: pie chart: etiologies

In teaching conferences as in reading room practice, the presentational device of contiguity—juxtaposition of two things—is exploited as technique and tactic. Comparison is one of the operations which contiguity affords—a rich and important one. Radiologists favor "correlation"—as in "imaging:pathology correlation"—but this term emphasizes the intellectual exercise over the surface of representation. Contiguity is a more materialist notion—the seam or edge of juxtaposition constituting a hinge, one that is quite plastic.[35] Contiguity can summon judgments of similarity or difference between two elements; it can suggest progress or regress over time; it can transmit a contagion, materialize a controversy, secure a consensus. Contiguity in imaging displays can produce coherence, but it also can produce instability, polysemy, dialogical exchange—with complex intellectual and aesthetic entailments—at the heart of radiologic method.

Been There—or Not

"One of those things that it's easier to understand if you were there," notes a resident, presiding over CT slices on the display-top, supposedly showing a bowel fistula. He pauses, and then adds wryly, "I wasn't there either." Scattered chuckles register this twist on a standard formula—the claim of special authority (one kind of radiological charisma) based on familiarity with the case.

In teaching conference, the presiding radiologist often brings cases with which she can claim firsthand involvement. She might have been "on service" when a study was performed—might have talked to the patient, or a clinical team, or dictated or "signed off on" the interpretation. She might have looked at the case over the shoulder of another radiologist—or been "curbsided" for a second opinion. Or: she might *not* have such firsthand knowledge.

One need not speak from a position of existential involvement in a case to read out a capsular history, guide a resident's interpretive efforts, and reveal the diagnosis written on the jacket. Cases culled from Teaching Files, or loaned by a colleague, have plenty of value. The presenter can speak *as if* familiar with the case.

If the presenter was not involved in the case, often someone else in the room was.

> Attending: Nora, let me show you this [*pulls* MRI *from folder, places on display*]
>
> Resident: Is this the one that I did on call?
>
> Attending: Ah, it must be the one that you did on call—so no, let's, let's get—Pauline, you look at it.

Others may have seen a case already because good cases have exchange value—in hallways, after work—with their associated tales of risk, error, skill: intrigue. It may be difficult to find someone who has *not* seen a study—as with one case of intussusception which *all* the residents present claim to have seen.

Conference cases are always haunted by someone's having been there. Fresh image interpretations are theatric simulations—requiring matching of resident to case, selective revelation of history, suppression of prior interpretations and pathological evidence. Disclosure of the diagnosis on the jacket restores a case to its archival status and dissipates any pretense that it is new. The veil of ignorance, and other devices constructing the temporality of conference interpretation, are also reminders that each case is a mimetic revisiting of a primal interpretive scene.

Testimony of someone previously involved—radiologist or clinician—can extend a case's relevance, compound its intrigues. She can reflect on acquisition of the images. "[The] junction of enterocleisis . . . [is] obvious on this film, thank God, but in fluoro, this was very difficult . . . The only time that this lesion was seen was when patient was in very steep RPO obliquity." Implied: in fluoroscopy one must be persistent, willing to place the patient in awkward positions; and it's a good thing I was, or we might have missed this. (Self-congratulation can operate even—perhaps best—in the passive voice of procedure-speak.)

Comparable discussions about CT image making often pertain to contrast administration and timing. "If I had known ahead of time [of the carcinoid], I'd have done a noncontrast, then a dual phase." Logistics of imaging "workup"—known best to someone previously involved—can be the most interesting aspects of a case:

Fellow: This was also a call case from awhile back—one week . . .

[*displays a plain abdominal x-ray*]

Fellow: Follow dilated colon to the region of the splenic flexure, and then it just stops . . . [*displays a barium enema (BE)*] . . . couldn't get any kind of contrast past distal transverse colon . . . area of narrowing . . . very sharp taper, I guess would be the best way to describe it. Think . . . adhesion or volvulus . . . much less likely to be a mass . . . We did a CT on this gentleman too . . .

[*displays slices of CT*]

Attending (previously uninvolved in case):   CT was pre-BE or post?
Fellow: CT was before . . . I did a 3D reconstruction through it [*pause, displays reconstruction*] and you can see it as a narrowing point [volvulus] . . .

[*enter three dark-suited residency applicants with resident tour guide*]

Attending: Pretty good call on CT scan.
Fellow: Actually it wasn't—it wasn't called until BE . . . For these last two cases, the clinical teams were right on top of things.

This case initially seems open/shut. Presented in CT conference, culprit lesion defined in 3D CT reconstruction: one could infer the Body CT team made the diagnosis. The attending seizes on workup logistics: which study enabled "the call"? His question and cagey compliment elicit the admission that this lesion was initially *not* defined on CT. The compliment is deflected toward clinicians. They (not radiologists) pursued the BE, on the basis of which the call of volvulus was made and a revised interpretation of the CT enabled. This thicker tale unfolds because the fellow's initial exercise of rhetorical prerogatives, based on prior involvement in the study, meets canny questions.

A resident comments, after a memorable case showing extensive infection: "Did you ever see this lady? On Vascular or anything? She was so sick!" The attending replies: "I'm impressed this lady is still alive. But she is." Most conference cases do not invoke the patient as person. Conventions about including personal identity (of patient or staff) in the radiologic case differ from conventions of clinical case presentation. In medicine and surgery, cases are generally presented by persons with firsthand knowledge, to others without—often in terms of ownership: "my service," "my patient." In radiology conferences, forefronting of personal involvement is selective. Radiology conference is not the expiatory ritual of the surgical M&M.[36] Perhaps because touching or speaking to the patient is so infrequently a radiological prerogative. Most radiologists have modest stakes in image-production craft (province of techs).[37] And privileges of ownership deriving from direct involvement are transitory—given rapid circulation of cases in networks of hospital exchange.

Conference Correlations

"Correlation" is a familiar and versatile term for radiologists. Its implications of *retelling* and *togetherness* have special importance—given the disparate forms of evidence, and ways of knowing, that radiological correlations seek to bridge. Correlations specify links to radiologists' clients (clinicians) and to their proof (pathology).

In conferences, cases are scheduled, tidied, and re-presented for collegial gatherings. Retellings aim toward pedagogical effect, or inter-disciplinary balance. Correlation can be between image and histology (rad-path), as well as among specialized viewpoints from inside and outside radiology, which might or might not have been brought to bear in a case's first pass across the viewbox (rad-clin).

The first form of clinical correlation for an imaging case is the indication on its requisition, or history offered by a client colleague. Radiologists feel these cues are crucial to interpretation. Their incompleteness is a recurrent source of frustration at reading room viewboxes—but in conference, history is elaborated just as the presenter wants it.

Clinical correlations devolve from a patient's "demographics"—aspects of actuarial status. Age, gender, "race or ethnicity," travel exposures, risk factors of various sorts. In one teaching case, concerning a cerebellar mass, the attending specifies: "Seven-year-old. Yeah . . . a—Mexican little kid." He prompts the resident for relative probabilities in his differential. The resident is unsure: "Hemangioblastoma is the most—ah—is NOT the most common." Hard laughs. Nor is cerebellar astrocytoma the most common. The resident cannot say which is. Attending: "The most common cerebellar tumor in children is? Nancy?" Nancy guesses: "Medulloblastoma?" The attending confirms this—and only then reveals that this child had a medulloblastoma resected five years previously. A key omission—for teaching.

Conferences teach that some findings cannot be interpreted *without* clinical context. An attending, displaying a series of Neuroimaging findings from the same patient, says to a resident: "You need to put the problem of the [narrowed] internal carotid artery, the 'puffs of smoke,' and the intraventricular hemorrhage [together] . . . It all happened at the same time, so they have to be—they have to be related." Explaining the association among these findings depends on the clinical datum of their temporal coincidence.

Sometimes information beyond history and demographics is required. "To make the difficult call of diffuse hepatocellular carcinoma, correlate [MR images] with alpha-fetoprotein." Radiologists increasingly consider laboratory data which have traditionally served correlative work of clinicians.

Time constrains rad-clin correlations. An attending reflects on limited knowledge of early CT findings in stroke. "The patient who has a stroke going into the CT scanner—now that would be the ideal patient." He laughs, both at the scenario's unlikelihood and the facetious callousness of such a wish. Rad-clin correlation is necessary, and in most instances necessarily retrospective, because of the irreducible temporal gap between a clinical indication and a radiographic study.

Radiologists in conference remark on friction with clinical colleagues—for instance, regarding clinical skepticism about their testimony. "This is a case [*two slides*] where it was a terrible struggle to get the clinician to admit . . . chronic Budd-Chiari." There is also anticipated skepticism: "If they don't believe that plain film, there's nothing you can show them."

Inversely, radiologists often register skepticism about what clinicians say. In Imaging-Pathology Conference the attending displays an abdominal MRI: "This is a patient who'd had E. coli sepsis . . . continued to run fevers, was imaged for that, and they [Body Imaging] found this lesion in the dome of his liver." The fellow who had been on service clarifies: "When we did this MRI we didn't have all that history." The Imaging service initially said the lesion was consistent with hepatoma—so the surgeons scheduled an operation. Had the Imaging service known about recent infection, they would have placed more emphasis on an alternative possibility, hepatic abcess. It is hard to tell these apart from imaging appearance.

Attending Sue:	Pete, what do you think? [*to a senior attending*]
Attending Pete:	Impossible.
Attending Sue:	What do you mean?
Attending Pete:	Impossible to distinguish infection from hepatoma . . . unless you put the pictures side by side . . . Nothing beats a little clinical suspicion.
Attending Sue:	We have great imaging tools but you're not going to get histology out of them . . . you can lean one way or the other . . .
Attending Pete:	And the other thing is, never trust the clinician's history.
Attending Sue:	I don't think anything grew out of him . . . unable to get tissue . . . they were using this five-month follow-up . . . [if better after antibiotics, must have had infection]
Fellow:	Biopsy *did* come back . . . consistent with necrosis—done in Interventional.

With a choice between hepatoma and abscess, deprived of the history of fever following bloodstream infection, the original interpreter/s of this study missed the call, and the patient almost underwent an unnecessary operation. Attending Pete says the missed call was forgivable: images cannot resolve the issue. His maxim, "Nothing beats a little clinical suspicion," captures the wish for fuller history. A companion precept, "Never trust the clinician's history," describes a reflex of suspicion—a form of radiological resistance to being steered wrong by fallible colleagues. The radiologist wants the clinician's clues but wants the freedom to declare some red herrings.

Attending Sue noted: "You won't get histology" from imaging. She highlights the nonequivalence of image and tissue, the unreasonableness of expectations of surgeons for definitive diagnoses based on imaging alone. She suggests that the current, revised diagnosis of abscess is based on the patient's response to antibiotics. But unknown to her (pointed out by the fellow), imaging tools of radiological colleagues have already been enlisted in procuring a tissue diagnosis.

Tissue remains, for the most part, the diagnostic gold standard.

Relations of radiological to pathological findings are typically those of conjecture to confirmation. Even radiologic findings made with relative certainty are stabilized, their evidential status augmented, by corroborative histology. The temporal vectors of this reverse those of rad-clin correlations. While imaging findings relate to clinical events through a lens of retrospection, in relation to pathologic diagnoses, they work prospectively. Tissue consummates radiological prophecy. Radiologists speak of findings as tissue-proven or not; pathologists do not speak of radiologic proof.

Resident:	This was biopsied and turned out to be a cystic renal cell CA [cancer].
Attending:	So this was biopsy-proven renal cell?
Resident:	Yeah.
Attending:	'Cause we were voting heavy on multilocular cystic nephroma.
Resident:	Well, there you go.

[*titters around the room*]

One attending comments on the limits of radiology in predicting tissue type: "As much as I like . . . [to finesse a call]: thirty percent of the time we can predict the histology; thirty percent of the time we're just very close; thirty percent of the time we just say, here are the findings . . . Otherwise, a tumor is a tumor is a tumor."

It is notable that, for most rad-path correlation in teaching conference, the pathologist is absent! The pathologist is represented by annotation, second-hand reference—not by acts of primary witnessing or evidential display.

The radiologist has become not only a consultant to the pathologist but also a representative of the pathologist's diagnostic capabilities. As described in chapter 2, "Cutting," the radiologist is increasingly a procurer of tissue. Handling the biopsy needle, the radiologist assumes traditional prerogatives of the pathologist or the surgeon.[38] The cutting edges of sectional imaging are aligned more and more with beveled steel.

Attending:	Can you put a needle in that [pericardial lesion] to make a diagnosis?
Resident 1:	No.
Attending:	Woe betide the radiologist and cytologist who try to make the diagnosis of mesothelioma from FNA [fine needle aspiration]!

Resident 2 (in audience):   So, Lou, how should they sample it?

Attending: Open biopsy . . . Thorascopically now.

Dire tones here imply that there are medicolegal stakes in knowing how to proceed toward a tissue diagnosis. The imaging finding summons, not only description and differential diagnosis, but also active participation in staging further workup. The putatively noninvasive study plots coordinates of an invasion: the radiologist's testimony indexes the acquisition of tissue and making of a pathologic diagnosis.

As with rad-clin correlation, rad-path correlation can involve interdisciplinary chafe, competition. An attending notes in conference that a formal pathology report described a lesion at a venous "bifurcation"—and remarks that the proper term should have been "confluence": "I'll talk to him: they sure look at our reports!" In another setting, a pathology resident describes a Morgue Conference case in which the radiologist (third-year resident) demonstrated pulmonary nodules on CT scan—while the pathologist's gross dissection revealed relatively homogeneous lung tissue. The pathology resident recalls shrill insistence by the radiologist: "They [nodules] have to be there!" He remarks that such rad-path inconsistencies arise "now and again."

Indications for Intrigue in the Reading Room

In conference settings, "the case" lends itself to both manageable repetition and productive variation, in a controlled theater of expert testimony and pedagogy. Conferences develop intrigue through theatric artifices: selection of naive readers, deliberate withholding of information, interrogation in the hot seat. Such choreographic tools intensify experiences of viewing and testifying, and modulate suspense, anxiety, and pressure among witnesses-in-training. But these artifices are also calculated to mimic conditions of "real-time" interpretation, as in the reading room, or contested interpretation, as in a courtroom. What is perhaps most artificial about conference case presentations, and differentiates them from cases as they unfold in reading rooms, is their relative temporal coherence. In conference, one case is moved steadily to some denouement before the next is begun—whereas in the reading room, cases lurch along, many overlapping, subject to interruption, recursion, contradiction. Cases at the reading room viewbox unfold less predictably.

There are sociological continuities between conference room and reading room: pedagogical hierarchies and cross-disciplinary correlations, upon which conference theatricality draws, also characterize reading room work. Divisions

32. Divisions of expertise at the back viewbox, Body CT reading room

of labor are as salient as they are in conferences: rad/clin divisions, foremost, but also specialty divisions within radiology, and among various clinical experts who come for consultation.

Similarities between conferences and reading room viewbox settings are useful in considering felicity conditions of intrigue and curiosity on front lines of diagnostic work. Under impress of clinical timelines, removal from conference formalities, what are the contours and ingredients of intrigue? What solicits, sustains, inculcates curiosity and detective acumen in "primary" diagnostic processes at the viewbox?

In a first encounter with a case, often the first thing the CT radiologist meets is the "indication" on the requisition. A Body CT attending expresses gratitude for one indication: "When they say 'rule out something,' it is *such* a clue!" But in Neuro CT, the fellow reads an indication and remarks on its comic inconsistency: "'Rule out infarct, history of syrinx.' These don't go together."

Later that day, the same fellow recalls another indication that was even more comical, and briefly mystifying:

Neuroimaging fellow: [*dictating:*] Chapman on Parsons, 1068263, MRI of
                         the brain dated 1—correction 2–11–97 period. New
                         paragraph.

History:	Twenty-four-year old male with bitemporal hemianopsia. New paragraph.
	[*stops dictation; aside, to resident:*] They [the techs] called me—they wanted a protocol for [*pause*] "bitemporal hemangiomas."
Neuroimaging resident:	[*laughing*]
Fellow:	I said "They have *what?!*" She says "Bitemporal hemangiomas" and I'm like—and I'm sitting on the phone thinking, what in the heck is that? "Bitemporal hemangiomas"—I'm thinking this is something I haven't seen before. This sounds really cool!
Resident:	[*laughing*]

This is high viewbox comedy. Unrepresented here are the clinician's bad handwriting, the clerk's translation error (hemangiomas from hemianopsia), the tech's dutiful propagation of the error—because "hemangiomas" has such a nice ring. But it is the fellow's willingness to play out this chain of mistranslations that is so funny: she forces all at the viewbox to imagine what such a lesion would look like—a unique and potentially grotesque prospect—"really cool!"

Humor here is sublimated annoyance. Problems with cryptic requisitions have been exacerbated by new clerical procedures—the Order Entry System. But often the target of annoyance is the client clinician. In University Hospital this clinician is sometimes a stranger, perhaps a resident or student. So requisitions are already hefted with latent suspicion. "Is this a *real* doctor's patient?" Occasionally suspicion galvanizes investigation, and an intrigue unfolds, perhaps having less to do with diagnosis than with sorting out a question of professional responsibility.

Even legible, articulate, and reasonable indications solicit conjectural efforts. A Neuroimaging fellow dictates: "History: rule out parotid mass"—and then breaks off to question the resident.

Neuroimaging fellow:	Tiff, why are they probably looking for parotid mass?
Neuroimaging resident:	Where?
Fellow:	[*laughing*] No, why?
Resident:	Ah, physical exam?
Fellow:	I suppose it could be. What else? What's in your parotid gland?

The fellow pushes the resident to guess that this patient presented with a facial nerve problem. The indication summoned an exercise in clinical-radiological correlation, in a hypothetical register—an abduction. The express utility of this inquiry is the anatomy lesson. But it is not clear that such utility is the sole reason for spinning hypotheses like this. Perhaps the Neuroimaging team will find out later today that their inference about symptoms is correct, and derive some satisfaction. It might redound to their credit if they can tell the clinician the patient's symptoms before the clinician tells them. And even invented patient histories might be aids to sharpened radiological attention.

A requisition in Neuroimaging reads: "recurrent pituitary tumor . . . radiation therapy treatment planning." A puzzle takes shape within seconds of mounting the film. The target lesion is missing.

Neuroimaging fellow:	I ah—don't see a pituitary tumor [*boardbuzz*] that they're going to irradiate. [*boardbuzzes*] All right.
Neuroimaging resident:	Hmm.
Fellow:	They should have done a, ah pituitary protocol. Anything that's pituitary anything should get a pituitary protocol.
Neuroimaging resident:	Is that the same as cuts of the sella?
Fellow:	Yeah, except they're big ones. I don't know what these are—they're pre- and post-[contrast]. Um. I'm not sure what's going on. [*boardbuzz*] [*sigh*] Here's the . . . pituitary recess, the pituitary gland is in here and then coming down you get into clivus, and I don't really see what is—[*boardbuzz*] I mean, she looks like she's a post-resection. Here's the pituitary—a little bit of residual something— pituitary gland . . .

The fellow finds other confusing features of the study. "I hate to be the one stuck with this. I like someone else to read these things." Regardless, he embarks on formal dictation. He interrupts himself several times with questions: "Now what are those there?" "Why would they be?" The resident asks: "Do you want to wait and see if we can get her outside films, or—?" In a tone of resignation, the fellow says: "No—I mean, I don't see anything." And continues dictating:

Findings colon: the midline structures are intact period. [*sigh, pause four seconds*] The sella is shallow and appears to be partially empty period. [*pause four seconds*] No large or enhancing tumors are identified period. The optic chiasm is intact period. The cavernous sinuses [*boardbuzzes*] are clear period. [*stops dictation*] I don't know what to say. [*long boardbuzz*] They're not going to like these images . . . [*dictating:*] No definite ah pituitary tumors are identified period. No other enhancing lesions are seen period. New paragraph [*pause*] . . . [*brief interruption*]

Impression one period. No definite evidence of pituitary tumor period. Patient should return for high resolution images through the sella if clinically indicated period. Signed.

Within this dictation, pauses and interpolated comments mark residual uncertainty. But the recorded version on RTAS, and that transcribed to paper, will betray none of this. This puzzle, about the absence of an expected recurrent cancer, has been constricted in the rhetorics of official testimony to a series of negatives, with only a few qualifiers: "no definite . . . tumors . . . no definite evidence." Unless or until someone else inquires, this brief bloom of a conundrum is restricted to the reading room. It devolved from dissonance between indication and images, within the structured informality of viewbox pedagogy.

Compare the requisition's role to that of other texts bearing on the instant images: reports on prior images. Radiologists can be standoffish about viewing old x-ray reports. One resident prefers to review images himself, new and old, *before* he looks at reports on any prior studies, "to see if I see what they saw." Many radiologists feel consulting prior reports before looking at images is cheating.

Perhaps reading old reports intrudes into the empirical purity of autopsis. But perhaps the fact that reading the indication first is not only permissible but preferable to "blinded" image review, also derives from the nonindexicality of the indication, from its polysemic capacity to prop up or produce diagnostic puzzles.

Encasement in Clinical Inquiry

8:40 a.m., Body CT: The resident receives a phone call from Mick, the resident covering last night. A patient overnight had a suspicious finding near his colon, but since his gut was not well-contrasted, may need rescanning. The resident hangs up the phone and announces this "add-on" from the night before.

Fellow: I wonder what happened. He probably refused the study or something.

Resident:	Mick said he got the study done—it's just that they can't differentiate between colon versus—he's got a fluid collection, and they're trying to differentiate between colon and fluid. And they're waiting for the—for it to pass.
Fellow:	Oh, the rectal contrast?
Resident:	Well, no, they didn't give rectal contrast.
Fellow:	Why not?
Resident:	That's a good question. That's a damn good question, actually.
Fellow:	I don't know why they wouldn't have.
Resident:	I wonder if I should call . . .
Fellow:	Ask Mick why he didn't get rectal contrast . . . if they were worried about pelvic or low abdominal fluid collection.

The resident cannot reach Mick (who has "signed out" to go home) but reaches the scanner room where the study was performed. The day tech knows no details. The ER Radiology desk doesn't know either. The resident is exasperated. "Ah, shit. Apparently the guy's an inpatient. So if we need it, we'll just bring him down, fill his bowel up with contrast and go."

Ten minutes later, a team of seven clinicians enters the reading room. Two pick up phones to answer pages. The attending asks the CT resident if he can "help look at a CAT scan" he has carried in. It is last night's scan of the suspicious colon.

Clinical attending:	Frank Smith: he's a man who, ah, has a history of pancreatic mass, liver lesions which are consistent with mets—no diagnosis on that ah mass—that's had multiple biopsies. [*boardbuzz*] He comes in clinically with small bowel obstruction and, ah, had an ultrasound done of his abdomen yesterday that showed a questionable abscess. In what area was that abscess?
Clinical resident 1:	Left . . . quadrant.
Clinical attending:	Left *lower* quadrant?
Clinical resident 1:	Left quadrant.
Clinical resident 2:	Left abdomen to the pelvis.
Clinical attending:	OK, we weren't sure if this thing was, uh . . .
Body CT fellow:	Bowel.
Clinical attending:	. . . continuous with bowel—and the problem is we weren't sure he received adequate contrast, so we're trying to . . .

CT fellow:	Yeah, we were just talking about this actually, and trying to figure out why—the rumor was—I haven't seen these—that he didn't get rectal contrast.
Clinical attending:	OK.
CT fellow:	Which would've been nice . . . Yeah, usually—they should—and really it's part of a routine abdomen here—we give rectal contrast unless there's some reason that somebody says not to.

[*boardbuzz, filmflap*]

| CT fellow: | Um, he's an inpatient—we can always—[*filmflap / paper shuffle*] There's two things that we can possibly do on him. One is wait until some of the contrast from up above gets through, and the other is, uh—[*board-buzz*] we can bring him back down and give him rectal contrast. |
| Clinical attending: | Um-hmm . . . Well, we'll probably favor the latter— just to try to help get a diagnosis about whether this is something the surgeons need to go in and drain. |

[*boardbuzz*] [*filmflap*] [*filmflap*] [*boardbuzz*]

There is a one-minute conversational pause while the fellow examines the images intently.

| CT fellow: | OK, well, startin' at the top: he's got a ah, moderate-sized left pleural effusion with some underlying atelectasis. [*sigh*] That hemidiaphragm on that side is, uh, elevated and, uh, probably from all the dilated colon pushin' up on it . . . [*boardbuzz*] See that? |
| Clinical attending: | [*murmurs assent*] |

[*boardbuzz*]

CT fellow:	There's air in the left biliary system—he must be status post some kind of biliary procedure like a choledocho-jejunostomy or something.
Clinical attending:	Yeah—yeah, he had that, that's right.
CT fellow:	And—he has the mets that you know about—

Clinical attending:	Um-hmm.
CT fellow:	—ring-enhancing. And—wow. [*pause*] There are certainly dilated loops of large and small bowel here. [*boardbuzz*]
Clinical attending:	OK.
CT fellow:	There's this—this all pretty well opacified small bowel. It's starting to look pretty big. [*paper flap*] It's not huge, but—at least over a few centimeters. [*measuring bowel with edge of paper*] This one's two and a half so—it's dilated. [*boardbuzz*] And there's a lot of large bowel dilatation too, and then you get into this—
Clinical attending:	Mm-hmm.
CT fellow:	Which is kind of suspicious for two reasons: one is the density isn't the same as the colonic stuff around it—
Clinical attending:	OK.
CT fellow:	—which might be getting some contrast into it already—
Clinical attending:	Mm-hmm.
CT fellow:	And two, it's enhancing. The margin of this thing looks like it's enhancing. The other thing is—that bothers me about it is that, ahm, there's little gas bubbles in it.
Clinical attending:	Mm-hmm.
CT fellow:	But—and you can get that in fecal matter [*boardbuzz*] but [*boardbuzz x 2*] these are little gas bubbles within this very uniform fluid [*boardbuzz*] and uh . . .
Clinical attending:	Mm-hmm.
CT fellow:	I think this thing's an abscess.

A Medicine attending and resident enter and engage the CT resident about a scan underway downstairs. The conversation between the CT fellow and clinical team continues.

Clinical attending:	Would it help you to do this rectal contrast study that'd make it more definitive?
CT fellow:	Um—

Clinical attending:	Because they may suggest, uh—I'm not sure if they've asked if the radiologist [i.e., Interventional/Vascular[39]] would feel comfortable trying to stick a catheter in there at this point.
CT fellow:	Have you asked 'em?
Clinical attending:	No, I haven't—I can bring the films down—
CT fellow:	I would ask 'em.
Clinical attending:	OK.
CT fellow:	They may want to give the rectal contrast or wait, or they may feel since they're the ones who are going to have to put the needle in it—I don't—I don't want to—I think it's an abscess.
Clinical attending:	OK. Alrighty.

[*boardbuzz*]

Clinical attending:	I'm not sure these have been officially read.
CT fellow:	One reason I think it's an abscess—is because, I can't link it up to bowel down underneath here.

[*exchange omitted:* CT *fellow redescribes his findings*]

CT fellow:	I think it's an abscess. And I think if you show it to uh, the Interventional people—
Clinical attending:	Yeah.
CT fellow:	—They may be ah, happy enough to say that they think it's an abscess too and that—
Clinical attending:	OK.
CT fellow:	They'll—put a needle in it.
Clinical attending:	Well, let's show it to the surgeons and—initially, and make sure they don't want to do anything, and then they're going to probably suggest—they're going to probably suggest Vascular Radiology to take a peek at it . . . OK.
CT fellow:	OK.
Clinical attending:	All right, do these films need to stay here?
CT fellow:	They do, actually.
Clinical attending:	OK, OK. Alrighty. Thanks for your help.
CT fellow:	How sick is he?
Clinical attending:	Ah, very sick . . . Great. Thanks a lot.
CT fellow:	Welcome.

The clinical attending and team depart. The CT resident slides his chair over beside the fellow. The fellow is still intently inspecting the images. The resident has some questions.

CT resident: [*sotto voce, leaning forward:*] Question . . . How did you know . . .

CT fellow: [*full voice:*] Why do I think it's an abscess?

Resident: Ring-enhancing—gas collection . . .

Fellow: Well—I don't like the fact that it's enhancing—the wall is enhancing—a lot more than the bowel wall is enhancing. There is some bowel wall enhancement [*boardbuzz*] developing [*boardbuzz*] but if you look at it this is certainly the brightest ring in his abdomen.

Resident: Mm-hmm.

Fellow: The other thing is if you [*boardbuzz*] go to the top of it—um—[*boardbuzz*]

Resident: Image 20—

Fellow: Uh, maybe it [*pause*]—I don't even know if that's the top—[*filmflap: film held up before hot light*] [*boardbuzz*] [*ten-second pause*] [*boardbuzz*]—I think the top of it is in here, just kind of starts here and I can't really hook it up to a definite loop of bowel.

Resident: Yeah.

Fellow: Then it comes down, [*boardbuzz*] and the thing that makes me the most suspicious of it is that I definitely can't hook it up to a loop of bowel down here. Right?

Resident: Mm-hmm.

Fellow: Here's rectum posterior to it—rectum, rectum [*in serial cuts*].

Resident: Are you above the bladder right there?

Fellow: You're above the bladder. Bladder is here, here, here, here, here. [*serial cuts*] Starts right about the bladder with this little bit of enhancement and then this thing takes off. OK. The sigmoid is back here. [*boardbuzz*] Back here. [*boardbuzz*] Here's the sigmoid [*boardbuzz*] coming around [*boardbuzz*]—sigmoid is still behind it—[*boardbuzz*] Sigmoid is behind it and then the bowel is all behind it. You know? I never see anywhere where the tube goes into it. It just kind of stops right there—you know it's not really—I just get this feeling it's an abscess.

Resident:	How did you know—you said something about the air in the biliary system?
Fellow:	[*ten-second pause, long boardbuzz*] Oh this air? See that air?
Resident:	But, how—did he give you a little bit of history about this area or something?
Fellow:	No, there's a collection there—
Resident:	[*whisper:*] Oh, man—
Fellow:	—and there's a loop of bowel that comes right up to that.
Resident:	. . . clues . . . was it a choledochojejunostomy, is that what you said?
Fellow:	Um-hmm.
Resident:	[*repeating hard word:*] . . . do-cho-jejun-ostomy. What was his primary for the mets—what does he have, colon cancer?
Fellow:	Yeah. Uh, actually I don't know—I'm sorry. Kind of an un-known primary.

[*brief exchange omitted*]
[*boardbuzz*] [*boardbuzz*] [*boardbuzz*] [*pause eight seconds*] [*boardbuzz*]
[*pause three seconds*]

Fellow:	I'm sorry I'm not listening to you. [*boardbuzz*] I'm—still thinking about this.
Resident:	. . . the viscosity of all this . . .
Fellow:	Bubbles.
Resident:	They're not going to the dependent—they're rising up to the surface of the . . .
Fellow:	Yeah, exactly—

The resident's line of questioning is interrupted by the CT tech, brandishing a yellow paper: "One more add-on." The fellow rises and removes the films from the board. "I'm actually going to show this to the Vascular guys and tell them that there's a potential procedure." He leaves with the folder under his arm.

The resident has been rapt with interest and admiration—and worry about his own inadequacy to the kind of skillful ad hoc consultation he has just witnessed. "Oh, man—. . . clues." The fellow has remained absorbed by filmic detail—concerned to make sure he has not overlooked something or conveyed to the clinicians a level of certainty unsupported by the images.

There is a conundrum here, but its contours and content have changed. The smaller puzzle with which this case opened a half-hour earlier—why did this patient not get rectal contrast?—has been rendered moot by the images, and the clinicians' reframing, and several rereadings by the fellow. Now the puzzle has assumed the form of a specific lesion and a well-padded hunch, a kind of bet, on the part of the fellow: this is an abscess. This hunch includes a prediction: Interventional Radiology may drain it.

On the radiological horizon, the puzzle calls forth a "potential procedure" that might answer all. The "very sick" Frank Smith—pancreatic mass, liver metastases, recent gallbladder surgery, fevers, bowel obstruction—is oddly incidental to the terms of this intrigue.

The fellow returns in a few moments, with the news that Vascular will probably drain the lesion without further imaging. He replaces the films on the board. He and the resident resume work at separate boards. Five minutes later, the Interventional Radiology fellow, a woman in scrubs, enters. She sits beside the CT fellow at his board to look at Frank Smith's images.

Interventional fellow:	It's a nice enhancing rim.
CT fellow:	Yeah . . . It's just too bright to be a loop of bowel—and the other thing that's pretty nice actually, even though we didn't give him rectal contrast, [*board-buzz*] he's getting enough stuff down from above that he is enhancing the colon a little bit . . .
Interventional fellow:	Mm-hmm.
CT fellow:	. . . because he had a little delay to his scan. Do you guys need this? 'Cause I'll dictate it and—you can have it.
Interventional fellow:	Sure—take your time, I can come back and get it.
CT fellow:	OK.
Interventional fellow:	Sounds good. Was he done last night—guess it was—yeah.
CT fellow:	Yeah.
Interventional fellow:	. . . last night . . . [*indistinct*]
CT fellow:	So, what size drain you going to put in this guy? Put a big one—
Interventional fellow:	Probably just an 8 French again will do—maybe 10 . . . From that—put to straight drainage—
CT fellow:	I got a feeling it's going to be viscous—got all these little air bubbles trapped in there.
Interventional fellow:	Um-hmm . . . looks like—wonder if it communicates with bowel . . .

CT fellow:	You know, it's possible—I don't see it but . . .
Interventional fellow:	Were you on here yesterday?
CT fellow:	Yeah.
Interventional fellow:	We drained that woman last night.
CT fellow:	Which one?
Interventional fellow:	Svenson.
CT fellow:	That lady I don't remember.
Interventional fellow:	We got seventy cc's of pus out of her—it was foul-smelling—
CT fellow:	Where was her abscess?
Interventional fellow:	In her anterior abdominal wall—she had all these thickened bowel loops going up to it and I bet she had a fistula.
CT fellow:	Svenson . . . oh yeah, the lady with the, uh, history of cervical, uh, cancer.

The plot centering on Mr. Smith's lesion has thickened with details of viscosity and drain size—and, by association with another case, with the prospect of malodor. After the Interventional fellow departs, the CT fellow puts some other films on the board, then writes on Frank Smith's film jacket.

CT fellow:	You know, I can never spell choledocho-
Investigator:	Choledocho—too many C-H's—
Fellow:	Yeah, too many C-H's.

Another Radiology resident enters, a second-year. She chats with the fellow about a recent graduate's new job at another institution. The conversation unfolds against the backdrop of the Smith images.

Visiting resident:	His [graduated resident's] senior roast gift was a dump truck, however.
CT fellow:	Why?
Visiting resident:	He's a dumper [i.e., of workload, on fellow residents].
Fellow:	Oh, really? [chuckles]
Visiting resident:	That's OK, we all loved him anyway. He's really sweet. This looks like a sick person.
Fellow:	[with Inspector Clouseau accent:] This person is not good.
Visiting resident:	No.

| Fellow: | Look at that—that's one of the biggest ones I've seen in awhile. Time to be drained. |
| Visiting resident: | Nasty. That should be easy to get though [*laughing*]. |

Suspense about the could-be abscess, and accessibility to drainage, has all but disappeared. The visiting resident, on the basis of casual inspection, remarks upon the "sick person" and his abscess quite matter-of-factly—"should be easy to get"—and the fellow's comments about its size and "need to be drained" belie the care and uncertainty that passed before. The case no longer calls for a detective: Clouseau will do.

The CT resident speaks up from the RadCare terminal: "Did the Vascular General people think that was going to be a uh, abscess?" The fellow confirms that they will be draining it under ultrasound guidance.

Fellow:	I think it's going to be thick stuff too, 'cause of the way those air bubbles are floating—the gas bubbles are floating in the middle of it.
Resident:	Yup.
Fellow:	So I would like put the biggest tube in I could put, right—
Resident:	Do they actually ever have to apply suction to those things?
Fellow:	Uh no, they usually—they can do gravity drainage. It's nice when you're doing it, if you can, to suck out as much as you can, you know, while you're kinda standing there, waiting for things to happen and . . . and then hook up to gravity drainage. There's—I don't know, you can potentially make the person feel better faster . . . if you get as much pus out as you can. [*pause*] I have no scientific proof for that theory. [*laughing*] It just kind of makes sense.

Last wisps of speculation here—about textures of goo—are structured by the persistent eager questions of a trainee.

The resident begins putting up films on his board, while the fellow dictates the case of Mr. Smith.

| CT fellow: | [*to RTAS dictaphone:*] Dr. Fleck dictating with Doctor . . . Biggers to cosign, the patient is Frank Smith 952745—12–6–96 is the date—522976 is the order number—new line. [*long boardbuzz*] |
| | Examination CT scan of the abdomen and pelvis . . . [*boardbuzz*] New line. |

Clinical Data colon: [*five-second pause*] [*click*] Fluid collection identified on ultrasound, question abscess, elevated white count and [*indistinct*] fever period. [*boardbuzz*] Technique colon: spiral CT of the abdomen and pelvis was performed from the lung bases to the pubic symphysis following the intravenous administration of 150 cc's of Omnipaque and the oral administration of contrast material period. No rectal contrast was administered period. The scans were formatted into 8mm collimated slices and, uh [*click, two-second pause, click*] comparison is made with prior studies of, uh [*click x 2*] 6–27–95 period. [*clicks*] New paragraph findings colon [*clicks*]—[*click*] There is a uh [*clicks*] moderate sized left pleural effusion with uh [*clicks*] associated left basal atelectasis period. [*clicks*] There is air identified within the intrahepatic biliary system predominantly within the uh [*clicks*] left lobe [*clicks*] of the liver consistent with the patient's history of pancreatic cancer and uh status post choledochojejunostomy period. [*pause, click*] There are uh multiple ring-enhancing metastases identified within the uh [*clicks*] liver in the anterior and posterior segments of the right lobe period. [*to self*] There's one in the left lobe too.

RTAS: [*click: playback:*] . . . segments of the right [*click*] the anterior and posterior segments of the right lobe . . .

Fellow: [*click*] . . . of the liver and the medial segment of the left lobe of the liver period. There are uh clips in the uh [*clicks*] porta consistent with patient's biliary surgery period. [*click, pause*] [*click*] The uh spleen and kidneys are unremarkable period. [*click*] The adrenal glands [*click, pause*] [*click*] are unremarkable period. [*clicks*] The uh pancreas uh is not well visualized—there are clips [*clicks*] present around the pancreatic head [*clicks*] [*boardbuzz*] period. [*boardbuzz x 2*] New Paragraph.

There is ah [*clicks*] marked dilatation of large and small bowel which is uh fluid filled [*clicks*] throughout the uh abdomen [*clicks*] period. [*clicks*] No uh [*clicks*] definite area of mechanical obstruction is identified period. [*click*] [*boardbuzz*]

There is ah, a large [*clicks*] fluid collection consistent with an abscess in the left lateral abdomen extending uh from the uh level of the midpole of the left kidney to uh the level of the superior aspect of the left acetabulum. This demonstrates uh wall enhancement and there are some central areas of gas collected within the abscess cavity. New paragraph.

Impression colon: 1 period. Large abscess in the left abdominal uh—[*click pause click*] cavity as described above [*clicks*] period. [*clicks*] 2 period. Multiple dilated loops of large and small bowel [*long boardbuzz*] without evidence of a definite site of mechanical obstruction period. [*clicks*] This may represent an ileus period. 3 period. Known hepatic ah metastases in this patient with ah pancreatic colon cancer [*clicks*] status post ah choledochojejunostomy with intrahepatic biliary air period. 4 period. Ah moderate left pleural effusion with left basilar atelectasis period . . . [*indistinct*] [*clicks*] Dr. Biggers has reviewed the above study and agrees with the above report period.

A series of overhead pages summon Anesthesia and Surgery to the ICU. "That's not good," the fellow comments. The resident gets off the phone and whistles "Your Love Put Me on the Top of the World" while he deletes studies from the MagicView.

CT fellow:	I just wanted to get that one done, in case the Vascular guys wanna take it . . . You saw the abscess, right? Got to see it?
CT resident:	Frank Smith—yeah, I saw the abscess . . . You want to keep that on the monitor [MagicView] just in case?
Fellow:	Ah, keep that one on the monitor in case Hank [Biggers, the attending] wants to look at it and Vasc—and Vascular takes the films.

The fellow's expedited dictation does not mention his discussions with the "Vascular guys" or the clinical attending—nor even the anticipated procedure—though it provides them all with testimony upon which their interventions can be formally hinged. Nor does this dictation indexing a lesion have any further use for the "index question" that started off the case this morning—namely, why the patient did not receive rectal contrast. The fellow's description of findings is matter-of-fact and systematic.

At this point, "abscess" has ceased to be a daring diagnosis; during the series of conversations here and in Interventional Radiology, the fellow's "I think this is an abscess" has been corroborated enough that it has assumed the form of a declarative "Impression" of confident syntax. The case's opening tones and postures of uncertainty have been whittled away. Initial ambiguity and conjecture have been shaved to a fine rhetorical edge: "consistent with."

This lesion's encasement in layers of radiological and rad-clin effort and expertise is both consummated and elided by this dictation's declarative voice. The dictating fellow's tone of authority formally updates and amends earlier, fuzzier interpretations of the clinical team—whose attending carried the films in just one hour before. Yet, notwithstanding the closing lines of the dictation, the official signature underwriting this authority—Dr. Biggers, CT attending du jour—has as yet not seen the films.

Frank Smith is not mentioned in the reading room for the next three hours—though over lunchtime the films disappear. Just before one o'clock, the Body CT attending du jour enters the room and encounters the Interventional resident in scrubs rummaging at the left viewbox.

Interventional resident:	[*filmflap*] [*paper shuffle*] . . . Fucking board—[*long boardbuzz*]—she must have grabbed it—[*paper shuffle*]—son of a bitch—[*mumbling*] [*noticing CT attending*] Are you on CT today?
CT attending:	Yah.
Resident:	Did Maureen come down and take Frank Smith's CT already?
Attending:	May very well have. I've been interviewing people this morning, Austin. So I've been in and out of here.
Resident:	Well I don't see it on the board, so I assume she must have grabbed it.
Attending:	Probably grabbed it. They usually put a note in a book though. Frank Smith, yeah. See four through—someone must have taken it.
Resident:	Yeah.
Attending:	You doing something on him?
Resident:	Draining his abscess. He's got a huge abscess, apparently.
Attending:	[*reading log:*] Yeah, left lateral abdomen with central gas and wall enhancement. [*throat clearing*]
Resident:	Thanks.
Attending:	Sure.

Later the attending and fellow sit together to review the morning's cases. The attending rewinds the board.

[*long boardbuzz*]

CT attending:	Someone did—it looks like they'd taken one of the cases—
CT fellow:	Yeah, Vascular took the one—
Attending:	And it looked pretty straightforward, with a big abscess—
Fellow:	Yup.
Attending:	I looked at it on the monitor.
Fellow:	I saw the pus.
Attending:	Yeah.
Fellow:	Actually they were walking down the hall with the tubes and said—and went like this to me—I was like OK, yeah—[*both laughing*]—I see it—
Attending:	In other words, you made the right call.
Fellow:	Yeah, it's, it's an abscess, we knew. [*boardbuzz*] Put the cap on it.

The making of certainty in this case was accomplished with only nominal contribution from the attending du jour. His contribution has been reduced to the thinnest veneer of ex post facto approval. What was briefly a viewbox puzzle was converted over a period of hours to near-certainty—to which the fellow now contributes autoptic proof: I saw the pus. With the attending, the fellow reasserts the diagnostic radiologist's prerogative to maintain a dignified distance from bodily fluids. Even though in the attending's eyes this diagnosis was "pretty straightforward," it was still creditable: "you made the right call."

This case was indeed pretty straightforward. It unfolded, from scan to capped pus, in a half day. Still, in its various testimonies and teachings, it generated uncertainties, postures and rhetorics of detection, moments of suspense. These varied with audience and timing. Certainty developed as performances of reading accumulated. A crust of cross talk and interlocutions, wrapped around a soft lesion: such is the constitution of the viewbox case. Its interest and complexity are often belied by declarative rhetorics of formal dictation.

Viewbox Correlations: Imaging-Imaging

Slicing summons comparison: CT cases with others . . . and within the instant case: one slice with the next, axial cuts with sagittals, contrast images with noncontrast.

Comparisons may establish "consistency." But consistency may not mean resemblance. One finding may be consistent with another with respect to causal mechanisms—for example, edema in separate locations. Two studies may represent consistency of the same lesion over time. A lesion's appearance in one modality may be consistent with appearances in others.

Consistency across studies: an exemplary case unfolds at the Body CT viewbox. The fellow has reviewed a "trauma abdomen" with the attending and has begun dictating. "The spleen comma, pancreas comma kidneys and visualized small bowel are normal period."

The fellow interrupts himself to pull down a film and carry it across the room to the attending (at the other viewbox). He has noticed irregularities in the spine.

Body CT fellow:	Vertebral bodies . . . she's twenty-seven years old—and the rest of them . . .
Body CT attending:	Wait a—hopefully we are not missing a vertebral fracture. That may need a plain film . . .
Fellow:	Probably question mid or—
Attending:	That's a good pickup . . .
Fellow:	There's some irregularity in the lower thoracic vertebral bodies.
Attending:	Yeah. There's no vertebral—pre-vertebral hematomas so . . . hopefully not acute, and also the back in spite of all else looks all right.
Fellow:	[*returning to his viewbox*] I will call them back and let them know that—
Attending:	Yeah, this may be—
Fellow:	—to get some thoracic/lumbar spines—
Attending:	Right. And—
Fellow:	T-spines. OK.
Attending:	This may be an old—you're absolutely right, [she's] too young to have those things.
RTAS:	[*dictation playback:*] . . . kidneys and visualized small bowel are normal period.
Fellow:	[*dictating:*] New paragraph. [*pause*] There is irregularity of a lower thoracic vertebral body parenthesis (at the level of the renal vessels end parenthesis) which [*pause*] may represent a vertebral fracture period. Plain films of the thoraco-lumbar

RTAS:	spine are recommended for further evaluation period. No prevertebral hematoma or—[*click*]
	[*dictation playback:*] . . . lumbar spine are recommended for further evaluation period.
Fellow:	[*dictating:*] No pre-vertebral [*click*] hmm [*click*] abnormality is noted [*click*] hmm. [*filmflap*]

Again, the fellow interrupts his dictation to carry a film over to the attending.

Attending:	You found something else, huh?
Fellow:	Is her vertebral—is her spinal canal . . . I mean, look at this one, this one here. [*boardbuzz*] The spinal canal there looks . . .
Attending:	Oh I would not be—that's going down . . . if the patient had—has—ah, if they are worried about vertebral fractures, they gotta get something else. This, this is an excellent pickup—this I'd want to take to my peers . . . hopefully she doesn't wind up with a compression fracture, with this little fragment . . .
Fellow:	I was just—do you think her spinal canal there was a little . . .
Attending:	Triangular, narrow.
Fellow:	Yeah . . . Compared to the other vertebral—I mean the other levels. I'll question it—if they get plain films we'll . . .
Attending:	Well, the only question there is—maybe she has . . . some impingement . . . [*filmflap*]
Fellow:	Exactly.
Attending:	Maybe she has a—
Fellow:	That triangulation is certainly—irregular.

The fellow returns to dictation.

| RTAS: | [*dictation playback:*] which may represent a vertebral fracture period. |
| Fellow: | [*dictating:*] Period. The spinal canal at this level is also somewhat [*pause*] irregular dash dash——plain films versus ah CT is recommended for further evaluation period. [*pause*] Period. There is no ah prevertebral soft tissue abnormality period. |

[*pushing phone buttons to page clinician*] [*dictating:*] New paragraph.

Impression. No ah visceral or soft tissue abnormality, period. Next number. Irregularity of a lower thoracic vertebral body with narrowing of the ah spinal canal which may represent a thoracic vertebral body fracture [*pause*] period. Further evaluation is recommended with plain films versus ah CT of the thoracolumbar spine period. End. [*phone ringing*] Actually new paragraph.

This study was reviewed concurrently with Dr. Bynum period. End.

[*picking up phone:*] Hello, is this Jeff [emergency attending]? Jeff, on Ortiz—do you still have an eye on her? Or is she already gone? Are you—are you in the ER? Taking care of her? OK, her—at the level of her renal vessels, probably lower thoracic spine—we went back and looked at this again. One of—she has an irregularity of one of the lower thoracic vertebral bodies. [*pause six seconds*] Now, did you know that before I—you already knew that from a plain film? Beautiful. And you—her spinal column is a little narrowed, so there may be some impingement. [*pause*] See ya. Beautiful. All right, I just wanted to let you all know. Thanks. Bye. Appreciate it. [*pause eight seconds*] Got it. OK. Thank you much. Bye-bye. [*hanging up phone*]

Attending: What do we know about that one?

Fellow: L1 burst fracture.

Attending: L1—that's a good pickup. That's what they usually—compression fracture . . .

Investigator: And Jeff already knew that?

Fellow: Well, they had suspected it, they weren't really sure yet. They were going to send her back over and get some other cuts, but they thought that that's what it was.

Attending: Needs some more . . .

RTAS: [*dictation playback:*] . . . CT of the thoracolumbar spine period. End. Actually new paragraph.

This study was reviewed concurrently with Dr. Bynum period. End.

Attending: That's a very good pickup though, Dale Smollett.

Fellow: [*dictating:*] Transcriptionist, could you go back to the body of
           the report where I said "lower thoracic," just say ah—substi-
           tute in there "approximately L1" period. And then in the—in
           the Impression, "questionable burst fracture involving the L1
           vertebral body," and then end that report, thanks. [*hanging up
           dictaphone*] [*sighing*]

This case is remarkable for several reasons. One is the fellow's "good pickup,"
noted by the attending, who had not noticed the vertebral fracture on first pass
through the films. Two is the consistency between two kinds of imaging study,
called "beautiful" by the fellow. But another, more unusual feature is that this
fellow, in service of beauty, has made an ex post facto adjustment of his formal
report. Not that revising a report is uncommon: radiologists often dictate ad-
denda, to reflect new information or discussions with clinicians; they correct
transcriptions' grammar and spelling. But amending a report like this after it
has been dictated, to make it consistent with other imaging findings, is uncommon. To be sure, the difference between "lower thoracic" and "L1" is small, and
the elapsed time between dictation and revision is trivial: this amendment is
almost beneath notice. But this radiologist has traded his autoptic privilege
away for a principle of cross-image consistency. He has not seen the plain
x-rays on whose account he has changed his report. Consistency here functions
not as a device of empiricist description but rather as a desire, a compulsion.

Consistency is a category of relation that accommodates notions of similarity,
stability, and metaphysical identity (of person, lesion, or disease). Observations
of consistency can aid in navigating, and sometimes solving, diagnostic puzzles.

Then there is inconsistency, difference, change—image/image correlations
that are highly productive of intrigue. There is unexpected change over time—
from a prior study to the current study. Radiologists often look closely at earlier
images to verify when a change really occurred. It is a radiological reflex to ask:
should I (or they) have been able to see this finding back then? This is the stake
in the following suite of exchanges at the Body CT viewbox. The first exchange
is among attending, fellow, and resident (third-year). They are discussing a
worrisome two-centimeter mass.

Attending: That's probably—does that look like a lung cancer? Does that
             look like a lung cancer . . .
Resident: Let's go back—let's go down. [*boardbuzz*] [*boardbuzzes*]
Attending: Is that the old study?

| Fellow: | Yes. |
| Resident: | Yeah. So it was there previously. [*boardbuzz*] |

[*pause*]

Attending:	How long ago was that?
Resident:	April 22. It's over a year—well, it's almost a year ago . . . They mentioned it previously [on the official report].
Attending:	Yeah. Why do—oh, you know what? I bet—
Fellow:	I think they said something—a granuloma—that something was stable and now it's bigger?

In the previous scan, a year ago, a much smaller mass had been identified in roughly the same location. The report had indicated it might be a granuloma— a benign scar. But the mass of concern on today's scan is not a granuloma. What makes it confusing is that there does seem to be a granuloma on today's scan, adjoining the mass.

Resident:	Well there's the—there's the granuloma.
Attending:	Right. Where's the ah—lesion?
Resident:	Here's the mass.
Fellow:	The mass starts like—below it.
Attending:	Oh—well—some people are unlucky. Some people—some people are unlucky . . . We said it's—we said it's a grannie. Where did we see that before? I'm trying to find the—[*boardbuzz*] I'm trying to find the, ah [*boardbuzz*] [*boardbuzz*] This is a different relation . . . It doesn't stay here. This is very dense. Oh, is this a—wait a minute. Is that a blood vessel—we think it's a . . . ?
Resident:	I don't think it's a blood vessel—I couldn't follow it.
Attending:	OK. All right. It's very dense—it's got to be calcium—but where was it before—when we called it?

From the details it emerges: a two-centimeter mass, probably a malignancy, has developed in a man's chest, in roughly the same area where the CT radiologists "called," one year ago, a small mass *or* a benign "grannie." The more definitive mass was subsequently noted on a routine chest x-ray, at the outset of the current hospital admission, for chest pain. The team can see a granuloma on the present study—a bit different in appearance—abutting the mass. Their intense focus on the edges of the tumor, their careful attention to adjacency and continuity, and density (characteristic of granulomas), is in part an effort to reassure themselves that their colleagues did not understate something a year ago. With "some people are unlucky," the attending registers his belief that the previous call was appropriate—that while this is, unfortunately, a cancer, distinguishing it from granuloma was not possible a year ago.

The next time the case is addressed, it is in dictation by the resident. (The attending is not present.) The pertinent finding is described this way: "There is a two-centimeter mass in the posterior costophrenic angle on the right, period. This mass previously measured six millimeters, period . . . There is a calcified granuloma adjacent to this mass [*pause*] which is new, period. [*pause*] No other pulmonary nodules are identified, period."

When the dictation is almost complete, the resident turns to the fellow: "Did, did we ever notify this doctor about the . . . ? I'll give him a page if not."

Fellow: I guess we could tell 'em—just let 'em know it's not the same, it's not the granuloma but . . .
Resident: [*dictation playback: "New paragraph."*] [*dictating:*]
              Impression colon: Interval growth of a two-centimeter mass in the posterior costophrenic angle on the right—[*pause*] Dr. Steiner is aware. [*aside:*] I'm not gonna mention that, they already know about it. [*dictation playback: . . .*
              "*angle on the right*"] [*dictating:*] Period. End.

The dictation differs slightly from the consensus developed at the viewbox with the attending. The dictation says there is an enlarged mass and a new granuloma—whereas, earlier at the viewbox, the discussion revolved around the call of either an old (and persistent) granuloma and a new mass, or an enlargement of a previous mass that was once indistinguishable from granuloma.

Soon the clinician paged by the CT resident calls the reading room. The resident informs him that Isaac Hardy has a chest mass that has grown from six millimeters to two centimeters in a year. This clinician questions a prior reference to granuloma; he seems to suggest the mass is a new problem.

Resident: What do you mean the mass was incidentally found? [*pause*] The mass was there a year ago. [*over shoulder to fellow:*] Did—did we call that a granuloma a year ago?
Fellow: Well, it says "granuloma versus neoplasm" on the dictated report.
Resident: Oh. [*to phone:*] They said "granuloma versus neoplasm," they weren't sure. Yeah.
Fellow: It was too small to tell at that point.

Resident: Yeah. OK. Bye. [*hangs up*] Um—

[*boardbuzz x 3*]

Resident: I can't—I can't believe they would—I mean, yeah, it could be a granuloma, but—
Fellow: Yeah, I'm surprised they raised that question, actually.
Resident: There's no calcification in there to suggest . . .
Fellow: You can see it [the lesion] on three cuts.
Resident: Yeah. There's no calcification in there.

There has been a bit of fending here. The clinicians do not want responsibility for having failed to follow up on a mass identified a year ago. The resident and fellow have clarified that Radiology said what they could at the time. But privately they question the previous reader's even having "raised the question" of a granuloma: "there's no calcification." They feel that the prior CT readers should simply have called a small nodular lesion, questioned possible malignancy, and left it as "too small to characterize."

A couple of hours later, a chest surgeon comes to the reading room. The Body CT attending greets him: "Dr. Samson, today's your heyday. I mean, we . . . we're finding lung cancer nodules left and right . . . unfortunately a nodule that we read a year ago—I guess he got lost to follow-up or something." He shows Dr. Samson the films.

Attending: The interesting thing is that that could be a granuloma— He has—in reality he did have a granuloma, he does have a granuloma—because just above this thing is this.
Chest surgeon: Mm-hmm.
Attending: I thought maybe this was the same old grannie, and this just popped up. But in reality, if you look at it, this is actually a nodule which has grown from before. If you look at it level by level—so this is . . .

This review with the surgeon places the emphases differently: on the previous scan there was a granuloma (more prominent now), and just above it a small nodule, which has in the ensuing year grown to two and a half centimeters. The size is important: this can be resected.

In this case, these differences among sequential interpretations are minor. They are consequent upon happenstance distribution of responsibilities, within the CT team, for engagement of surgeons, clinical attendings, and the dictation system. No radiologists will lose sleep over this. The formal report on the previous CT raised the question of neoplasm versus granuloma, a frame that still works for today's reading, and absolves the radiologists of responsibility for delay. The patient's return to medical attention is fortuitous. He will undergo a biopsy soon, then surgery.

The specter that haunts several of these discussions is the worry about having missed or misstated something. This shapes expressions of conjecture and surprise which attend different versions of the CT interpretation.

If one imaging study fails to corroborate another, intrigue may ride the coattails of one of these questions: Which is right? What went wrong? Who made a bad call? What is the best modality? Comparisons between modalities, like comparisons between readers, inevitably produce dissonance and embarrassment along with corroborations and triumphs. This agonistic structure of the diagnostic enterprise is not unique to radiology—but in radiology, systematic incorporation of comparative methods confers a kind of generality, and banality, upon its agon. Yet mere "pitting against" is an oversimplification. If that were all it were, the radiologic agon might only reproduce binaries like superiority/inferiority and correctness/error. Its performances of nuanced comparison also produce intrigues that solicit curiosity and judgment.

Viewbox Correlations: Rad-Clin

At the viewbox, some clinical correlations are gratifyingly compact. After a clinician's one-minute visit to Body CT:

CT Fellow:	OK, see. He wants the CT for a specific question. Boom. That's what it's for.
CT Resident:	That's what it's for. Not for "We want to rule out full body lesions."
Fellow:	Yeah.

Radiological interpretation is plagued by sprawl. CT radiologists have an interest in keeping investigations focused—a matter of time and intellectual effort, and sense of proper fit. They do not like using CT for screening. "So *much* information . . . We find so many incidental things. So then it's 'what do you do with them?' And 'Who's going to pay for that?'"[40]

Sometimes clinical correlation is summoned by an absence of findings.

Neuroimaging fellow:	Did we get in touch with this person's intern?
Neuroimaging resident:	Yeah, I already called.
Fellow:	Paged him?
Resident:	Yeah, and talked to him.
Fellow:	What did he say?
Resident:	Um, he just said the clinical history was she had like a several month history of an increasing fullness in her neck—he couldn't palpate anything on exam, um, he couldn't—I mean the exam was basically normal.
Fellow:	OK. Did you tell him we didn't see anything?
Resident:	Yes. That's what I told him. [*pause*] *After* I elicited that.
Fellow:	Yeah, he said: "Oh, there's nothing there."
Resident:	"There's nothing there." [*laughing*]
Fellow:	[*laughing*] That's not fair they can know all that stuff.

These radiologists understand full well how their own negative search corroborates the negative search of a clinician—how congruent absences can become a factoid of positive knowledge—and also how the withholding of a clinician's exam was necessary to the independence of their investigation. Their laughter, and comment on fairness, acknowledge the game. They are victims of a kind of wild goose chase (here and daily): their jocularity is a standard device in the return-to-senses of the dupe.

A more complicated kind of correlative engagement turns on the occasional skepticism of clinicians—as when some imaging finding does not agree with patient symptoms or signs.[41] This may summon work of negotiation and persuasion—as in the following case.

The Body CT attending sits at the MagicView reviewing a new chest/abdomen CT—of an outpatient with Hodgkins. "Uh-oh: is this a saddle thrombus? See what they've said before . . . That sure looks like thrombus." This is alarming, because clot in pulmonary arteries can be fatal. There is no mention of thrombus on old reports, nor is it evident on a CT from several months ago. "I think we need to call the doctor." When the clinician, an oncologist, answers his page, the resident takes the call.

Body CT resident:	[*on phone:*] Hi, Dr. Farley. We're looking at Flanders . . . We think he may have a large thrombus, a saddle

	thrombus in his pulmonary artery, and we're thinking that he may need a pulmonary angiogram . . . Yeah . . .
Fellow:	[*to Body CT attending:*] You'd think he'd be symptomatic.
Resident:	Maybe I'll let you talk to Dr. Baum. [*hands phone over*] They're not clinically suspicious.

The attending speaks with the oncologist. "He doesn't have any symptoms?" She is surprised. "Gosh—this sure looks like a saddle embolus . . . You don't usually see this as an outpatient, you know." Patients with large pulmonary artery clots are usually very short of breath. Off the phone, the attending says: "I think it's gotta be, but you'd think he'd be more symptomatic." She pages a colleague, a chest specialist. Then she inspects the films again. "I think he's got the best saddle embolus I've ever seen."

The chest specialist answers his page. Again the resident fields the call: "We'd just like to get your opinion." Then he calls the scanner to make sure the patient has not been sent home.

In moments the chest expert enters and proceeds to the MagicView. He is emphatic: "Doesn't need an angiogram. It's pulmonary embolus. Pulmonary thrombus." Dr. Baum explains to him that the oncologist "doesn't believe it." The chest CT expert replies: "He'd better believe it. I'll talk to him." Dr. Baum also explains that the patient has been on a particular drug regimen for his Hodgkins—that the clinicians "thought he was short of breath because he was on BCNU."

Chest CT expert:	He's short of breath because he has pulmonary embolus . . . This is an extensive thrombus . . .
Body CT attending:	I didn't think that you see them as walking/talking outpatients.
Chest CT expert:	I have . . . They really aren't clinically silent . . . there's usually something . . .
Fellow:	Are we going to see more of these because of our spiral technology and the nice boluses . . . ?
Chest CT expert:	Yeah, I think that's part of it.[42]

So far the patient's paucity of symptoms has weighed in against the radiological reading of pulmonary thrombus. A chest radiologist who has "seen this before" seems less swayed by dissonant clinical information. Competing opinions have been mobilized with impressive rapidity. The oncologist is paged again.

| Chest CT expert: | [*on phone:*] Hi, Owen, this is Hank. [*pause*] As Crista said . . . he indeed has very extensive thrombus within his pulmonary artery. He basically has a saddle embolus. It not's quite . . . [obstructing flow] . . . You can actually follow these thrombi down to both lower pulmonary arteries. [*pause*] To me there's no question . . . Crista said something about angio. I really feel very strongly that doesn't need to be done . . . He's had some shortness of breath, is that right? [*pause*] I see . . . I see. [*looking at screen as he speaks and listens*] I have . . . it's not frequent . . . but it doesn't knock me out of my shoes . . . In terms of looking at lung parenchyma in lung windows . . . He doesn't look like BCNU lung . . . [He *does* have] . . . rather extensive pneumatosis in colon . . . hepatic flexure is just *full* of air in bowel wall . . . [*pause*] I've seen . . . when you see pneumatosis, one of the things that can cause it is ischemia . . . Could be related to infection, thrombophlebitis . . . can be a benign thing in patients on steroids. Has he been on steroids? [*pause*] Uh-huh. [*pause*] Right . . . the Hodgkins: he's got para-aortic adenopathy as well . . . two to three in a chain. |

This conversation convinces the oncologist. He knows the chest CT expert and attending well—is himself no stranger to the reading room. The assertion that the patient does not need an angiogram—the gold-standard test for pulmonary embolism in the early 1990s—is a strong one. The oncologist cannot come see the study firsthand—but he asks that the patient be sent to the Heme-Onc Clinic to make plans for admission.

Upon hanging up, the chest CT expert reviews the hardcopy films—just arrived—with the CT team. They discuss the oncologist's questions of BCNU pneumonitis, for which the patient has been treated for months. The bowel wall air remains disturbing. "Has he thrombosed anything else? . . . See all this air? . . . zoster . . . did he thrombose his mesenteric vessels?" This last question is not answerable from images.

I walk over to the Heme-Onc clinic, in another building, and introduce myself to Dr. Farley. I explain my study of CT—and that the source of my interest in Flanders's case is the unexpected, "incidental" finding of a significant lesion. Dr. Farley discusses the patient's experience with Hodgkins and many complications. Since a marrow transplant six months ago, Mr. Flanders has been "on and off steroids." A month ago he was admitted to the hospital, with abdominal pain that was eventually attributed to a zoster infection. In addition, he has had dizzy spells attributed to "some sort of autonomic neuropathy." There were elevations of liver enzymes, raising a question of viral hepatitis (perhaps zoster) or recurrent Hodgkins. The reason Dr. Farley ordered the CT was to "rule out recurrent Hodgkins." "He has had intermittent shortness of breath . . . we attributed it to BCNU." The oncologist admits that since his conversations with the CT radiologists, the clinic nurse has said Mr. Flanders called yesterday to report more difficulty breathing.

These comments are offered mostly without reference to the present imaging findings. Dr. Farley does remark that Hodgkins disease "produces clot, more than in other cancer patients." But Flanders has not had "frank thrombus" before. Dr. Farley welcomes me to accompany him in his consultation with Mr. Flanders.

En route, I am introduced to the clinic nurse as a physician following up on the "incidental finding" from CT. She shakes her head with exasperation and sympathy. "Stephen's had every incidental finding you could imagine." Dr. Farley also meets up with an oncologist colleague. "Did you hear about Stephen?" The colleague says yes and conveys his astonishment. "Has he had any workup for hypercoagulability?" Dr. Farley says: "Not yet . . . but you can bet he will from Stanbridge." (Stanbridge, the Heme-Onc attending who will supervise Flanders's hospitalization, has a research interest in hypercoagulability.)

In the consultation room: Mr. Flanders is in a wheelchair and wears a helmet. He is with his wife. Dr. Farley says: "I'm not going to tell you your Hodgkins is back . . . a few small nodes, but you've always had those." He expands on "good news" from the CT. "But there is something wrong . . . You have a clot in the pulmonary artery . . . Usually patients have more symptoms . . . so I need to check how you've been feeling . . . and also we need to admit you to the hospital, to put you on heparin . . . A big deal, this can kill you."

Mr. Flanders says he has been dizzy and winded, even when walking to the bathroom, since starting a new pain medicine a few days ago—so he stopped that medicine. He has felt better today. His zoster pain is significant. Neither Mr. Flanders nor his wife questions the diagnosis of pulmonary thrombus or the need to be admitted. Mrs. Flanders asks: "What could have caused this?" Dr. Farley speculates about hypercoagulability related to Hodgkins and possibly to prednisone, about clotting factors, about "recent work from Sweden" implicating a protein called Factor V Leiden. The wife's questions (she is a nurse) are astute: "Is heparin therapy OK with his low platelets?" She also asks about "low molecular weight heparin." Dr. Farley thinks "not for initial treatment."

The clinicians' initial surprise and puzzlement about the lesion are not shared with the patient. Questions in the clinic are no longer versions of "Is this real?"—but, rather, "Why did this happen?" and "What is to be done?" And there are further rad-clin disconnects. The finding of air in bowel is not addressed (even given the patient's recent hospitalization for abdominal pain). Nor is a possible relation between pulmonary thrombus and dizziness. There will be further opportunity for clinicians to consider the CT. But at this moment there is a remarkably "incidental" quality to the relation between clinical and radiological discourses.

Back in the reading room, the team is pleased to hear the patient is indeed short of breath, more than the clinician conveyed. Their pleasure derives from pride in radiologic diagnosis—a muted "I told you so." It is as if they predicted something—or at least beat the symptom to the consultation.

Radiologic diagnosis can prompt and rectify clinical information gathering. And puzzles which bloom around CT lesions in the reading room can have a rather contingent relation to puzzles of the clinic and the lives of patients.

Another case exemplifies how some image interpretations depend crucially on clinical information. In the Body CT reading room, a clinical fellow in a long white coat enters and asks the CT attending: "Can I be the fiftieth person to ask you about Murray Pardo?" The attending replies, "Sure, of course," and slides his chair across the room to the right viewbox. He summarizes the history he knows: "He has lupus and he's twenty-four, and he gets a [renal] biopsy . . . and now he has abdominal pain."

The clinical fellow interrupts to add that the onset of the pain was "coincident with hearing the click of the biopsy gun."

The question that develops before them is whether certain abnormal findings in the mesentery could be blood or bowel contents (a complication of biopsy), or could represent something else. The CT attending speculates: "Could be blood, not clear-cut blood . . . at the lower pole of kidney . . . The clear thing here is infiltration around mesenteric vessels." After commenting on this "clear thing," he invokes Occam's Razor—ancient logical device prized by diagnosticians: "Now, you ought to have one disease [not more] . . . [But] I will tell you, though, that the problem here is twofold . . . That's a long way off from that kidney, and you'd have to go through spleen to get there." So, against the principle of parsimony, there is a topological issue: those mesenteric findings are a bit far from the biopsy target to attribute to a local procedural complication.

The clinical fellow adds: "I just examined him, and he has peritoneal irritation on the left side." Then she asks: "Now, you guys thought he had lymphoma?"

The CT attending denies this: "Nah-nah-nah-nah-nah-nah-nah . . . thickening of mesentery from fibrous tissue." But, he asserts, what she has just said changes things entirely: the most important prospect is no longer pursuit of mesenteric fibrosis. "You tell me he's got peritoneal signs. No one else has offered that history . . . He needs a surgeon." The radiologist has modulated his position and now endorses the theory of a procedural complication. "They're seeing him now," says the clinician.

Circumstances of onset of pain, and the evolution of clinical findings, have made a determining contribution here to this review of imaging appearances. What became key to this radiologist's thinking is the last datum "offered" by the clinician—pain when the examiner jostles the peritoneum. These clinical signs now overshadow speculation about mesenteric imaging changes, and even override objections to the biopsy complication theory based on spatial geometry.

It is not clear whether this clinician is previously known to the CT attending—but he seems to trust her physical examination, enough to hinge his interpretation upon it. This trust is worth hefting alongside the skepticism of the radiologists in the case of the saddle thrombus—and skepticisms registered in conference settings. The nexus of trust/skepticism provokes the question: why do radiologists not interview and examine patients more often themselves? (Some do—but they are exceptions.) The hinge of collegial reportage—from the history on the requisition to these viewbox conversations—is, like that between two films brought into contiguity, extremely plastic. Rad-clin correlation is founded not only on the conventional exclusion of patients from the reading room but upon the removal of the radiologist from the clinic—specifically, from practices of interviewing and physical examination. These exclusions and abstractions are commonplaces of professional specialization, produced by institutional layouts of diagnostic spaces and persons' trajectories, and enforced by new technologies, including CT scanners. On one hand, these mutual and reciprocal exclusions make necessary the discursive stitchwork of "correlation." On the other hand, they also found many of the gaps and dissonances, suspicions and leaps of faith, that are constitutive conditions of diagnostic intrigues and judgments at the viewbox.

Viewbox Correlations: Rad-Path

This book opened with questions about the status of the cadaver in reading diagnostic images. Throughout, I have noted how CT interpretations are reconnected with dead tissue, cores or slices of flesh—through which imaging findings can be "tissue-proven." The much-touted "noninvasiveness" of CT scanning is hinged to the possibility, if not the eventuality, of this retrieval of cadaverous chunks.

Neuroimaging fellow:	All right, so this is a thinker.
Medical Student 1:	Gonna biopsy it?
Fellow:	Biopsy it—definitely.
Investigator:	[*chuckling*]
Medical Student 2:	That's our recommendation.
Fellow:	For everything—[*chuckling*] everything we don't know what it is—biopsy it, definitely. [*chuckling*]

Biopsy always looms as a possible solution to a viewbox puzzle. But it is invasive, and expensive, so its very mention sometimes provokes a recursion: return to the puzzling images, renewal of efforts to finesse a diagnosis.

A Body CT fellow has finished dictating a case. I ask him why he is looking at the films again. "I'm just double-checking myself because of the findings of metastatic lesions." He was not able to find a primary cancer—lung or kidney would be the likely source—so the patient will need to have one of the mets biopsied to clarify its type. "It's always frustrating to me when you see all these mets and you can't find any—any source. It's just strange."

Likewise in Neuroimaging, a prospect of biopsy compels reinspection of images:

Neuroimaging attending:	I think we're going to have to biopsy it, what else?
Fellow:	Well, what do you think it could be?
Attending:	What else could it be? It doesn't look like PML. It's in the—in the—could it be? Could it be PML then?
Fellow:	Yes.
Attending:	It doesn't look like classic PML. Could it be an astrocytoma, a low grade astrocytoma—white matter, the right age, has a little bit of mass effect.
Fellow:	You think it does?
Attending:	Just looking at the cortex here looks a little squashed, and it extends into the corpus callosum.

Fellow:	Right. [*boardbuzz*]
Attending:	Maybe we should do spectroscopy, you know, just to see if it's a—it looks like it—
Fellow:	If it's got tumor spectra?

In this case, deliberations of fellow and attending do not produce greater diagnostic confidence. Here, MR spectroscopy is invoked as one way of finessing tissue typing through imaging. It is still an investigational technique in 1996—so the Neuroimaging attending's mention of it remains properly equivocal (notwithstanding enthusiasm among others on the team for recruitment of more "MR spect" cases). Biopsy is the option of last resort—marking a limit, and in some ways a failure, of radiologic means.

Though radiological diagnosis with "tissue specificity" is a desideratum, sometimes the limits of radiologic diagnosis become grounds to be defended—against presumptions of colleagues who ask too much. One morning a CT attending returns to the Body CT reading room enthusiastic about a case from Tumor Board conference. She describes it to the fellow: a chest CT of a patient with "known colon cancer" showed infiltrative changes of the mediastinum. The changes were "low attenuation . . . with calcification . . . measuring 20–50 [Hounsfield units]." Her voice is animated describing some "interesting" locational issues: "*middle* mediastinum . . . *bounding* the SVC." But, she remarks, the surgeons "wanted answers." She sounds indignant and disdainful: "As if we could do the biopsy through our x-ray vision." She says to the fellow: "I was hoping you weren't hit with it"—for the initial reading, during one of his shifts: a tough case to be put on the spot about. Both attending and fellow chuckle about the fact that, because there were questions about heterogeneity of the liver, one of their Body Imaging colleagues will surely "get an MRI." They expect this will happen whether or not this patient "goes to biopsy." Implication: though they think the limits of radiologic inference have been reached in this case, it will be difficult to fend off this last, probably superfluous, episode of radiologic workup.

There is disagreement about limits of radiologic diagnostic specificity. There are jokes (with a foothold in accusatory seriousness) about the limitless capacity of radiology for "self-referral": one radiologic test leads to another, ad infinitum. Radiologists are expert in knowing the gaps in the catch-net of each diagnostic modality, and the best imaging technique to fill those gaps with.

Final diagnoses may be bad for business—but they are helpful in culling cases for conference presentations. A Body CT resident asks: "What do you think about my showing a microcystic adenoma [at CT conference Monday]?"

Attending:	Has it been excised?
Resident:	No.
Attending:	How do you know it's a microcystic adenoma?
Resident:	I thought that was an MR diagnosis! If the cysts are small enough, then it's . . . [*pauses*]
Attending:	[*shaking head*] There's enough overlap . . .

The conversation turns to a more general issue that plagues selection tasks like this. The resident is supposed to bring recent cases, encountered on his watch. But: "It's hard to get path proof on recent cases." If not the adenoma, considers the resident, "I can get appendicitis . . . and I've got a lymphoma on a non-con CT . . . I may be able to get a cystic renal cell CA."

As this resident laments, the relation of reading room viewbox work to "path proof" is anticipatory. Current diagnostic puzzles rarely come to a tissue denouement in the same day, or even the same week. Many diagnostic puzzles never come to biopsy or autopsy at all. And some cases that do come to biopsy are not resolved.

A neurologist enters the Neuroimaging reading room to check two different MRI scans. He views them firsthand, scans the logbook, and uses the wall phone to review one of the studies' dictations on RTAS. The case was dictated the day before by Dr. Morrison. Today's attending advises discussing the case with Dr. Morrison. Then, as the neurologist is on his way out, the attending warns: "He's going to say, 'I dunno.'" The neurologist chuckles and responds: "Trouble is, then we do a biopsy—and the pathologist says, 'I dunno.'"

This jest is predicated on a caricature—and it is understandable that diagnosticians, with their nuanced relationships with uncertainty, should joke in this way about pathologists. But radiologists also have a subtle understanding of what pathologists can do. This is partly because radiologists procure material for pathologists.

During a viewbox lull, the Body CT fellow expresses concerns to the attending du jour about his recent difficulties obtaining diagnostic samples on radiographically guided biopsies.

Fellow:	We get "atypical cells suspicious for malignancy," but nothing else.
Attending:	They [Pathology] won't call it?
Fellow:	One of the things I might have done is selected a place where I got the easiest window [aimed the needle to minimize collateral damage, not necessarily to access the best part of the lesion].

Attending:	When you first plated it, on the wet prep, did they tell you you got cellular material?
Fellow:	Thick, viscous stuff . . . four mls of fluid . . . they said, "We'll spin this down, and we'll do a cell prep" . . . The pathologists thought it was enough.
Attending:	All I'll say is, I wouldn't feel quite as bad with pancreatic biopsy . . . you're pretty dependent on the expertise of the cytopathologist . . . Pancreas is kind of a funny actor, if you will, largely because of desmoplastic response . . . The yield in terms of how good they are in making that call is pretty limited . . . The only thing I'd try is go to an eighteen gauge [needle]. I wouldn't go higher . . . It [pancreas] is probably the most troublesome of CT/ultrasound guided biopsy . . . The other one, the paraspinal mass, I don't know . . .
Fellow:	I made three passes . . . didn't want to go too medial . . .
Attending:	I think everything was done appropriately . . . I don't know any . . .
Fellow:	Did that adrenal lesion today: cytology was "necrotic material"—so I did a core, sent it to Surg Path.

This earns a chuckle of approbation from the attending. Cytopathologists work with liquified material, often aspirated through a needle, and cannot view intact tissue architecture—whereas surgical pathologists work with chunks of solid tissue. Cytopathologists' occasional "nondiagnostic" interpretations frustrate radiologists—even as radiologists feel implicated in providing inadequate samples sometimes. The other side of the coin is that radiological targeting finesse, combined with good cytopathology, can become a point of pride—obviating the need of some patients to undergo more invasive surgery.[43]

Radiologists register misgivings about pathologic diagnoses for various reasons. A resident in Body CT says: "This is Ms. Marlin . . . we saw her outside scan last Friday . . . here thickening of rectal wall . . . she was scoped, they biopsied an ulcer, lots of teams came down . . . They [Pathology] mentioned ischemia." The attending listens thoughtfully, then comments that the rectum is an "unusual place to get ischemia—because there's a dual blood supply." This exemplifies an objection based on anatomy—of which radiologists are proud of their comprehensive grasp.

Another CT attending explains rad-path tensions in differentiating neurofibroma (benign) from neurofibrosarcoma (malignant):

Body CT attending:	And ah, when they take it out, histologically, [it] can be very benign.
Body CT resident:	But they grow back.
Attending:	They'll grow back. But—so, when they come back, [pathologists will] say "neurofibromatosis, no evidence of neurofibrosarcoma"—trust me.
Resident:	OK.
Attending:	Because, ah—I have seen cases which literally grew, ah, double in six weeks, ten weeks—and, based on radiologic evidence, we've gotta call it neurofibrosarcoma. And the surgeon's going to ask you, "Is this sarcomatous degeneration or is this neurofibroma?"
Resident:	Yeah.
Attending:	And you've got to say: "Grows like a weed, got to say this is sarcoma." Take the histological evidence.
Resident:	Neurofibroma.
Attending:	Because they base it on mitosis, anaplastic pleomorphism and junk like that . . .
Resident:	Is there a chance they maybe didn't cut it in the right position?
Attending:	Maybe they didn't cut it, maybe—no, some—some of them they cut it right, and they say—I usually call them up and say, "This thing's growing like weeds."
Resident:	Yeah.
Attending:	Morphologically and histologically, still looks benign . . . grows like that—there's a discrepancy. When the surgeon comes calling back: "You called sarcoma, this is a neurofibroma, you're wrong"—we're not wrong, because we're going by the growth rate, they're looking at mitosis.
Resident:	Yeah. We're both right.
Attending:	We're both right. We haven't reconciled the difference yet.

Here the objection to the pathologist (or surgeon) from the radiologist is based on a lesion's gross morphology, and change over time—as opposed to lesion-internal criteria like "mitosis . . . and junk like that." A lesion which grows "like weeds" ought to be classified as malignant, according to this radiologist—regardless of its cellular architecture under the microscope. The radiologist-in-training needs to understand how different diagnoses can collide, even coexist, in rad-path correlations—partly to defend the specificity of his viewpoint against presumptions of colleagues. The idioms of settlement here ("we're both right") belie the extent to which rad-path correlation is able to resuspend a diagnostic dilemma—and thereby renew prospects for engagements by, and disagreements among, specialized experts.

This exchange not only exemplifies dissonance in rad-path correlation but does so in a hypothetical register: the attending is advising the resident not so much on this case as on prior and future cases-like-this. As mentioned above, histopathologic findings in a particular case are usually not available at the time of primary radiologic interpretation.

An exception to the temporal disjunction between primary radiologic and pathologic readings is the CT-guided biopsy. A CT biopsy places CT interpretation in an indexical relation to tissue interpretation. That the biopsy is being performed means the limits of CT interpretation have been reached: the scanner is targeting a lesion whose diagnosis remains elusive.

One morning in the Body CT reading room, the fellow mounts on the back viewbox some staging CT films for an upcoming biopsy. These films are reviewed with the attending a few moments later.

Body CT attending:	Now, this is the one that we're biopsying?
Body CT fellow:	Well I haven't quite figured out what we're biopsying yet, so I thought maybe that might tell us.
Attending:	We need to talk to Dr. Len Marcus [Interventional radiologist] . . . So, looks like he, looks like he did a portal-caval shunt, a TIPS? . . . Looks like they did a shunt thing. This is ah—We need to invite Dr. Marcus . . . maybe he wants us to biopsy this.

They continue to inspect the films. There are two tubes apparent, a shunt (a vascular connection) and a stent (holding open the biliary tract). The attending marks the film to facilitate recognition.

Attending:	So this is shunt and this is stent. So one is "SH" and one is "ST." So presumably they want us to biopsy this, Mona, to document this ah—cholangiocarcinoma . . . So hopefully we don't hit this thing. I want to just stay away . . . right idea. Aim for this stuff.

Fellow:	Aim for that biliary tube.
Attending:	Yeah, right. That's a nice artery—that's the gastroduodenal artery, that's the anterior pancreatoduodenal. When you see red stuff, you know you don't want to . . .
Fellow:	Should go in . . . [*indistinct*] . . . straight.

This film review with the fellow—a meticulous slice-by-slice examination—has enabled some plausible conjectures about procedures recently undergone by the patient, and the probable biopsy target. To seek clarification, the attending pages Dr. Marcus, the Interventional attending, who has some prior familiarity with the case. In a few moments he has him on the phone: "Hey, Len. Question for you. You, presumably, are the one who referred us this biopsy of Lisa Conner—and they're looking for histological proof of cholangiocarcinoma? [*pause*] Right, right. TIPS, right. Uh-huh. I've got the CT scan in front of me . . . Yeah, it's more, more—sort of a periductal—there's a whole bunch of junk in there—soft tissue junk in there. [*pause*] I see. So if it's—so if it's a positive they ain't gonna do it." Though the CT attending's tones are more definitive, and his question focused ("they're looking for histological proof of cholangiocarcinoma?"), he receives important new information from Dr. Marcus: this patient is a candidate for a liver transplant, but she will no longer be a candidate if the biopsy reveals cancer. Dr. Marcus, for his part, has reviewed some prior studies, but not the instant CT. A few moments later he joins the CT attending at the rear viewbox.

Interventionist:	Yeah—what do you think this—what do you think this . . . ? See, that's the stent. What do you think all that stuff is right around there?
CT Attending:	Yeah, that's what we're going to biopsy.
Interventionist:	I guess it could be inflammation.

The CT attending looks through a folder of old studies but cannot find anything to compare present images with. The rest of their exchange works through several questions beyond the principal one, of whether "this schmutz" could be a cholangiocarcinoma, or simply inflammation, perhaps related to prior procedures. Could there be associated lymph nodes? Maybe. Is there a retained ductal stone? Doubtful.

The CT attending and the interventionist leave, and the resident and fellow busy themselves with the day's workload. They do not reconsider the biopsy case until a tech enters to announce that the patient has arrived. The resident has not participated in the prior discussions or yet had the opportunity to engage the images—even though he will be the person performing the biopsy. The images remain on the back viewbox.

CT tech: How would you like Ms. Conner positioned and scanned?
CT resident: Um—on her back.
Tech: On her back.

[*resident and fellow move to back board; the tech remains temporarily*]

Fellow: So about right here . . .
Resident: I'm not even sure . . .
Fellow: You're sticking it right into this, this stuff.
Resident: That low density stuff?
Fellow: This low density stuff right in front of that stent . . . so it's just gonna go—straight down.
Resident: Just lateral to—that vessel.
CT tech: 8's are OK? Do you want 5's? [mm slices]
Fellow: Why don't you do 5's.
Resident: The only thing is I'm not going to be able to see that vessel since we're not giv—since this is not contrast.
Fellow: Right. So we'll just stay lateral, I guess.
Resident: I guess as long as I'm—fairly, uh—parallel—or—as long as I'm fairly vertical—
Fellow: Actually you could just—as long as you're real—just right vertical, you'll be fine.
Resident: I'm gonna um—and aim for the stent and just go proximal to the stent by one millimeter . . . right?
Fellow: 'Cause you'll be able to tell at the level of the SMA, with the stent—just aim right for the stent.
Resident: Yeah.
Fellow: About a centimeter short. [*filmflap*] [*filmflap*]

Most of the remainder of the biopsy staging takes place in the scanner room (recounted in chapter 2, "Cutting"). Within a few hours, a pathology resident conveys biopsy results back to CT. They are posted, on a sheet in the CT Biopsy Log, on the back wall of the reading room:

Date _____   MD taking booking _____
Referring MD: <u>Carpenter/Marcus</u>   Beeper <u>5717</u>

Date Bx scheduled: <u>4/13/97</u>   Organ/structure <u>peripanc tissue</u>

Pt's name <u>Lisa Conner</u>   Unit # <u>10498375</u>

History, Imaging findings & biopsy desired

_____

Prior CT or US scan (Y/N) _____   Location of prior scan _____

(If possible outside scans should be retained in CT/US area until biopsy)

Coag parameters PT <u>4/13 13.2</u>   PTT <u>4/11 34.6</u> Platelets <u>4/13 90K</u>

(Coags should be available prior to date of study)

Comments _____

Bx performed by: (resident/attending) <u>Alpers/Bynum</u>

Date of Bx <u>4–13–97</u>

Needle type   <u>20</u>   Number passes <u>2</u>   Result <u>Inflamm cells</u>

____ ___ _____

Technique notes _____

Complications _____

So the archival consummation of the CT biopsy in the reading room is terse: "Inflamm cells." Without complication or annotation, this is simply a lesion well targeted, an arrow quivering in the bulls-eye. There will be other, more elaborated reports about this biopsy—official reportage in the patient's chart, and perhaps discussion in the reading room, conference room, or in the halls. But this official record retains little or no trace of the evolution of pre-biopsy curiosities in the reading room. Much of the process from "schmutz" to "inflammatory cells" remains unrepresented: the suite of modest viewbox intrigues, partly constituted by gaps in communication and distributed across differences in expertise, comprising questions like: What could it be? What else could it be? How do we get there? What happens if I miss? Nor are key remaining questions represented: Could this needle aspiration still have missed a cancer? Should this patient undergo surgical biopsy?

Or I Will Feed the Tech the Film

In the Neuro CT control room, a few minutes before 8:00 a.m., just after the techs' change of shift: a Body CT attending enters and greets the tech.

Attending: All right, you're getting ethnographically studied—as well you should be. God knows, they've needed to study you guys [techs] for awhile.

Investigator: [*laughing*]

Attending:	Do you want to do me a [favor]—[print] one while you're waiting here?
CT technologist:	What's that?
Attending:	It's one done last night, on Tucker . . .
Tech:	The entire case?
Attending:	Yeah, you—everything you can get, pretty much. You can cut out the lung windows, I don't need that, but I do need the rest of the case.
Tech:	OK—done.
Attending:	It's a good teaching one for the residents.

While the tech prints the case, I ask the attending what interests him about it.

Attending:	Well, there's colonic wall thickening involved in the rectum and sigmoid—it's a little difficult the way it was tricky and it fooled the resident last night—so whenever I see something that's—I like to collect the cases that—[*pause*] are, one, classic for something kind of unsuspected, two, if I know that it's been tricky enough for a resident this good to miss it . . .
Investigator:	Right.
Attending:	Then I might collect it because it just tells me that it deserves more emphasis.
Investigator:	Sure . . .
Attending:	Well, this is, this is a case of, uh, probably pseudomembranous colitis. Now, pseudomembranous colitis is really tricky, you'd think it'd be obvious, because classically it presents with fevers and diarrhea, right? But it often presents in these funny, ah, patients who are getting antibiotics for abscess in the hospital, start running a fever. You start looking to see whether the abscess has been correctly drained or not, you see another problem. And it might be totally clinically unsuspected—I mean, the patient may have just fevers—and it's the kind of thing that radiologically . . . 'cause I mean, this is just classic—if this turns out to be anything other than pseudomembranous colitis, um—you know, I would—I'd feed Zeke the films.
Tech (Zeke):	[*laughing*] So . . .
Investigator:	I didn't realize that was a classic radiological diagnosis. I think of it as—it's a [toxin] titer diagnosis.

Attending: It's a classic colonoscopic diagnosis. But when it's fully developed, it has massive wall thickening. There are very very few things that will give you the degree of wall thickening you see in pseudomembranous colitis. Um, there are a few things: neutropenic colitis . . .—but the setting there is completely different. Those are patients who have AML, or somebody's . . . immunocompromised.

The attending receives his films from the tech and departs. The tech is faintly pleased to have retrieved and printed the study faster than the attending expected. We speculate on how long it took the resident to call back the clinical team to amend her prior report and make sure they know of this serious but eminently treatable condition.

A month later, I see the resident who was covering that night. I ask whatever happened about that colitis case that Dr. Samuelson overread with her. Six hundred cases ago: the resident knows it instantly. She is concerned that I know about her "missed call"—but I explain my interests are elsewhere, in what people think of as interesting cases. "Oh, yes, that one. Well, that *was* interesting. It turned out not to be what he said it was. That was biopsied and turned out to be a case of eosinophilic colitis, *not* pseudomembranous colitis. It was presented at one of the GI Imaging conferences."

I ask: "Will you wind up, by virtue of your relationship to the case, writing it up? Or will Dr. Samuelson?" "Well," she says, "he originally asked me to do it but—Sarah Mounier is going to do it for one of her AFIP cases." And yes, she says, the case still has a sort of "fascinoma" quality.[44] I pass on the attending's compliment—"a good resident"—and she is appreciative.

Later, her resident colleague does indeed bring films and pathology slides to the AFIP: the Armed Forces Institute of Pathology—in Bethesda, Maryland. This is for her required six-week course in radiologic-pathologic correlation. The AFIP is *the* site in the nation where radiologists-in-training engage formal study of rad-path correlation. Bringing two "tissue-proven" cases is one of the prices of admission. "Fascinomas" are especially welcome.

So here is a CT reading which becomes correlated with stained tissue, the gold standard: radiological and pathological evidence are brought together—in a national archive—to constitute a new element of positive knowledge.

But what of the institutional chains of custody of these different orders of evidence? Does the ballast of the pathology slides account for the interest generated by this case across the radiology and GI services? The marvelous trope of eating film weighs in the materiality of images—ultimately *against* the counterevidence of biopsy results. The locution establishes a radiologist's relation to filmic lesion like that of a gambler to his champion: a sure bet, a good hand.

Though I did not attend any conference which discussed this case, it is likely that tales of two radiologic misses were made a part of its frame. The memorability and exchange value of such a case are vastly enhanced by a cumulative narrative of missed calls. It is also likely that the politics of face-saving will entail further attributions of "trickiness" to this lesion, colonic wall-thickening—its capacity to fool not only a good resident but also a good attending.

Reprise

Testifying and teaching are twined practices, in conferences and at the reading room viewbox. In both settings they are part of a theater of expertise, intrigue, and proof.

Conferences are more ritualized in their structuring of roles and performances. In both settings there is pressure of a docket—but in the reading room there are added pressures of frequent interruption, urgencies of client demands. Each setting draws upon the other: dictation at the viewbox, interrupted as it often is, quotes in its recorded form the ideal rhetorics of formal case presentation; professorial inquisitions in conferences reproduce the missing links and pressures of viewbox testimony.

Sociologically, testimony and teaching in conferences accomplish more than the finessing of diagnoses. Conferences dramatize hierarchical divides (teacher/trainee, specialist/generalist) at the same time that they develop hinges and exchanges among them. Conferences reconstitute departmental and professional identity: performing and learning to perform expertise is crucial for a discipline whose primary clients are other physicians.

Though conferences vary in genre and topic, there are predictable sequences laced through the calendar that establish rhythms of engagement for particular disciplines. There are also pedagogical genres that recur throughout the conference series: slide/lecture presentations; case presentations from particular services; question-and-answer exchanges organized around images.

One of the constitutive ritual structures of resident teaching is the hot seat. The hot seat is a pedagogical device which intensifies experiences of seeing and saying, teaches rhetorics of testimony and rules of evidence, develops poise under pressure. Hot-seat dialogues emphasize performative dimensions of expertise in an atmosphere of cross-examination.

In conferences as at the viewbox, the key performative and epistemological unit is the case. When a case is presented in conference by a resident who has not seen it previously, there are conventional phases: a discovery phase of solo image inspection (autopsis), then a presentation phase. This latter is parsed further into a period of orienting display for colleagues, a reconnaissance of findings, a focus on key findings, and a differential diagnosis, ideally leading to a consummating diagnosis. Thus the radiological case presentation is not a narrative of events and circumstances, like the clinical case presentation. It narrates a reconnaissance of images, then a reconnaissance of a table of diagnostic prospects. When a final diagnosis is hazarded, it is usually on the basis of best fit with known morphological features, and epidemiological predilections of morbid entities.

Framing a case and offering a clue produce whiffs of intrigue. Hazarding a provisional diagnosis in face of withheld proof intensifies intrigue. Yet case presentation in teaching conferences is not calculated to produce unified effects. It is often cross-laced with interrogations—as are so many interactions at the reading room viewbox. Moreover, radiological teaching and work are assembled as case series. In conferences, though cases are selected for interest, teaching value may not inhere in each case per se but in its juxtaposition with others. This is one version of radiology's use of the presentational device of contiguity. Juxtaposition works among cases, and also for elements within a case—images, image/text, image/diagram. Contiguity provides a hinge of great flexibility, shifting radiological thought and pedagogy across and through the manifold of visible materials to which it addresses itself. The device of contiguity, and the considerations of similarity and difference that it facilitates, may not generate a narrative trajectory, a causal or chronological development. Contiguity and the series of cases can lend themselves to a more Cubist enterprise—montages of multiple views, associations, correlations.

Still, this Cubist approach, among or within cases, and with or without pathologic correlation, often stages, or compounds, intrigue.

The reading room viewbox is for the most part a more intimate and insular setting than the conference room. Many intrigues that bloom here are founded on this insularity. Patients are excluded, and clinical evidence and testimony— patient experience, exam findings, details of colleagues' curiosities—are often missing. Radiologists prize their abstraction and yet need clinical contributions to their interpretive work. Sometimes attempts to reconstruct missing context produce a bloom of intrigue.

What is hidden and locally intriguing in any given imaging case, in the reading room as in the conference room, is historically and socially structured—by hierarchies of authority and expertise, by divisions of labor and training, and by correlation with other evidential modalities.

Intrigues help to secure the prestige of agents as well as the stability of specific truth claims. Acumen and judgment, crucial faculties for diagnostic work of the hospital, are too often theorized as private skills, matters of individual psychology. They are performed in the theater of intrigue—and their performances structure thought, affect, and belief. The suite of spectatorial irritabilities, curiosities, and astonishments which unfolds around the CT viewbox is no mere by-product of the diagnostic economy. It unfolds, within pedagogical theaters, as a positive phenomenon—as device to solicit engagement of trainees, as "fun," as solicitation to the emulative learning of judgment.

While the economy of intrigue has clearly changed since the days of "plain films," it seems, at the CT film viewbox, that curiosity has not quite succumbed to the superficial starkness of the section, to Taylorist regimentation, to the proliferation of information, or to the exhaustive "transparency" of embodiment.[45]

Yet intrigues that bloom at the CT viewbox are not centered on patient experience; they are not like the holistic, life-historical puzzles prized by clinicians. Viewbox intrigues interrogate motives and expertise of colleagues, the inadequacies or unavailability of previous studies, the local duplicities of radiographic signs. These complications and intrications of the diagnostic puzzle sometimes produce versions of intrigue more consistent with early modern connotations of the term—trickery, plots, waylaid messengers. In any case, the powerful nexus of the medical mystery and the decryptive interest it solicits, born around the autopsy table and nurtured in more than a century of CPCs, is still undergoing transformations.

There are important pedagogical issues, radiological and extraradiological, embedded in collegial viewbox practice. The historical fact that the mechanical viewbox is disappearing—already gone, from University Hospital—should be giving medical educators pause. It is not that anyone should be nostalgic for filmflap and boardbuzz, for the sleight-of-hand of film shuffling. But the theater of interdisciplinary and correlative colloquy which has historically devolved around a viewbox array of "original" films, site of clinicians' pilgrimages, is a powerful device in the teaching of diagnostic judgment.

Judgment cannot be taught except through its exemplification.

And should medicine be nostalgic about the cadaver? Many commentators are concerned about what is missed with the decline of postmortem autopsy. Some think medical imaging and other diagnostic tests have been accorded too much authority—that imaging is no substitute for tissue.

To approach (once more) this question about the authority of medical imaging relative to histopathology—and ways in which medical imaging extends its prestige and authority as the trademark of modern medicine (the doc at the viewbox)—I turn to another kind of setting, outside University Hospital: the Expo.

6 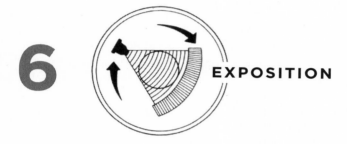 EXPOSITION

A mild, or what artists term a cool light, with its consequent warm
shadows, will do wonders for even an ill-furnished apartment.
• • • **EDGAR A. POE,** "The Philosophy of Furniture"

Exposition trades on exposure: setting out, placing in view. There are two venues
of exposition where radiographic projects, including CT, are set out comprehensively, for broader polities. Both are outside University Hospital.

One is a training venue. Each radiology resident, in the third year, attends
a six-week course in Radiologic Pathology at the Armed Forces Institute of
Pathology (AFIP) in Washington, D.C. The same course is attended by most
radiology residents in the United States and some from other countries.[1] The
AFIP Department of Radiologic Pathology also teaches shorter courses, in the
United States and around the world.[2] These courses disseminate knowledge and
method throughout the radiological profession.

The other is an annual event involving faculty—and some residents, administrators, and techs. It is the Scientific Assembly and Annual Meeting of
the Radiological Society of North America (RSNA), in Chicago: major showcase

for radiological research and technology, nationally and internationally. Special arrangements are required to ensure that services in University Hospital are covered during the week of the meeting, in late fall.

The AFIP Rad-Path course and the RSNA meeting are sites of pilgrimage for nearly every young U.S. radiologist. Both are singularly grand. The AFIP houses the world's largest tissue archive—and its Department of Radiologic Pathology is unique. The RSNA is the world's largest scientific trade show: only Chicago has (in the 1990s) the convention space to host it.

These metropolitan expositions are dedicated, like many historic precursors—the Exposition Universelle in Paris in 1867, the Columbian Exposition in Chicago in 1893—to technological progress and specimens of the Other.[3] They form a crucial part of the cultural backdrop of the CT Suite, at University Hospital and throughout the civilized world. Radiological Expos, representing radiology's marketers, purchasers, and publics, constitute arenas of intensified trade in the commerce of detection, intrigue, and judgment whose home turf is the local hospital viewbox.

The AFIP Rad-Path Course

The AFIP's Radiologic Pathology (Rad-Path) Course cycles all year—six-week sessions, with one-week breaks. Each six-week course is held in a building adjacent to the main AFIP building ("Building 54," on the Walter Reed Army Hospital campus): the Radiologic Pathology Education Center. A plaque in the foyer names donors: American College of Radiology, American Roentgen Ray Society, RSNA, among others.[4]

The auditorium of the Radiologic Pathology Education Center seats 232 persons. Concentric tiers of curved desktops slope down toward a large projection screen at center front. Seats are padded, desktops wired. In anterooms are lockers and a coffee dispensary. Coffee is a staple for lectures in a darkened hall.

Most of each course day is devoted to slide lectures on rad-path correlation.[5] Many are delivered by faculty of the Department of Rad-Path, the remainder by experts from other academic hospitals. The syllabus is sectioned by body-region/system, mostly: Chest and Mediastinum; GenitoUrinary; NeuroRadiology; GastroIntestinal; Musculoskeletal; Pediatric.

After lectures, a seminar addresses "Unknown Cases" displayed on ten fixed viewboxes on the side walls of the auditorium. The last half-hour of each day is devoted to discussion of five Unknowns on one side of the room—after which

33. Columbian Exposition stereoscope card, 1893

they are removed and a new series hung. This Unknown course unfolds inter-actively. According to Dr. Rosado de Christenson, department chair in 1997, it began as a weekly event—but "the residents liked it so much, we decided to do it every day."[6]

Unknowns are approached through residents' questions. Some are reluctant to guess at diagnoses in the large group. "If they nail it, they feel great"—but many are "afraid to say something stupid." Dr. Rosado de Christenson tells residents, "If you get it wrong, I can *really* teach you." She likes "puzzling over things": "I remember every case that I stayed up at night with as an intern." She also remembers mistakes.

There are no hot-seat interactions in the Rad-Path Course—largely because of the problem of scale. Some wish for more interactivity in lectures—the Education Center has recently been wired for "audience response"—but lectures and large question/answer sessions presently remain the teaching formats of choice.

In the foyer of the Education Center is a glass case exhibiting AFIP and Rad-Path mementos—T-shirt, sweatshirt, medal, cup—given to course participants, as well as a volume of the department's important *Atlas of Radiologic-Pathologic Correlation.*

A Medical Museum

I first visited the AFIP thirty-some years ago, with schoolmates. My seventh-grade class visited the Medical Museum—now the National Museum of Health and Medicine. I remember our fascination and horror at bone saws and pickled fetuses, old microscopes and waxen pustules.

The public Museum seems far from the military and scholarly missions of the AFIP. (It is a long walk around to the Institute's front entrance; there, one encounters a security checkpoint.) But the Institute began as a museum. From 1862, when the Surgeon General directed medical officers to forward "specimens of morbid anatomy" to a new Army Medical Museum, until 1946, when the Museum was subordinated to an Army Institute of Pathology, museological functions were central, and the public was an important constituency.[7] In the words of recent AFIP directors: if the Center for Advanced Pathology is the "heart" of the Institute, then the National Museum of Health and Medicine is its "soul."[8]

The first curator of the Museum sought Civil War specimens for a "grand national cabinet"—particularly projectiles and the body parts they damaged. Field surgeons were induced to comply through the curator's direct supervision at hospitals and battlefields. Soldiers parted with severed limbs and Minié balls. "All the specimens collected by the medical officers belong to the national museum," asserted the curator.[9]

Scientists at the Museum undertook research in pathology—including tissue staining and photomicroscopy. Pathologists from the Museum conducted Lincoln's autopsy—and indeed Ford's Theater briefly became the Museum's home. The pathologist Joseph Woodward remarked: "What nobler monument could the nation erect to his [Lincoln's] memory than this sombre treasure house, devoted to the study of disease and injury, mutilation and death?"[10]

Military and civilian physicians consulted the Museum daily. "Officers and soldiers who had lost a limb by amputation would come to look up its resting place, in some sense its last resting place"; and also "the public came to see the bones, attracted by a new sensation."[11] Early specimens included examples of military surgery, plaster casts, skulls and skeletons, comparative anatomic specimens, "microscopical" specimens, and specimens showing "morbid conditions of the internal organs in fever, chronic dysentery and other camp diseases," as well as "morbid anatomy of the diseases of civil life."[12] The Museum also compiled anthropometric data. European commentators remarked in 1870 that the Museum's published catalogs (surgical, medical, microscopical) contained more specimens than in all similar European museums combined.[13]

34. Army Medical Museum Great Hall, 1890s

The Museum hosted Washington's Philosophical Society meetings. Among its European visitors was the director of the Vienna World Exposition of 1873.[14] The Army Medical Museum was itself a War Department exhibitor at the Philadelphia Expo of 1876, the Columbian Expo in Madrid in 1892, and the Columbian Expo in Chicago in 1893.

The Museum's missions were broadened by the second curator, John Shaw Billings, also curator of the Army Medical Library (precursor to the National Library of Medicine).[15] As its pathological collection grew, along with techniques of pathological analysis, the Museum offered more direct service to medical professions. The Army Medical Museum was the first institution in Washington to obtain an x-ray machine (used in 1896 to aid in the removal of a bullet from a young girl).[16]

World wars generated an influx of new pathological materials, requiring more pathologists and new systems for cataloging.[17] The Surgeon General again requested materials from postmortem autopsies in field settings. Curatorial staff dispatched to the field still often obtained the best specimens.[18]

The Museum's expanding mission of pathological consultation coincided with early development of specialty organizations. In concert with specialty academies, it established "Registries" of specimens and cases: a new form of cooperation between archive and clinic, and between military and civilian professionals. Eventually, twenty-seven registries were consolidated as the American Registry of Pathology, administered by what became the AFIP.[19]

A Surgeon General's Report (1937) remarked that the Medical Museum was "one of the show places of the Nation's capital."[20] Yet it became necessary to justify collecting morbid anatomy in a *military* archive. Advantages of military administration included extra control over specimen procurement and implementation of research findings—via "contract with bodies of troops under discipline."[21] These issues remain important to the present arrangement: a National Museum of Health and Medicine as a core division of the AFIP.[22]

The history of the AFIP exemplifies the importance of a museological ethos, a patriotic one, in genealogies of contemporary biomedicine. This ethos has comprised natural historical and even ethnological concerns: not only morbid pathological specimens, but also weapons, comparative anatomic specimens, insects, instruments. Twined in the curatorial practices of this early museum were complementary classificatory systems: taxa of artifacts, animals, ethnic varieties, and deranged human tissues.

These curatorial histories conjure up several bioscientific projects cited in this book. There is comparative anatomy, with its debt to Cuvier, and work of conjectural reconstruction. There is anthropometrics—referencing Bertillon's work on identification of criminals, and "galleries of culprits." There is craniology, historically important to brain imaging. And there is pathology, pursuit of morbid phenomena in fixed tissue.

The loop from Victorian museological sensibilities, and their scientific apparatus, to the Expo is exemplified in the Army Medical Museum's installations at the Philadelphia Exposition of 1876. In one room, set up like a tropical field hospital, mosquito netting over each bed, the walls held a series of Joseph Woodward's large framed photographs of stained sections of pathological material, along with Thomas Eakins's painting of surgical teaching, *The Gross Clinic.* Display of Woodward's magnified slices of pathological material publicized the virtues of the sectional image in biology and medicine on an unprecedented scale. Some of these images were also among those displayed at the Columbian Expo in 1893.

If today the cadaver seems to have disappeared from the teaching hospital, and from the ritual of the CPC, then one of the places it has gone to is the AFIP Museum.

Yet the cadaver is there not merely as scientific specimen. Or rather: its scientific values include affective engagements of the patrons of "this somber

35A AND 35B. Army Medical Museum exhibits at (a) the Philadelphia Exposition, 1876; and (b) the Columbian Exposition, 1893

treasure-house." The Museum became a hallowed space—apotheosis of military discipline (without whose impress so many body parts and war mementos could not have been accumulated). The collection inculcated in a nation's civilian citizenry a "new sensation"—a mixture of morbid curiosity, reverence of sacrifice, and pride in progress of (American) medicine.[23]

In this museum of horrors of warfare and camp diseases, there has been a certain privileging of the projectile. Finding and removing bullets also provided impetus for the development of tomography.

Recently I brought my own children here. The Museum is modernized: more levers to pull, buttons to push. There are exhibits of x-ray apparatus, a simulated fluoroscope, even a CT gantry, with explanatory videos. But the most riveting exhibits are still body parts in glass cases.

Archives of Radiologic Pathology

It is not so much as parent and citizen but as professional and scholar that I am granted a tour of the AFIP by Dr. Rosado de Christenson. She escorts me through locked doors and ID card readers. She points out photomicroscopy, and the general photography section, which serves her department and twenty-plus others. Some hallways are lined with cabinets with small drawers, and posters from recent conferences.

We enter the Rad-Path Case archive: a large brick-walled room crowded with file cabinets. Dr. Rosado de Christenson notes that this same room once housed the teaching operations of the Rad-Path Course as well. A projection screen on one wall is now partially obscured by file cabinets; on sides of the room, steps lead to a balcony gallery, once for lecture seating, now containing shelves and cabinets.

I have asked to see the case of eosinophilic colitis brought to the Rad-Path Course in 1997 by one of University Hospital's residents (as noted in chapter 5, "Testifying and Teaching," every participant in the Rad-Path Course submits two cases). I recount the tale of the attending who said that this was pseudo-membranous colitis or he would feed the tech the film. This elicits chuckles from Dr. Rosado de Christenson and one of her chiefs: the phrase it plays on—"if it is not X, I'll eat the film"—is a pet expression of her own.

A computer problem prevents our locating the case I seek—and I have omitted to bring the patient's first name. "It's a *real* puzzle now, I guess," chuckles Dr. Rosado de Christenson. But she shows me other cases. Each Rad-Path case is kept in a colored cardboard folder, filed by organ system and diagnosis and bearing a unique patient "accession number." The folder contains the case abstract, a page of clinical history, and a report on the histopathology.

We open a chest case. In a 5 x 7 manila subfolder is a handful of miniature x-ray films, just over 4 x 6 inches. Dr. Rosado de Christenson explains that original images were (until 1997) refilmed on this "Log-E" film—a fine-grained aerial reconnaissance film. "You have to get used to looking at it," she says, noting that AFIP radiologists often use loupes. The miniaturized images for this case contain, in addition to conventional x-rays, sectional images: ultrasound (2 x 3 matrix) and CT images (3 x 5 matrix).

Dr. Rosado de Christenson explains the accession process for cases brought by residents attending the Rad-Path Course, or cases arriving through the consultation network. All materials—histopathologic slides, photographs, gross specimens, radiographs—are sorted by a case manager. The radiographs are, since 1997, no longer kept on Log-E film but are instead archived digitally.[24] Photographic transparencies, sometimes refilmed under the old system, are also archived digitally. Images are stored in the Rad-Path archive. Histopathology slides, tissue blocks, and gross specimens are sent to the Tissue Repository, and most are warehoused off campus.[25]

All cases submitted by residents are "reread by AFIP pathologists": diagnoses by pathologists at originating institutions are not considered. Cases "are reviewed in divisional conferences as unknowns." AFIP pathologists' reviews may result in overturning of original diagnoses. "Sometimes we actually contact the originating institution to notify them we think a lesion might not have been well-sampled."

For residents' cases submitted with tissue blocks, specimens may be sent directly by departments of pathology at originating institutions; some are carried by the residents themselves. A story is recounted about a foreign resident held up in U.S. Customs with an amputated limb.

If there is no gross tissue specimen, the course directors require high-quality photographs of gross pathology—as 2 x 2–inch slides. Histology alone usually does not meet AFIP criteria for proper rad-path correlation. "A case with fabulous radiology is reduced to mediocrity by poor gross photography."[26] Photographic images should show the specimen "in the same orientation as at least one of the diagnostic images": residents are enjoined to "work with your surgeons and pathologists so that optimal sectioning of specimens can be achieved."[27]

I ask Dr. Rosado de Christenson whether it really matters whether residents attend the surgery or autopsy to supervise photography of resected tissue. She affirms that this is often significant. Involvement in custodianship of tissue and photography results in "better cases." I find it fascinating that such explicit instructions—not only about seeing for oneself, but also about imposing a kind of military discipline on tissue procurement (like the curator gone to the battlefield)—are required for good rad-path correlation. The AFIP's version of rad-cath correlation is more rigorous than what passes for "tissue-proof" in a case's home institution.

Of the cases brought by residents attending a rad-path course, only the best—by vote of section chiefs—are presented to course participants. Dr. Rosado de Christenson explains other ways good rad-path cases can be exhibited. The AFIP has long contributed a staple feature of the journal *Radiographics* (published by the RSNA)—entitled "From Curating of the AFIP"—which incorporates cases into topical reviews. Moreover, the editor of *Radiographics* is now welcoming AFIP's "Best Cases" as submissions for publication. This reflects a wish to "give back something to the institutions that have given their cases to us."[28] Dr. Rosado de Christenson notes authorship conventions: the resident who submits a case is listed first; a staff member who assisted preparation is second; the pathologist at the home institution is third; the AFIP is not listed.

Beyond this publication cycle—cast as gift and countergift between institutions, and credit due to procurers of images and tissue—Dr. Rosado de Christenson hopes to make other "best cases" available online. She demonstrates the format on PowerPoint slides. The cases are first presented "as Unknowns" correlating radiographs and gross photographs, then photomicroscopy, then (sometimes) historical correlations, then a diagnosis, and last, attributions (naming the preparer and institution). As Dr. Rosado de Christenson clicks through cases, she waxes enthusiastic: "echinococcus in the kidney . . . from a Belgian resident . . . a great case"; "beautiful, beautiful case of a mesenteric cyst."

The World Wide Web has enabled ambitious imaginings of rad-path educational outreach. Dr. Rosado de Christenson is developing reference materials for diagnostic radiologists "at the point-of-care,"[29] textbooklike but cross-referenced via hyperlinks. I addressed some of these materials late in chapter 3, "Diagnosing," comparing tables of pathological appearances with early police photo archives. Dr. Rosado de Christenson demonstrates pursuit of a diagnosis of thymoma. A click brings up a Short Description of thymoma; next a Long Description of thymoma illustrating rad-path correlations. One can then click to the telltale formats (shown in figures 26a and 26b): "Range of Disease"—a variety of appearances assumed by thymomas—and "Differential Diagnosis"—a range of other diagnostic entities that can become confused with thymomas. A final linked format is the Unknown—an image which can be offered to the trainee as a puzzle.

The key contribution of this AFIP resource is its exemplary completion of the epistemological circuit of cadaver reference. Each image on its field-guide tables represents a numbered specimen in its Tissue Repository. The footsteps (or gurney-wheelmarks) of the missing cadaver—gone from the clinic and scarce in the CPC—are herein reconnected to the work of the viewbox—at University Hospital and elsewhere. Admittedly, the AFIP is not so much a repository of whole cadavers as of small parts—slices and cores of dead tissue. But it is rigorous, more so than any institution in the country, in its correlations of radiographic images with pathology. The ballast of the cadaver, indeed of many cadavers, is returned to the scene of image reading and pedagogy. The cadaver was never gone: it has always been in the background, accessioned and registered, in Washington—and even online, only a few clicks away. It underwrites the epistemological authority of medical images, including CTs, which remain in a fundamental sense derivative: radiographic claims to "noninvasiveness" are at best temporary and provisional—or they are parasitic (via magics of resemblance and contiguous display) upon someone's handling of tissue, in Washington—or beyond.

Some disease entities are better represented in AFIP archives than others. Concerning GI cases, instructions to residents read (in 2000): "Please *do not* bring a case of leiomyosarcoma unless the case is unusual in some way. (We already have hundreds of these cases.) . . . In order to diversify our cases we request: Inflammatory Disorders with CT (pancreatitis, cholangitis, diverticulitis, colitis, etc.); AIDS and its complications."[30] What of the prospect, faintly implied, that Rad-Path case holdings of the AFIP might achieve more or less full archival coverage of some varieties of disease? Would the need for further pathologic correlation of imaging appearances then be obviated?

The scale of archival ambitions of the AFIP is conveyed by a previous chair of its Rad-Path Department, Dr. Madewell, in an interview retained in the Museum's Oral History collection:

Academic health care centers and radiology departments in those centers have teaching files. The American College of Radiology has an exquisite teaching file. But, in those files, you get five or six or ten of something. And they are selective; these are ten cases that illustrate the example of a disease process. That is very helpful, and I think that is certainly available throughout the country. But I think the person who really wants to understand in depth, to look at the raw material, not selective material that someone else says here's the way osteosarcoma or bronchogenic carcinoma should look like, should go to the database and look at a thousand of them. And there may be new perceptions in there that the person who has selected those cases to go in the teaching file may not have understood at that time. I don't think all institutions have to have this raw database. But I really feel, as a society, we need to have at least one of these medical databases, so that we can accumulate this wealth of material, both in longevity, going back to the Civil War, and also in the breadth of the process in numbers. Because it's only in numbers that we can see the occasional or the less-common expressions of disease states.[31]

These reflections link basic radiological values discussed in this book—especially *autopsis*, seeing for oneself—to the master archive, a repository of "raw" and less-selected material—a lot, a thousand cases, a "breadth . . . in numbers." What Madewell calls "longevity" is partly about accumulation of "wealth" and a kind of archival immortality. But it is also about historical change: the possibility of "new perceptions" suggests that disease taxonomies are neither rigid nor final. The historical archive constitutes both a repository of specimens and a diachronic sequence of classification schemes—a suite of "new perceptions."

Dr. Rosado de Christensen is concerned about both numbers and history when she laments the decreasing autopsy rate nationwide. I ask if pathologic "ground-truthing" of radiologic images will be an enterprise of lasting importance—or whether radiographically imaged morphology might one day assume an independent stability comparable to that of pathological morphology. Her response is that pathology, particularly histopathology, provides access to "reasons for the morphology"—enhances "understanding of the physiologic process." "Otherwise it [reading radiographs] is just pattern recognition." She is emphatic about benefits of tissue correlations for her own viewbox work: "I understand better—I read better" in the shadow of pathology. She can offer more refined diagnoses: "I think it is better to give a differential diagnosis of one or two things, not ten things."

"Reasons for the morphology" engages Morgagni's old notion of "seat and cause," transposed to the cellular and molecular level. This radiologist's embrace of pathology is founded not precisely on morphological correlations but on microscopic process which underwrite them.

What of CT? I ask Dr. Rosado de Christensen. Does she agree with the senior radiologist from University Hospital that something is lost in the survey of slices—that CT makes it "too easy" for residents? She will not say things are easy; but she agrees that "plain film is much harder than CT" to interpret. "I still shiver in my boots when I have to read a plain chest x-ray . . . There's so much you can miss." She quotes a colleague who refers to the CT scan as "the truth test." But, notes Dr. Rosado de Christensen, "not everyone can have a CT."

In a speculative mode, Dr. Rosado de Christensen remarks that the current director of the Institute (a medical examiner by training) has suggested that radiology and pathology may be converging on a single specialty. This prompts me to inquire about her sense of identity as a radiologist at the Institute—where the Department of Radiologic Pathology is somewhat oddly subsumed under the Division of Environmental Medicine—alongside Veterinary Pathology and Infections/Parasitic Disease. Dr. Rosado de Christensen indicates this recent "streamlining" has some nonobvious rationales—including emphasis on teaching (which Rad-Path shares with Vet-Path, for instance).

Dr. Madewell situates the issue of radiologists' roles and identities at AFIP within a broader ecology of work in other institutions—for instance, at Walter Reed, a "workaday hospital, where pathology is not so concentrated":

> To accumulate . . . experience and skill . . . you need to look at a lot of normal radiologic imaging. You also have to understand the normal anatomy and the normal physiology, to create that normal image. That's the foundation . . .
>
> So I think it is critical that the radiologist keeps his/her footing in the normal study group, because a lot of what we do in medicine is study normal people to make sure that they don't have an abnormality . . . A lot of medicine is the practice of screening or detecting subtle abnormalities. So that normality base is a very important piece for the radiologist.
>
> But you also have to see a lot of abnormal pieces. One of the real advantages I think the AFIP provided me and other people who have been there, as well as students who have come through the program, is the ability to sit down and look at thousands of carcinomas of the lung, or any entity you want to talk about. You can sit down and accumulate a life experience in a matter of four hours.[32]

"A life experience in a matter of four hours . . . focused, concentrated observations . . ." Value of the Exposition. I turn to the other key Expo that University Hospital radiologists attend: the annual meeting of the RSNA.

RSNA Technical Exhibits: Specimens

Chicago, McCormick Place North, December 1996. The RSNA meeting's largest Technical Exhibit area is an immense steel-girdered room, perhaps 300,000 square feet of floor space, with aqua-green carpets.[33] It is filled with kiosks, banners, booths, tables, signs, corporate logos, and dark suits. Every important radiological vendor is here. I sit in a corner to review the layout in the program, *On Display*—but even with numbering system and color-coded maps, I cannot rationalize a navigable path. No place to stand above and survey the scene. I loosen my tie and plunge in.

One of the first exhibits—Baxter—displays what appear to be bones, under a banner that reads "Solid Medical Sciences." There are skulls: some pediatric, some clearly pathologic, one holding a brainstem tumor; there are mandibles, a pelvis, feet. It turns out these are not bones but solid models of CT datasets: slices, laminated together to reconstitute a scanned object. One full-sized femur is made of laminated paper. One of three men staffing the exhibit, a programmer, is talkative. I fill out a questionnaire. Its questions are mostly for surgeons, about possible applications of these models—for patient teaching, resident teaching, surgical guidance: how close should tolerances be, how transparent should a model be, what color? These men are here to find out what use these products might have; previously they have been to orthopedic and maxillofacial surgery conferences. The staffers' attentions fade when I explain I am an anthropologist. They send me off with a faux-bone keychain, copied from their paper-laminated femur.

Around the corner is an exhibit by VOXEL,[34] enjoining passers-by to "Experience the 3rd Dimension." On a wall are a half-dozen anatomic holograms. Each "VOXGRAM," sized like a standard x-ray film, offers a shimmering "sculpture of anatomy," which seems to have depth and can be inspected from different angles.[35] Below each hologram are sectional images (CT, MR), representing, in more conventional format, data from which the hologram was constructed. Captions list virtues of a holographic gaze—"Transparency . . . Perspective . . ."—and brief scenarios of utility: "Unwinding tortuous blood vessels and characterizing complex aneurysms more quickly." The VOXEL staff, purveyors of virtual 3D, are convinced of the value of their *materia* for the surgical theater.

Down the aisle, I find PIXY, a "Tissue-Equivalent Teaching Phantom." She (the gender is clarified by a staffer) is a somewhat skeletal creature, seated casually in a director's chair, ankles crossed. On closer inspection, she turns out to have clear plastic skin, pink-and-tan organs, and spheroid lesions suspended in her abdomen. She is a teaching tool, and a test object, for diagnostic imaging and radiation therapy both.

Three exhibits, three models of specimen embodiment: from laminated solid, to shimmering hologram, to tissue-equivalent phantom. My head spins with the glossy finesse of these simulacra. All of them are creatures of/for CT, and yet none of them is familiar. They are utterly commodified bodies, surrounded by well-dressed hucksters—yet none of them is quite for sale, either. I cannot buy a laminated-paper femur, a shimmering VOXELGRAM, or a PIXY

36. PIXY, Tissue-Equivalent Teaching Phantom, RSNA Technical Exhibits, 1996

phantom—not here, now. The customers for these commodities are institutions; and what is being exhibited in each is not so much the instant specimen as a way to make, to simulate, more specimens.

My head is still spinning when I find, around another corner, a man in a long white coat demonstrating wonders of liquid nitrogen. I have missed the part of his talk about MRI technology; for the moment he is Mister Wizard, exhibiting amazing things one cannot do at home—and I realize I have become a gawking fairgoer and should not proceed to the CT scanner exhibits until I can feel more poised among the confidence men of imaging. I leave the Technical Exhibits, to find my way to Scientific Sessions.

RSNA Scientific Sessions

I head to McCormick East, to a midafternoon session I flagged in my Scientific program, on CT/Pathology Correlations. The room holds several hundred chairs. About 150 persons are in attendance—90 percent of them men. The session consists of six "papers," given as slide presentations (nine minutes apiece), all addressing CT or MR-guided biopsy. All PowerPoint talks, shown in double-projector, double-screen format. Every slide in all six talks conforms to the same formatting conventions: blue background, white writing. This turns out to be true for the entire conference: 1,676 papers in nine time slots over six days: nearly all conform to these presentational standards. The consistency is oddly reassuring, in view of the variegated booths and signs in the Technical Exhibits. It helps demarcate science from the carnivalesque.

The first paper is about "inconclusive" CT biopsies. The presenter and colleagues reviewed 116 needle biopsies performed with a semiautomated device and found that only one of several factors (a "paraxial approach" on a second pass) gave more conclusive results. This is important because a third of lesions whose biopsies were inconclusive turned out, on follow-up procedures, to harbor malignancy. One of the members of the audience queries the effect of the cytopathologist. This has "not yet been analyzed." The presenter studiously focuses on technical/geometric factors, and statistics, and does not engage matters of interpretive judgment.

The next paper, delivered with a German accent, concerns the "precision and clinical value" of MR-guided biopsies of lesions of the liver and head and neck—lesions not visible on CT or ultrasound. The paper involves closer description of both the biopsy device and its handling. The presenter tells of a tissue core from a liver lesion in a woman with a prior malignancy. Though the liver lesion was suspected to be a recurrent cancer, its "histopathology revealed focal nodular hyperplasia"—a benign diagnosis. This presentation offers fewer statistics and more narrative examples.

The third paper concerns angled needle approaches to "difficult" lesions. The results include a list of "structures avoided" as well as one of sites biopsied. The author advocates unorthodox approaches and patient positions, and tolerance for greater needle insertion depths (12 cm is his average), in the interest of reducing organ crossings. This paper is more anecdotal than the last. Yet even anecdotes problematize a standard distinction between "invasive" and "noninvasive" by displacing invasion from skin to organs.

I retreat from the last three papers—and wonder how the rest of the registrants distributed themselves for this wedge of the day. This session I'm leaving is attended by fewer than 1 percent. There are twenty-three other Scientific Sessions underway simultaneously, throughout McCormick Place East. Over two more days at the RSNA, I attend another four sessions like this—some with bigger audiences (CT in Evaluation of Pulmonary Embolism), some smaller (Emergency TeleRadiology).

Late afternoon: good time to visit Scientific Exhibits—"posters"—while crowds are thinning. These are on another level, in another enormous hall hundreds of feet long, with fifty-foot ceilings and even greener carpets. Here

37. Scientific Exhibits, RSNA Meeting, 1996

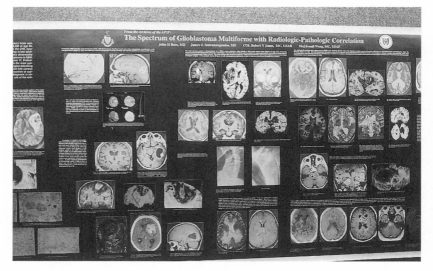

38. Rad-path correlation, AFIP Scientific Exhibit on brain tumor, RSNA, 1996

one can stand on a balcony and look down—on rows of posters receding into the distance, under specialty banners: Neuroradiology Head & Neck; Breast Imaging; Pediatric . . . At a few posters, authors linger, but most posters are now unattended. Some are illuminated from behind—images and text displayed as transparencies, one or two 3 x 5–foot panels. Other posters are opaque, under fluorescent lights. RSNA judges of scientific merit have made rounds, because some posters bear ribbons—"Best in the Division"; "Honorable Mention"—like pies at the fair.

Most of these posters have been produced with help from medical illustra-tors, elements arranged for visual and narrative impact. Disparate pictorial genres are controlled with background colors and borders: each poster aspires to artistic coherence. Yet each is distinctive compared to its neighbors. A col-lective effect of these same-sized, compositionally distinctive exhibits, side by side, row after row into the distance, is an oddly variable regularity—a cross between the conformity of the blue/white slide shows and the carnival of Tech-nical Exhibits. These posters materialize a vision of how science unfolds: in similar-sized increments, with a modicum of aesthetic conformity. And at a regular pace: physician registrants earn Continuing Education credit for walk-ing among the posters—by swiping their "Expocards" in a card reader every half-hour. There are card readers every few aisles, so one can travel from aisle to aisle, establishing rhythms of critical appraisal across the landscape of or-gans, and store up credit for one's serial attentions.

The variety of poster topics is a lesson unto itself. CT is everywhere—along with other sectional modalities. In the Genitourinary Section, there are exhibits titled "3D Model of the Male Pelvic Floor Muscles Based on CT, MRI and Sheet Plastination"; "Perinephric and Renal Sinus Diseases: Helical-CT and Pathologic Features." In the GastroIntestinal Section: "CT Virtual Reality: Incorporation of Inexpensive Virtual Endoscopy." And in the Neuroradiology Section, from Radiologists at the AFIP: "The Spectrum of Glioblastoma Multiforme with Radiologic-Pathologic Correlation."

The rows of posters are absorbing: I forget to swipe my card when I should. They are also exhausting: I leave feeling as if I have been too long in a museum—or a shopping mall.

RSNA Technical Exhibits: Scanners

In the Technical Exhibit halls where the big imaging machines are found, booths become buildings and signs become billboards. All major manufacturers of CT scanners have massive exhibits: GE, Siemens, Toshiba, Picker, Elscint, Imatron, Philips.[36] Each of these claims a large swath of the Exhibit floor plan. Each fills its space with towering architectures, large company logos. Walls and kiosks are covered with slogans, image panels, monitors. Here and there are the big white machines themselves: CT scanners, MRI scanners, ultrasound machines, nuclear medicine c-arms—in each corporate family, the latest generation. Around these institutional fixtures circulate dark-suited executives, prepared to testify about corporate versions of the cutting edge.

Mere squinting at a panel of images attracts a rep's attention. I ask what's new in CT this year. I learn that the previous few years were important for scanner innovations: "spiral" scanning technology, primarily. This year, all major manufacturers have spiral-capable scanners. Their innovations are software modules and toolkits—for rapid reconstructions, contrast timing, biopsies. The other focus is "networks" for storage and transmission of images.

The GE exhibit is typical in many respects: "Faster, Smarter, Better." Signs boast ease of use of GE's helical scanner, "Ensuring High Throughput with Consistent, High Quality Results." GE has many other machines on display—ultrasound, MR, and more—so CT occupies less of the GE floor plan than it does some other companies.' Elscint devotes a larger proportion of its exhibit to CT. Its signs tout "Affordable Spiral CT" and "State-of-the-Art Multi-Slice Applications."

At the Siemens exhibit, one of the showcased scanners is the machine in service at University Hospital: the "SOMATOM Plus 4." Siemens' slogan is

"Improving Your Outcomes." Beneath this, it celebrates its twenty-two-year "History of Clinical Leadership" in CT innovation; its "Global Reach" (colored dots on a world map); and its range of expertise as "Not Just a CT Manufacturer . . . Your Partner in Healthcare Offering a Full Portfolio of Solutions."

At the Picker exhibit, I observe two women in identical business suits flanking a large projection screen, coordinating a presentation, for twenty men, on image enhancement and reconstruction. The hand gestures of these twinned mistresses of ceremonies quote stagehand revelations of TV game-show prizes and flight attendant demonstrations of air-travel emergencies.

In all these exhibits I am permitted to make photographs—though I must negotiate permission at each location. Some exhibitors scan my RSNA ID for the company database. Occasionally I must explain what an anthropologist wants with a photograph of a trade show scene. At the Toshiba exhibit, although permission is granted from on high, vigilant staffers wave their arms at my camera, one after another, each time I turn a new corner. Finally one staffperson, who watched me photograph wall displays, absolutely refuses to let me photograph a cutaway/*écorché* model of their latest scanner, exhibiting its inner workings.

I have heard of anthropologists taken for spies—but the episode surprises me. It teaches me about high stakes of technical innovation in the CT market. It also says something about sensitivities within this corporate Expo complex: beneath the skin of the scanner, one may peer but not aim a camera. I come to understand that corporate espionage has plagued past RSNA conferences—compromising a competitive edge held by one or another scanner manufacturer.

At the Imatron exhibit, Alphonse Renato, International Marketing representative, has interesting things to say about the history of CT competition. He joined the GE Service division in 1978, shortly after the eclipse of the EMI scanner.[37] He recalls a classic Harvard Business Review case concerning EMI's product management detailing early neglect of scanner service which "the industry" did not quickly forget. GE, conversely, emphasized service: "When streaks showed up on scans in the afternoon . . . we worked all night and had the machine up again the next morning."[38]

Mr. Renato offers a telling anecdote about refinements in early machines around the time of the RSNA Meeting. In 1978, "flaws" with the GE 7800 scanner were addressed in the GE 8800. For a time in 1979, GE had only "one working detector in a region"—and they had to shuttle it among five regional scanners "so that we could present data from five institutions at RSNA." In this tale, the migratory fixer-detector serves, not patient care or client satisfaction,

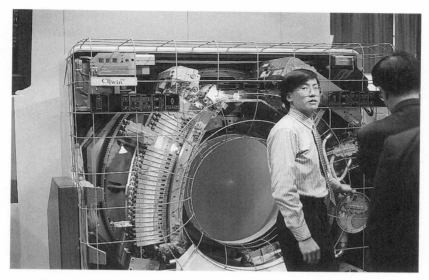

39. Exhibit staffer protecting exposed scanner, RSNA Technical Exhibits, 1996

but rather a convention of scientific reportage, the multicenter dataset. Yet this multicenter dataset, piety of experimental replicability, exaggerates "diffusion" of CT machines bearing a particular logo.

I ask Mr. Renato about relations between CT and MRI. In the early 1990s, he observes, CT was thought passé—and GE projected conversion of sectional imaging platforms to MRI within a decade. When "spiral CT came along" and GE had no spiral scanner in the pipeline, they "had to scramble" to avoid losing market share. But quite apart from GE's product line: "calcium saved CT," meaning that calcium in tumors—and for Imatron, in coronary arteries—is an important target of higher-resolution CT. "MRI can't see calcium."

Imatron's principal showcased technology is electron-beam CT. This high-speed technology seems promising for cardiac imaging. Imatron has 10 percent of the "functional cardiac imaging market," Siemens 90 percent. Much of Imatron's client base is Asian ("Bangkok does seven hundred scans a month"), mostly patients with coronary disease risk. Mr. Renato puts this in terms of an "HMO internist's quandary": which patient can be treated with medicine, and which should undergo instrumentation of coronary arteries? Quantifying plaque in coronary arteries can lower "costs per life saved."

I am impressed—not so much as internist, by the truth value of this formulation, but as ethnographer, by its jewel-like compactness, its vision of more imaging machines, diagnostic finesse, and cost savings, in one brushstroke. The

HMO internist scenario places imaged lesion within risk-profiled patient, and the expense of its treatment within the actuarial concerns of a health plan.

This conjures up the doc-at-the-viewbox scene discussed in the introduction. The viewbox supplies at least the backlighting for this imagining: a diagnostic quandary, technology to the rescue, and corporate solvency (of health plan and a publicly owned imaging company). The internist conferring with the radiologist at the HMO viewbox is multiply engaged: she exercises clinical judgment; she is managing resources; and she is constituting a niche for Imatron technology. Unfortunately, this viewbox scenario does not accord with traditional divisions of labor in contemporary hospitals. "Radiologists are not allowed to look at the heart," Renato notes,[39] "and cardiologists are not allowed to buy CT scanners." Thus an institutional deadlock, a "turf thing," prevents the development of this Imatron "application"—this wishful viewbox scene.

My tour of CT vendors teaches me little more, by way of technical information about scanners, than the techs did at University Hospital. There is more in the books on the scanner room shelf than in glossy vendor brochures. But about the general status of CT within the imaging industry, and within the exhibitionary complex of the RSNA, it has spoken volumes.

Reprise

Expos attract crowds: polities.

At the AFIP: elite scientists, professionals, and their trainees are distinguished from the lay public, via policing of IDs. There is also a distinction between national and international clienteles—with emphasis on privileges of citizenship, alongside missions embracing outreach and global epidemiology. There is the frame of military discipline, structuring tissue procurements and chains of custody, more systematic than unregulated civilian efforts, in hospitals as on the battlefield. For all these polities, there is tension between practices of custodianship and of exhibition: conservation of specimens in a repository—a "wealth" of materials in a "grand national cabinet"—and then their display: locally, among elite scientists, using microscopes and loupes; and more broadly, before public and professional audiences, for pedagogical purposes, in glass cases, in lecture halls, on the pages of professional journals and conference posters, and on the World Wide Web. For the public looking at tissues in the National Museum of Health and Medicine, there is the patriotic appreciation of battlefield sacrifice yoked to progress of national, and international, science. And there is also voyeuristic delight and morbid interest, bound up with aesthetics of violence: gleaming surgical instruments; fleshy surfaces and depths whose exposure exceeds what one sees everyday; specimens fixed by formalin and taxidermy, mementos mori.

The annual RSNA meeting also serves a range of polities and purposes. It is both tradeshow and scientific conference. Its attendees include vendors and purchasers, radiologists, techs and nurses, hospital administrators and consultants, residents and students, "other" physicians, software developers, military intelligence analysts, and more.[40] The technical and the scientific exhibitions are somewhat segregated but also interwoven: CT scanner exhibits are replete with academic clients and study findings; scholarly references and posters are funded by vendors of scanners and contrast.[41] There is no clean distinction to be drawn between the austerity of scientific method and the *flânerie* of the arcade—nor between the scientific fact and the commodity. In the trade show and in scientific sessions, various forms of embodiment—laminated femurs, holographic anatomies, tissue-equivalent phantoms, biopsy results—are exhibited alongside techniques and apparatus which produce them. This hybridity and eclecticism flaunt the Nature/Culture divide just as the *Wünderkammer* did before modernity's outset.

The RSNA and the AFIP's Rad-Path Department are collaborators in this techno-corporeal hybridity and eclecticism. Rad-path correlation itself is perhaps the most radically hybridizing practice, in cross-weaving specimens in glass or formalin with the increasingly complex productions of imaging technology. The RSNA helps fund the project of Rad-Path education; AFIP is a Technical *and* a Scientific Exhibitor at the RSNA conference; and the journal *Radiographics*, where Rad-Path papers "From Curating of the AFIP" are a regular feature, is "the only journal in the field of radiology dedicated to scientific exhibits, evolving technology, and continuing education"[42]—a uniquely hybrid mission.

The AFIP's Department of Radiologic Pathology serves a special role in "ground-truthing" medical images in the *materia* of dead, fixed tissue. This work goes on in hospitals around the world; but the AFIP does it more, and more systematically, than anywhere else. Moreover, it teaches both U.S. and many international radiologists the standards of rigor and principles of spatial and sectional registration that should govern this endeavor. In doing so, the AFIP Rad-Path Department reifies and extends classificatory systems which organize the variability of the world's diseases. It correlates disease taxa with imaging appearances by interlacing them with tables of morphological resemblance. And it aspires to comprehensiveness—an old archival fantasy.

One of the contributions of the RSNA conference is to incorporate the AFIP's archival and correlative ambitions—along with those of other academic and corporate (and military) partners—in an exhibitionary complex also without rival. Yet the RSNA's message of technological triumph and promise of a healthier (because better-imaged) future are not open to all.

At University Hospital, discussing the RSNA meeting, one of the CT techs (who has "never been to the RSNA") recounts with enthusiasm his recent visit to Disney's Epcot Center. He was impressed by GE's exhibit there of an MRI scanner with "the side of it cut away," and a mannequin inside, showing "what the images would look like on the monitor." He also was impressed by a GE jet engine with interior exposed. I am impressed that this tech, so matter-of-fact about the workings of his state-of-art scanner console, should be so responsive to the charms of Epcot.

Epcot is a prime heir to the mantle of the international Expo. It is Disney's vision of technological progress and corporate ingenuity—spun into utopian, triumphal scenes, for the entertainment and education of visitors and the inculcation of consumer desires. Of course medical imaging machines are salient among its exhibits. And of course the innards of the machines, like innards of simulated patients, should be exposed.

But is it odd that professionalization has not conferred on this Expo-going tech some immunity to the enchantments of Disney's MRI?

Perhaps the tech is not enchanted so much by GE's machines as by the contexts in which Epcot places them. Epcot embraces the ambience of the carnival. Perhaps the tech delights in seeing these machines delivered from the serious business of the hospital. At Epcot's ultrasound machine, his child "could come and take the transducer"—a privilege unclaimable in the hospital. No children attend the RSNA. (Busloads of children do, however, visit the AFIP's museum.)

Neither the RSNA meeting nor the AFIP Rad-Path Course is open to the public. Though the RSNA meeting is hugely attended, widely publicized, and occupies a substantial wedge of Chicago real estate for a week each year, it is restricted to healthcare professionals and exhibitors. Likewise, most of the AFIP, including the Rad-Path Course, is restricted to professionals with IDs.

The exclusivity of professional versions of the Expo is familiar enough. Scientific societies have long been exclusive. They were founded on aristocratic manners, on proper rituals and rhetorics for witnessing of experimental practices, on maintenance of calm spaces for scientific inquiry. With radiology, there is the additional precedent of the exclusivity of the viewbox: the "sacred" space of the reading room, protected by convention and occasionally a Do Not Disturb button. Some of the rationales for this exclusion of the laity by the priesthood of the viewbox—care for the patient's sensibility—might be applicable to RSNA scientific papers and the grotesqueries of AFIP rad-path correlation.

But there are other reasons for insulating the radiological Expo. Confidences at issue are not merely the patient's. There is the trade secret: engineering details of cutting-edge advantage, like slip-ring configurations in a new spiral scanner; or exaggerated representations of machinic utility, as by the sharing of one detector among five crippled scanners—trick of the confidence-man.

There are other aspects of professional confidence at stake across these domains of Expo. Proscription of photography at the RSNA Toshiba exhibit echoed a proscription I encountered at the outset of my fieldwork—that of videotaping in the University Hospital reading rooms. There was concern that videotape of specific fingers pointing at specific lesions on specific scans could amplify medicolegal exposure—could be subpoenaed by malpractice attorneys if ever a case wound up in court. Radiologists are accustomed to oversight—but not full and public exposure. Expo must be carefully structured so as to enhance, not threaten, their authority. They are willing to make their work public when they select the cases to be published. Quite apart from the issue of patient protection—the Hippocratic injunction to secret keeping—there is a sense in which the daily work of tomographic exposure requires a measure of insulation from prying eyes.

Thus there is a paradox at the heart of radiological Expo: the commitment to exposure, the ostentation and the comprehensive ambition of its exhibitionary complex, is wrapped around some carefully protected secrets. Where precisely, this year, is to be found the apparatus's cutting edge? And where precisely are the wobbles, the uncertainties, the errors, in radiological reading?

 **IMPRESSION**

For the more frequently we interrupt someone in the act of acting,
the more gestures result.
• • • **WALTER BENJAMIN,** "What Is Epic Theater?"

A historical scene that launched this book is the cadaver on the autopsy table—anchoring the diagnostic gaze. I wondered if the continued prestige of lesion hunting today, in the apparent absence of the cadaver, is paradoxical.

I asked how the cadaver could have become a lost object for contemporary medicine. Much lesion finding in today's hospital is organized not around scalpel-cuts of the pathologist but around virtual cuts of the scanner. CT scanning unfolds the primal scene of the cadaver on the autopsy table into two scenes: that of a patient's body on the scanner table (patiently emulating the cadaver); and that of docs at the viewbox (paging through a corpus of slices). The doc-at-viewbox scene became a stereotypical picture of doctoring: a logo of professional expertise, a fetish of healthcare consumerism.

But the CT Suite is a more intricated set of scenes than just these two. I have represented suites of practice distributed through a suite of rooms. Here and there I have juxtaposed tomographic practices with historical scenes that they quote, resemble, or otherwise conjure up: machinic cutting and the metalworker's lathe; film filing and the police prefecture; viewbox attention and the camera obscura, observation tower, and museum; teaching conference rhetorics and the courtroom witness stand. I have tried to suggest that the contemporary CT Suite remains in dialogue with—if not haunted by—the nineteenth century.

Perhaps it is best to reserve *haunting* for the continuing roles of the cadaver—that which the imaging revolution purports to have transcended. The cadaver is, in the CT Suite, not entirely lost. Rather, it has been turned inside-out and—like a purloined letter—concealed in plain view. Dead tissue remains in continuing dialogue with the diagnostic imaging enterprise—in the immobility of its patients, in CT biopsies, and especially in the fully registered cases at the AFIP. Yet death is also changed in this new suite: its presence and authority are redistributed, in time and space; its material traces are handled in longer and more intricate chains of custody; its visibility is coded in new ways.

A crucial aspect of the CT Suite's relation to death is its investment in, amplification of, *discontinuity.* I have flagged this in various ways: in my own representational genres; in my claim that CT cuts up the hospital just as much as it cuts up bodies;[1] in my attention to practices of juxtaposition and rhetorics of contiguity—including those which radiologists call "correlation"; in my consideration of the aesthetics of what Benjamin called shock.

It is not that the CT Suite is without flows. But the knitting of parts into flows, of segments into a suite, summons further practices. Let me review briefly some specific practices, some of them congealed in objects, that are important to translations of discontinuity into followings.

There is slicing, and the section. Slices cross between rooms and facilitate comparisons across media. The slice is a mode of representation: a conventional form, a genre. It has a history tied first to conjectural, then to empirical, projects in geology and anatomy. The slice has some experiential entailments: it presents certain perceptual opportunities ("quick and easy," "a lot of information") and challenges (for example, to reintegration into a volume), at the same time as it stands in contradistinction to other perceptual experiences of spatial positioning (for example, stereoscopics). The radiological slice is also a kind of imaginary, a boundary object with a particular kind of hybridity, in that it defines a plane of substitutions, a topos within and across which one can switch among contiguous representations, "correlations." It is the virtual slice that connects representations across imaging modalities. The slice is where image and tissue are registered, in rad-path correlations, at the AFIP and beyond: teaching conferences, pathological field-guide tables, Virtual Human software packages. Slices are materialized through practices of cutting—in CT, via mechanisms of spinning—and thus they inscribe a mechanical violence and extend an aesthetic of "posthumous shock." To attend carefully to the career of the slice—even the "virtual" slice—is to call into question some of the presumptions of the epithet "noninvasive" so often used to describe medical imaging.

Then there is casework, the case. The case in the film folder represents a congealing of viewbox conversations, correlations with nonradiological information. As a unit of diagnostic thinking, the case represents a particular patient (a sad case, a threatening case), but its cultic status trades further on its being an instance of disease (a beautiful case), insertable in a case series. The case as produced in the CT Suite, and reviewed on the reading room viewbox, does not privilege chronologies of patient experience, or the teleology of life-toward-death, as much as do the cases presented in the ritual of internal medicine's CPC. This reflects the radical estrangement of patients from reading room, the pressure of the docket, the complicated nature of viewbox correlations, the desire to ward off tissue procurement. However, the relative elisions of patient and cadaver have not diminished the tomographic case's affinity with puzzles, with mystery. The reading room viewbox remains a locus of intrigue. Yet this is not merely a matter of solving Nature's puzzle: the viewbox is a place where images are invested and divested rapidly of supporting information, where gazes see that others do not see: pedagogical interrogations (as in the hot seat) or exchanges of supplemental or contradictory information (as in consultation) also produce and amplify intrigue. Intrigue is sustained by, even as it knits up, technical and social forms of discontinuity.

Registration: another practice that recurs throughout the CT Suite. Registration of persons with hospital resources is mediated by clerks. Registration of body with imaging device and standard anatomic axes is managed by techs, with headrests and velcro straps. Various registrations of images on the viewbox are undertaken by radiologists—with old studies, with clinical history, sometimes with biopsy findings, with landmarks of anatomical designation. Images are registered by curators: for example, by Film Management in their circulation around the hospital ("if it moves, sign it out"), by the national archive of the AFIP when new cases are accessioned. Registration is a tool for translating among media as they circulate through institutions—body/image, body/text, image/image, image/text. Registrations are constitutive links in the making of collections, and in custodial chains of evidence. Registrations of professionals (the badge, the Expo card), are constitutive links in careers and institutional polities. Registrations are archival practices that are both banal and utterly crucial.

Perhaps the practice which best knits up discontinuities in the CT Suite is that of display, exhibition. Display in the CT Suite trades on some historical precedents: it quotes museal exhibition of natural historical specimens and objets d'art—as well as more generically bureaucratic exhibitions on computer screens and bulletin boards, in many CT workspaces. At the viewbox itself, the principal structures of display, grid and section, have utilitarian rationales but also a range of aesthetic entailments. And a wide variety of combinatory displays—like the double-barreled slideshow and the field-guide table—exemplify the flexibility of contiguity and resemblance in radiological exhibition. The grand exhibitionary complex of the Expo materializes the exorbitant potential that resides within radiological display, even among professionals.

• • •

One of the frames for situating my emphasis on viewbox practices, and their roles in *constituting* diagnostic evidence and expertise, is the prestige and power of imaging in contemporary biomedicine and for burgeoning global markets. The massive scale of the RSNA Expo exemplifies this power for professionals and marketers. The widespread pictorial genre of the doc-at-the-CT-viewbox exemplifies this for the public. I called this scene of doc-at-CT-viewbox a cultural "fetish." Some of its standard features—miniaturized detail available to close inspection; the luminosity and the systematicity of a grid and the regular seriation of its images—are a kind of pictorial formula for truth. The scene is not only invested with value—heavily cathected—but it is also differently invested by multiple parties. For the patient, the doc@box scene is where the plight of the individual meets a fantasy of divinatory insight and archival comprehension. For the institution, the viewbox is a locus of efficient throughput, a structure of disease management. For the doctor, the viewbox is a backdrop and prop in the performance of expertise and authority.

In Marx's analysis of the rise of factories and mass-produced goods, "fetishism" signaled the *unreality* of relations that commodities could develop with each other, and with those who desired and acquired them, in contradistinction to the real human labor relations that produced them.[2] Consumers of the doc@box scene may, in valuing its representation of expertise and evidential authority, overlook the suite of labors that underwrites them. But I do not wholly trust this realist epistemology. Fetishism is implicated throughout in the suite of labors I describe. Lesions in the scopic field of the CT suite are held to resemble one another, to mimic one another, to be tricky, to be elusive. These relations summon curiosities and anxieties of interpreters and desires of collectors. Such relations, of culprit diseases with each other and with their pursuers, are also fetishistic—if and to the extent that they are naturalized, seen as independent of practices that make them.

Back to the doc@box: "fetish" signals that the viewbox advertisement does not bear its superfluity of value and promise without tension, anxiety. Fetishes relate to bodies through routes other than direct intention.[3] The viewbox references carefully handled embodiments, but it remains a site from which patients are largely excluded, where knowledge of disease is expropriated. The viewbox is also a site where an image's diagnostic use value can be parlayed into, and perhaps eclipsed by, other "cultic" exchange values: advancement of research projects, professional careers, and health plan profits. For the professional, the viewbox scene stages not just sight but also oversight—thus poses anxieties about professional expertise (even, perhaps, its potential displacement by machinic sight, heralded already by various forms of computer assistance). For the institution, the repetition-compulsion materialized at the viewbox sometimes exceeds rational control: one study calls for another, and each study becomes more extensive (more images, finer slices). Moreover, the implied claim to institutional distinction which is staged in the scene of doc-at-the-viewbox has a dark underbelly: the fact of its maldistribution. Not every person can afford the scan she needs.

. . .

In describing casework and intrigue in the CT Suite, I have emphasized that intrigues inculcate modes of thought related to *judgment*. This is in part to offer an account of judgment embedded in social practice rather than in private cognition—especially in its teaching. I suggested that the CT Suite's economies of diagnostic intrigue—discursive and affective economies, not just monetary—are being changed in the shift to digital imaging: the transition from film to computer monitors, from mechanical viewbox to PACS workstations.

The crucial change is not one of native human faculties, or even of technology itself—but of the ritual settings, suites of practice, that follow upon technological change.

It is not that "intrigue" is an intrinsic good. Like curiosity, intrigue solicits both praise and blame. In early modernity, curiosity was "the mark of discontent, the sign of a pursuit." Moreover, curiosity attracted attention to itself: "Curious spectators became simultaneously subjects and objects of inquiry."[4]

Under the sign of intrigue in later modernity, this book has documented a related dynamic.

Copresence of radiologists and clinicians at the viewbox reproduces social conditions constitutive of late-modern intrigue since the invention of the detective story: divisions of status, viewpoint, and labor. These were conspicuous in all of Poe's detective tales, including the one Benjamin called an "x-ray picture" of detection, "The Man of the Crowd." To borrow Poe's formula from that tale, "the type and the genius of deep crime" (in the CT Suite: the culprit situation of the lesion) devolves not merely from "refusal to be read" but from "refusal to be alone."

In the efficient image distributions of PACS, which allow radiologists to work uninterrupted in the reading room and clinicians to access images elsewhere, the lesion may be a bit more lonely. Prospects for rad-clin intrigues, for cross-disciplinary performance of abductive reasoning and judgment, are fewer and farther between. No doubt lesions will continue to be tricky, to disguise themselves in crowded arrays of other culprit lesions. No doubt radiology residents will still emulate abductive craft of their mentors. But if a broader spectrum of human agents is implicated in culprits' "refusal to be alone"—if we take seriously Poe's and Benjamin's emphasis on cityscapes and human throngs in their sociologies of detection—it is not clear what follows in the more "suburban" and telemetric conditions of a hospital of PACS.

• • •

In retrospect, this book is a case study concerning case study. It has described a suite of practices, a suite of persons, and a suite of furniture, distributed throughout a suite of rooms, in an academic hospital of the 1990s. Through ethnographic fieldwork, occasional historical comparisons, and a montage of representational genres—quotation, description, narration, analysis—I have endeavored to complicate and particularize notions of what has gone on at, and around the corner from, the CT viewbox. Features of contemporary medical triumphalism—diagnostic truth, evidential rigor, professional expertise, archival comprehension—do not need to be fetishized: they can be engaged as crafted products in particular historical circumstances.

NOTES

Introduction

1 CT suite—or CAT (computer-assisted tomography) suite—is a common term in large English-speaking hospitals around the world. In French, "suite" is an everyday, nontechnical word, with no special association with either *tomodensitométrie* (TDM) or *scanographie*.

2 Foucault, *The Birth of the Clinic*, 146, quoting Xavier Bichat. The delayed rediscovery of Morgagni by the Paris Clinic is discussed at 125–35. Foucault is not the only historian to emphasize the revolutionary status of pathoanatomy and clinical-pathological correlation. See Maulitz, *Morbid Appearances*, for accounts of movements of French practices across national lines; Ackerknecht, *Medicine at the Paris Hospital*.

3 Foucault's formulation. More precisely, tracings of a disease in flesh are its "secondary" spatialization—distinguished by Foucault from "primary" spatialization, position in a classificatory table, and tertiary spatialization, summation of "all the gestures by which, in a given society, a disease is circumscribed, medically invested, isolated . . . or distributed." *The Birth of the Clinic*, 16.

In calling cadaver-medicine a "platform," I engage, if glancingly, Keating and Cambrosio's fine analyses of "Biomedical Platforms." They combine social, technical, and administrative connotations of this architectural term more comprehensively than I attempt with "suite." (They would point out that pathology, work with fixed tissue, is specifically marginal to the *functional* concerns of *biomedical* platforms.) But I reject their concern about "reductions" in addressing techniques, tools, and other materials under an "increasingly vague, all-encompassing notion of practices" (341). I endorse the "mangle" of practice (Pickering, "The Mangle of Practice") even if this also requires specifying forms and categories.

4 The decline of postmortem autopsy is not viewed by all commentators as salutary. Nor is it new. Emerging disciplines had vexed relations with necropsy—e.g., hematology, circa 1916, according to Wailoo, *Drawing Blood*, 86–88. Autopsy's decline became acute after the 1940s: rates dropped from 50 percent of hospital deaths to 7–14 percent in the 1990s. In 1970, the U.S. Joint Committee on Accreditation of Healthcare Organizations withdrew its requirement that hospitals maintain postmortem autopsy rates above 20 percent. See King and Meehan, "A History of Autopsy." Studies continue to assert the relevance of autopsies—for example, Burton, Troxclair, and Newman, "Autopsy Diagnoses of Malignant Neoplasms," where 44 percent of autopsies revealed undiagnosed or misdiagnosed cancers, most among causes of death.

5 Evidence is a risen star in medical practice, research, and education today. But its usages are overwhelmingly positivist—even among some critics of Evidence-Based Medicine (EBM). That is, much is presumed about the robustness, timelessness, and countability of evidence, while historical specificities of procurement, custodianship, circulation, and framing of facts are omitted or undervalued. Critiques of received notions of quantitative evidence include Goldenberg, "On Evidence and Evidence-Based Medicine"; Gordon, "Clinical Science and Clinical Expertise," in Lock and Gordon, eds., *Biomedicine Examined*; Porter, *Trust in Numbers*; Timmermans and Berg, *The Gold Standard*. On images' status in hierarchies of evidence, see Stafford, *Good Looking*, 21–40; Beaulieu, "Images Are Not the (Only) Truth," 61–63.

6 In this respect my agendas mesh nicely with those of Mol in *The Body Multiple*, concerned with discontinuities and "distributions."

7 Benjamin, "The Work of Art in the Age of Mechanical Reproduction," in *Illuminations*. See also "Theses on the Philosophy of History." Benjamin's understanding of montage drew on visual displays in shopping arcades, on Dadaism, on methods of film construction, on the riveted assemblage of the Eiffel Tower—and on a literary device to which he dedicated himself, juxtaposed quotations. Translators of *The Arcades Project* refer to montage's "philosophic play of distances, transitions, and intersections, its perpetually shifting contexts and ironic juxtapositions" ("Translators' Introduction," xi). See Buck-Morss, *The Dialectics of Seeing*, esp. 72–77.

8 My transcripts are primarily lexical, with some pauses and gaps indicated; I have not supplied supplemental markings of pace, emphasis, or other features that are sometimes offered in meticulous line-by-line discourse analyses—as in, for example, Morris and Chenail, *The Talk of the Clinic*, or Goodwin, "Professional Vision." Some transcripts have been spliced for continuity (with gaps marked with ellipses). Longer, more inclusive transcripts of selected passages, along with audio recordings, are reproduced at http://www.ibiblio.org/ctsuite/ctsuite.html. Conventions of capitalization in this book mostly follow hospital usage: titles (of departments, conferences, teams) are capitalized, but generic roles or concepts (resident, surgeon, rad-path) are not, except where a specific usage warrants notice or emphasis.

9 Salient exceptions: Reiner et al., "Impact of Filmless Radiology on Frequency of Clinician Consultations with Radiologists," and commentary by Baker, "PACS and Radiology Practice."

10 Film is not yet obsolete. Many hospitals' transitions to digital imaging have not occurred at the pace of those of academic and metropolitan centers. But trends are clear.

11 Simmel, "The Stranger." Simmel links social interactions of the stranger with those of the "*potential* wanderer," with the condition of having "not quite overcome the freedom of coming and going" (402). I was an intermittent guest in the Radiology Department for a year.

12 See Rosenberg, *The Care of Strangers*; Tyson, "A Definition of Religious Studies in the Context of Health Professional Education." One introduction to hospital ritual studies which notes religious origins of the hospital is Fox, "The Hospital: A Social and Cultural Microcosm," in *The Sociology of Medicine*, 142–80. Many works on hospital rituals have addressed activity in the operating room, e.g., Felker, "Ideology and Order in the Operating Room"; Goffman, "Role Distance." Another classic, also on surgical rituals but extending to settings beyond the operating room, is Bosk, *Forgive and Remember*. One study addressing CT scanning through the lens of ritual is Barley, "The Social Construction of a Machine."

13 See Star, "Epilogue: Work and Practice in Social Studies of Science, Medicine, and Technology."

14 Beliefs/intentions are often complicit with dualisms that hold mental contents apart from embodied action; further, they often presume the (unhistoricized) individual to be the privileged locus of agency. While this book does not adhere rigorously to symmetries and "irreductions" (Latour, *The Pasteurization of France*) regarding nonhuman agency, it aspires to self-consciousness about its humanist residues.

15 Bourdieu, "The Specificity of the Scientific Field and the Social Conditions of the Progress of Reason."

16 This last, somewhat medicalized term can be problematic: medicine's embrace of humanities to solve crises of "dehumanization" has at times entailed warming and fuzzing of humanities pedagogy, as well as reification/universalization of notions of the "human." Medical humanities include important historical, literary, and philosophical studies (among others) of medicine. Classic studies particularly relevant to this book include Canguilhem, *On the Normal and the Pathological*; Fleck, *Genesis and Development of a Scientific Fact*; and Howell, *Technology in the Hospital*. For examples of medical anthropology texts, see Lock and Gordon, eds., *Biomedicine Examined*; Comaroff, "Medicine: Symbol and Ideology"; Farquhar, *Knowing Practice*; Hahn and Gaines, eds., *Physicians of Western Medicine*; Rapp, "Accounting for Amniocentesis"; and Young, "The Creation of Medical Knowledge" and *The Harmony of Illusions*. For some examples of STS resources, see Biagioli, ed., *The Science Studies Reader*; Brodwin, ed., *Biotechnology and Culture*; Knorr-Cetina and Mulkay, eds., *Science Observed*; Latour, *Science in Action*; Rudwick, *The Great Devonian Controversy*; Pickering, ed., *Science as Practice and Culture*; Shapin and Schaffer, *Leviathan and the*

Air-Pump; Traweek, *Beamtimes and Lifetimes* and "Discovering Machines: Nature in the Age of Its Mechanical Reproduction"; and Woolgar, *Science: The Very Idea.*

17　On visuality and representation, see Crary, *Techniques of the Observer*; Debord, *The Society of the Spectacle*; Elkins, *The Domain of Images*; Friedberg, *Window Shopping*; Jay, *Downcast Eyes*; Karp and Lavine, eds., *Exhibiting Cultures*; Krauss, *The Optical Unconscious*; Mirzoeff, ed., *The Visual Culture Reader*; Mitchell, *Colonising Egypt*; Mitchell, *Iconology*; Sturken and Cartwright, *Practices of Looking*; Taylor and Saarinen, *Imagologies*; and Tagg, *The Burden of Representation.* On scientific representation per se: Lynch and Woolgar, eds., *Representation in Scientific Practice*; and Lynch, "Discipline and the Material Form of Images." On embodiment, see Barker, *The Tremulous Private Body*; Crary and Kwinter, eds., *Incorporations*; Elkins, *Pictures of the Body*; and Gallop, *Thinking through the Body.* On performance, see Mol, *The Body Multiple*; Pollock, *Telling Bodies Performing Birth.* On narrative, see Foucault, *Death and the Labyrinth*; and Hunter, *Doctors' Stories.* Important materialist/constructivist studies (historical and ethnological) of relations among biomedical technology, specialized experts, and biomedical objects like diseases include Berg, *Rationalizing Medical Work*; Berg and Mol, eds., *Differences in Medicine*; Latour, *The Pasteurization of France*; Latour and Woolgar, *Laboratory Life*; Rosenberg and Golden, eds., *Framing Disease*; Sandelowski, *Devices and Desires*; Wailoo, *Drawing Blood*; Wright and Treacher, eds., *The Problem of Medical Knowledge.* I have particularly benefited from works on medical imaging technologies, including Cartwright, *Screening the Body*; Barley, "The Professional, the Semi-Professional, and the Machines"; Dumit, *Picturing Personhood* and *Minding Images*; van Dijck, *The Transparent Body.*

18　See Stafford, *Body Criticism*, among other works. See Reiser, *Medicine and the Reign of Technology*—though I attempt to qualify some of this book's humanist presumptions and notions of technological progress. See Montgomery's *How Doctors Think*; though too recent to have shaped this book, it is the best available description of tacit knowledge and judgment in the broader ecology of modes of clinical knowing.

19　Benjamin, "Theses on the Philosophy of History," in *Illuminations*, 255.

20　Bourdieu, *The Logic of Practice*; Farquhar, *Appetites*; Lave, "The Practice of Learning"; Pickering, "The Mangle of Practice."

21　*Habitus* is a key term of Bourdieu's—a term whose genealogy comprises usages of Weber, Elias, Mauss, others. Bourdieu names affine terms ("from Hegel's *ethos* to Husserl's *Habitualität* to Mauss's *hexis*") in Wacquant, "Towards a Reflexive Sociology," 43. Bourdieu's work is crucial to my notions of practice. On habit, skill, and the tacit, see also Polanyi, *Personal Knowledge*, esp. chaps. 4–5; Schaffer, "Experimenters' Techniques, Dyers' Hands, and the Electric Planetarium"; Grasseni, "Introduction" to *Skilled Visions.* On mimesis, see among others Taussig, "The Golden Bough."

22　In invoking aesthetics I do not address myself to beauty as such, or to Romantic problematics of genius and symbol.

23　Aries, *Western Attitudes toward Death*; Benjamin, "The Storyteller," in *Illuminations*; Porter, "Life Insurance, Medical Testing, and the Management of Mortality"; Agamben, "Politicizing Death," in *Homo Sacer*; Landecker, "Immortality, In Vitro."

24 Similarly, on functional imaging and changing concepts of life, see Cohn, "Increasing Resolution, Intensifying Ambiguity."

25 Poe's detective tales preceded and influenced those of Arthur Conan Doyle, the standard patron of literature-and-medicine studies of casework.

26 Latour, *We Have Never Been Modern.*

27 Dodier, "Expert Medical Decisions in Occupational Medicine"; Engelhardt, "Clinical Judgment"; Feinstein, *Clinical Judgment*; Gordon, "Clinical Science and Clinical Expertise"; Hunter, *Doctors' Stories*, esp. chap. 2.

1. Reading and Writing

1 The Hippocratic Oath includes the promise (before Apollo, Asclepius, and other gods and goddesses) not to divulge "holy secrets."

2 Privacy in viewbox activity is more about protection of the princely collection than protection of naked subjectivity. Images on the viewbox are objectified. The distinction is exemplified in this cautionary tale with many versions: Radiologist: "I'd sure hate to be the guy with that in his head"; Unannounced visitor in the reading room: "Excuse me, sir. I'm the guy 'with that in his head.'"

3 Nicolson, "The Art of Diagnosis," 801–25.

4 See chapter 2, "Cutting," on "black-boxing" of patient-machinic distance in CT.

5 Cf. Polanyi, *Personal Knowledge.* On history of practices and pedagogies of physical examination in relation to new devices, see Reiser, "The Science of Diagnosis"; and Reiser, *Medicine and the Reign of Technology*, 1–44.

6 Foucault discusses the ritual character of medical examination, and its transformation by successive regulations since the late seventeenth century, in *Discipline and Punish*, 184–94.

7 In reading-room conversations, "see" and "know" outnumber other action verbs. Next is "think."

8 I do not engage directly in this book with important and expanding scholarship on instrument recordings of physiological processes. See Cartwright, *Screening the Body*; Borck, "Writing Brains."

9 Parsons, *The Social System*, 428–79.

10 On the coroner's inquest, see Burney, *Bodies of Evidence.* Direct viewing of the body of the deceased was, in the early nineteenth century, a responsibility of both the coroner and a lay jury. "The view" served populist traditions as well as official powers. Around the turn of the century, rising prestige of medical experts, coupled with lay juries' senses of incompetence or offended decency, led to diminished expectations for semipublic viewing—"disengaging the body from its lay interrogators" (98).

11 The golden age of autopsis, according to Anthony Pagden, was the sixteenth through the eighteenth century, when contacts with the New World first pitted the veracity of travelers/eyewitnesses against the textual authority of the scholastics—and before the full materialization of the (Baconian) empiricist/objectivist project in, for

instance, the work and writings of Alexander Humboldt. Pagden, "The Autoptic Imagination."

12 This exchange pertains to an MRI scan, not a CT. Because neuro MRIs are read on the same viewboxes as neuro CTs, and because CT and MRI are both "sectional" modalities, I occasionally include such MRI references in this book.

13 Merleau-Ponty, "The Intertwining—The Chiasm," in *The Visible and the Invisible*, 130–55.

14 Older CTs were sometimes "set up" in a tidier array (fewer slices per film) in order to make available detail clear to the reader. But nowadays many slices can be displayed on one film—e.g., in a high-resolution CT of a relatively small structure (a cervical vertebra).

15 Amann and Knorr-Cetina discuss pointing at images as a discursive prosthesis, serving "optical induction," in "Thinking through Talk."

16 On radiological and military discourses of targeting, see Lerner, "Fighting the War on Breast Cancer"—in part echoing Sontag, "AIDS and Its Metaphors." See also Cartwright's discussion of DOD funding of U.S. breast cancer research in "Women, X-rays, and the Public Culture of Prophylactic Imaging."

17 "Saccade" can refer to any gesture (e.g., violin bowing) but it often describes minute ratcheting movement of eyes tracking an object.

18 Krauss, *The Optical Unconscious*, 2. Krauss also addresses this modernist stare as a "withdrawal from purpose"—at home in the art museum—thus perhaps less relevant to an instrumentally determined setting like the reading room.

19 See Geertz's discussion of artifice, parody, and conspiracy among schoolboys, "Thick Description," *The Interpretation of Cultures*, 6–7. The difference between "thin description" ("rapidly contracting his right eyelids") and "thick description" ("practicing a burlesque of a friend faking a wink to deceive an innocent into thinking a conspiracy is in motion") is important to considerations of intrigue (chapter 3, "Diagnosing") and ethnographic representation. Yet I do not engage eyelid gestures at the viewbox. And I expect there were plenty of winks among radiologists at my expense.

20 Beard et al., "A Study of Radiologists Viewing Multiple Computed Tomography Examinations Using an Eyetracking Device."

21 See chapter 4, "Curating," on Taylorist "scientific management" and hospital efficiency.

22 See, e.g., Barrett et al., "Unobtrusive Evaluation of Mammographer's Eye Movements during Diagnosis of Mammograms." This study used video to record eye movements ("gaze dwells") and pupillary dilation. The authors note that prompted re-viewings of some lesions may actually increase false positive diagnoses. Another instance of this genre of research is Kundel and La Follette, "Visual Search Patterns and Experience with Radiological Images." The authors compare "scan patterns" and "fixation patterns" of viewers at different levels of training, and conclude that scan habits develop in students' clerkships, prior to residency. An example of trade-journal reporting of similar research is Nodine, "Eye Tracks Lead Sleuths to 'Invisible' Lesions." For a current research program on radiological perception, see

Krupinski, "Medical Image Perception and Performance," http://www.radiology .arizona.edu/krupinski/eye-mo/index.html (accessed 2006).

23 For a general overview of this topic as a problem of aesthetics, see Gandelman, "The 'Scanning' of Pictures."

24 Benjamin, "The Work of Art in the Age of Mechanical Reproduction," in *Illuminations*, 236–37. Benjamin's usage partially underwrites the title of Krauss's *The Optical Unconscious*. His earliest use of this term (1931) was in "A Small History of Photography."

25 Benjamin, "The Work of Art in the Age of Mechanical Reproduction," 248–49.

26 Cartwright and Goldfarb, "Radiography, Cinematography and the Decline of the Lens."

27 Descartes diagrammed the cow's eye in *La dioptrique*, first appendix to his *Discourse on Method* (1637). His privileging of visuality, and its relationship to reflexive self-certainty, are discussed by Jay in *Downcast Eyes*, 21–82. On relations of Descartes's philosophy to dissective explorations, see Sawday, *The Body Emblazoned*, 146–58.

28 Jay, *Downcast Eyes*, emphasizes (52ff.) the "de-narrativizing" function of realist codes. His account of perspectivism is developed from 50–62. Alberti's 1435 treatise *Della Pittura* is usually referenced as the basis of Renaissance perspectivism. Perspectivism, with its reciprocal geometries of viewpoint and vanishing point, is properly distinguished from the "retinalism" of the camera obscura. See Hughes, "Coming into Sight."

29 Jay, *Downcast Eyes*, 56 n115. This does not mean viewers cannot glance or scan— but that perspectivist representational codes privilege one mode of viewing. On "the glance," see Foucault, *The Birth of the Clinic*, "Seeing and Knowing": Clinical-pathological correlations no longer favor the "gaze" that "records and totalizes." "The glance, on the other hand, does not scan a field: it strikes at one point, which is central or decisive; the gaze is endlessly modulated, the glance goes straight to its object . . . this is no longer the ear straining to catch a language, but the index finger palpating the depths" (121–22).

30 Mitchell, "Orientalism and the Exhibitionary Order," 290. One example of complicities of mastery and the viewpoint: Jeremy Bentham, erstwhile architect of the Panopticon, frustrated by the tangled streets of Cairo, wrote of "taking a peep at the town from a thing they call a *minaret*" (306). Mitchell's work critiques not simply imperial exhibitionism but associated epistemic assumptions, ingredient by now to commonsense representation: "It is not the artificiality of the world-as-exhibition that should concern us, but the contrasting effect of a lost reality to which such supposed artificiality gives rise. This reality, which we take to be something obvious and natural, is in fact something novel and unusual. It appears as a place completely external to the exhibition: that is, a pristine realm existing prior to all representation, which means prior to all intervention by the self, to all construction, mixing, or intermediation, to all the forms of imitation, displacement, and difference that give rise to meaning" (301). Mitchell's *Colonising Egypt* develops this further. See Heidegger on the modern *Weltbild*, grasping of the world *as* picture, in "The Age

of the World Picture." Also, on empire and its codes of representation, see Pratt, *Imperial Eyes*. She comments, in regard to writers like Burton and Humboldt, on representational genres of "monarch-of-all-I-survey" (200).

31 See discussion of history and subject-positions of the camera obscura by Crary, *Techniques of the Observer*, 25–66. His account somewhat overstates insular and static aspects of camerae obscurae—some of which admit multiple observers, moving apertures (Carol Mavor and elin o'Hara slavick, personal communications and demonstration).

32 Jay, *Downcast Eyes*, 44.

33 Ibid., quoting John Phillips (*The Reformation of Images*), 45. On Reformation iconoclasm, see also Besançon, *The Forbidden Image*, 185–90; Miles, *Image as Insight*, 95–126.

34 Critiques of perspectivist realism include Goodman, *Languages of Art*, 10–19; Wartofsky, "Rules and Representation."

35 Berger, *Ways of Seeing*, 109.

36 On fallacies of photographic realism, see Snyder and Allen, "Photography, Vision, and Representation."

37 The centrality of southern European perspectivism to early modern realism and scientific worldviews has been questioned by Alpers, who points to different, still highly rationalized, northern painterly traditions of representation. She discusses the Dutch "mapping impulse" and depiction of a world of flat surfaces—a kind of "retinal" perception. Alpers, *The Art of Describing*.

38 Relations between Cubists and radiography are discussed in Henderson, "X-Rays and the Quest for Invisible Reality in the Art of Kupka, Duchamp, and the Cubists"—though more concerning artistic treatments of "transparency" and "fluidity" than aesthetics of multiplicity and juxtaposition.

39 Photography with light-sensitive emulsions had been in the public domain for over half a century. Technical developments—flexible film, small cameras—had provided "untrained masses the means to picture themselves" (Tagg, *The Burden of Representation*, 60, 66). X-rays were received as one more extension of photographic technique. According to Kevles, in 1896, Parisian department stores displayed an x-ray machine alternately with Lumière's new motion picture apparatus (*Naked to the Bone*, 25); Alfred Russel Wallace's 1901 retrospective *The Wonderful Century* characterized x-rays as a photographic trick (which Röntgen resented).

40 In Röntgen's lab, use of x-rays on the human body was secondary; the impermeability of bones to the beam made the body a convenient substrate for investigating the x-ray—not the converse. See Kevles, *Naked to the Bone*, 22 and 33ff.

41 The history of radiology is an enormous topic, even prior to its exuberant development around the Röntgen centennial (1995). This book engages only limited portions: tomography (chapter 2, "Cutting"); recordkeeping (chapter 4, "Curating"); courtroom settings and rise of a specialty (chapter 5, "Testifying and Teaching"). For more comprehensive histories of radiology, see especially Kevles, *Naked to the Bone*; Eisenberg, *Radiology*; Brecher and Brecher, *The Rays*.

42 For historical discussion of technological influences on perception, especially the place of x-rays, see Lerner, "The Perils of 'X-Ray Vision.'"

43 Nor is this problem specific to "experts": it also belongs to common sense. Enlightenment writers addressed education of the senses, relations of sight to language—crystallized in the famous problem of Molyneux: the blind man suddenly awakening to sightedness, and what he could see in a dawning of naive vision.

44 See Schaffner, ed., *Logic of Discovery and Diagnosis in Medicine*. For a less cognitivist approach, cf. Brown, "Logics of Discovery as Narratives of Conversion." One prototypical tale of discovery is that of Röntgen and the x-ray. An international organization, the Radiology Centennial, conducted commemorative events throughout 1995. One centennial essay is Evens, "Röntgen Retrospective."

45 Formerly this final section was called "Interpretation." These two terms have, in informal radiological parlance, very similar valences.

46 Encounter with the original is discussed in chapter 4, "Curating," in the context of museological exhibition and the cult value of images. See chapter 5, "Testifying and Teaching," for discussion of film review as *staging* the encounter with the original.

47 See, e.g., Berbaum et al., "Role of Faulty Visual Search in the Satisfaction of Search Effect in Chest Radiography."

48 One exception is "scatter" artifact from very dense objects—e.g., "metallic artifact."

49 See Traweek, "Discovering Machines" and *Beamtimes and Lifetimes*, on key roles of particle detectors—"beasts in the research yard"—in organizing research and structuring knowledge in physics, and how these roles are systematically elided in production of scientific facts.

50 Squire, *Fundamentals of Radiology*, 21.

51 The common term "normal limits"—as in "Adrenal pancreas spleen kidneys within normal limits"—signals a quasi-statistical mode of perception, edges of normality defined by cutoffs at the tails of the bell-curve. An indispensable source for philosophical consideration of the "normal" in biomedical registers is Canguilhem's *On the Normal and the Pathological*. Canguilhem notes derivation from Latin *norma*, a carpenter's square. In his account, "normal" is dissociated from modern connotations of a usual, natural, and ontologically prior state, and resituated as an artifice of the tool-user and a condition extrapolated from—derivative of—the abnormal. Evidence for derivative status of the normal includes the historical priority of pathology over physiology. Social histories of statistical normality include Hacking, *The Taming of Chance*, and Sekula, "The Body and the Archive."

52 This observation is a shop-floor commonplace, confirmed by the Radiology administrator who supervises transcriptionists. "Minutes drop with training."

53 See Law and Lynch, "Lists, Field Guides, and the Descriptive Organization of Seeing": "'Perception' is list-driven" (270).

54 Bonitzer, "Hitchcockian Suspense."

55 Barthes, *Camera Lucida*, 49.

56 Ibid., 55 (emphasis in original).

57 Benjamin, *Charles Baudelaire*, 188–92.

58 Lacan, "Of the Gaze as *Objet Petit a*." For a nuanced review of Lacanian (among other theories of) visuality, see Krauss, *The Optical Unconscious*. On "hauntings" of radiographic images in popular sensibilities, see Grove, "Röntgen's Ghosts." Simon Cohn addresses agency of brain images in "Seeing and Drawing."

59 Krauss, "Grids."

60 Krauss's redux of this "etiological" situation of the grid in twentieth-century art: "Behind every twentieth-century grid there lies—like a trauma that must be repressed—a symbolist window parading in the guise of a treatise on optics." "Grids," 17.

61 The history of anatomic representation in the West—as distinct from *patho*anatomy—is an extensive topic. See Sawday, *The Body Emblazoned*; Stafford, "Dissecting," in *Body Criticism*, 47–129; Crawford, "Imaging the Human Body"; Bruno, "Spectatorial Embodiments." Roberts and Tomlinson's *The Fabric of the Body* is a great-man history of anatomic illustrations judged by present standards of accuracy.

62 Rudwick, "The Emergence of a Visual Language for Geological Science," 164. The following discussion of geological sectional representation is distilled from this and his fine case study, *The Great Devonian Controversy*.

63 Ibid., citing example of John Strachey (1719, 1725), *Sections of an English Coalfield*.

64 Rudwick, "The Emergence of a Visual Language for Geological Science," 167. An important French mineralogical survey in 1780, "Atlas . . . minéralogique," was notable because it did *not* include sections depicting distributions of resources.

65 Ibid., 167.

66 Ibid., 169.

67 Ibid., 170. Cuvier and Brongniart's conventions were adopted by the Geological Society of London.

68 Rudwick, "The Emergence of a Visual Language for Geological Science," 179–80.

69 Stafford, *Body Criticism*, 47ff.

70 For brief historical overview, see Eycleshymer and Schoemaker, *A Cross-Section Anatomy*, ix-xii. Examples of eighteenth-century sectional images are works by Soemmering and Vicq d'Azyr on the brain, and by Camper and Scarpa on the pelvis.

71 Hopwood, "'Giving Body' to Embryos." An extensive history of microtomy is Bracegirdle, *A History of Microtechnique*.

72 Woodward, a pioneer in photography of histological sections, was the microscopical curator of the Army Medical Museum, forerunner of the Armed Forces Institute of Pathology (addressed at length in chapter 6, "Exposition"). The image seen in figure 12 and others like it were displayed poster-sized at the Philadelphia Exposition in 1876 and again at the Columbian Exposition in 1893 (see figures 35a and 35b).

73 Eycleshymer and Schoemaker, *A Cross-Section Anatomy*, x.

74 Ibid., xi.

75 Sections from the late nineteenth century in the collection of the National Museum of Health and Medicine include German plaster sections of an abdomen (4–5 cm thick); papier mâché models of sectioned limbs (Auzoux); and serial arrays of 1 mm longitudinal bone sections (Hodge)—the latter exhibited at the Columbian Expo.

76 Gal and Cagle, "The 100-Year Anniversary of the Description of the Frozen Section Procedure."

77 This passage and the one following are from interviews; investigator's brief interjections removed.

78 For cultural history of the stereoscope, see Crary, *Techniques of the Observer*, 116–36. Crary suggests the stereoscope signaled a limit to the unitary "point of view" of earlier display techniques (especially the camera obscura) and enrolled the physiology of the viewer in more explicit and self-conscious ways. For brief consideration of Crary's work with respect to radiological perception, see Lerner, "The Perils of 'X-ray Vision,'" 385–91. An early review of stereoscopy, by Oliver Wendell Holmes, physician, archivist, and inventor of a popular stereoscopic apparatus, is "On the Stereograph and the Stereoscope." For a radiological perspective, see Keats, "Origins of Stereoscopy in Diagnostic Roentgenology." Stereoscopy is only briefly discussed in an otherwise extensive treatment of neuro-optical and cognitive substrates of radiological vision: Jaffe, "Medical Imaging, Vision, and Visual Psychophysics."

79 Stereoscopic devices are discussed in Webb, *From the Watching of Shadows*, esp. 103–7. See comments on the Watson apparatus in chapter 2, "Cutting."

80 For example, at the 1996 RSNA Convention, displays of VOXEL technology (now Holorad). Skolnick, "New Holographic Process Provides Noninvasive, 3-D Anatomic Views."

81 Radiologists discuss aesthetics of medical imaging, but not so intensely as, for instance, astronomers interviewed by Lynch and Edgerton for "Aesthetics and Digital Image Processing."

82 Interjections of investigator removed.

83 Interjections of investigator removed.

84 The following dialogue is reported from adapted fieldnotes (not audiotape); some dialogue is quoted verbatim, with ellipses indicating unrecorded portions.

85 This term invokes the craft of the lens-grinder—*prior* to polishing: a surface unfit for optical transmission. A brief Medline history: before it became a radiological term, "ground-glass opacity" described a light-microscopic appearance of pathological liver cells (1974). Its use in radiology dates from 1977, describing x-ray appearances of diffuse lung processes. In 1982 the term entered the CT lexicon, describing chest x-ray findings in pneumocystis pneumonia—just prior to the naming of AIDS. Chest CT now accounts for 85 percent of the term's usage.

86 Though I speak of Body CT fellow and sometimes Neuro CT fellow, officially these fellowships are in Body Imaging and Neuroimaging. "Imaging" comprises other diagnostic modalities besides CT. The Body Imaging fellow reads Body CT, MR, and ultrasound; the Neuroimaging fellow reads CT and MR of head, face, and spine. So even "Neuro" is sometimes a misnomer: the Neuroimaging team reads most studies above the neck, and spine studies below the neck.

87 On medical professional as shopkeeper, see Goffman, "The Medical Model and Mental Hospitalization: Some Notes on the Vicissitudes of the Tinkering Trades," in *Asylums*.

88   See also the discussion of interdisciplinarity in chapter 5, "Testifying and Teaching."

89   Radiologists have long been concerned about subordinate roles in diagnostic work—procurers of images rather than full partners in the diagnostic enterprise. See "Consultants or Pairs of Hands?" Radiologists express concern that clinical specialists may not interpret images properly or comprehensively. Though some extraradiological imaging enterprises are well established (e.g., obstetricians and prenatal ultrasound; cardiologists and coronary angiograms), others are more contested (e.g., general physicians reading chest x-rays).

      All CT scans at University Hospital, indeed at most hospitals, are interpreted formally by radiologists. This does not obviate contesting of expertise. For instance, many neurologists insist, especially in training contexts, on the need for their own direct review of images.

90   Specialization accompanied the rise of scientific medicine—especially in Germany, where departmentalization of intellectual labor in universities was acute, and in France. See Weisz, "The Emergence of Medical Specialization in the Nineteenth Century." A briefer overview is in Rosenberg, *The Care of Strangers*, 169–75. He notes that specialization in America began in outpatient clinics and was imported into hospitals in the early twentieth century, with the exception of specialties founded on hospital facilities. Also see Starr, *The Social Transformation of American Medicine*, 355–59. He discusses divisions of labor in hospitals at 220–25 (e.g., relations between pathologists and surgeons). See chapter 5, "Testifying and Teaching," on the rise of radiology as a specialty. For a view of specialization suggesting that negotiability of commodified products may be more important than complexity of intellectual tasks, see Gritzer, "Occupational Specialization in Medicine."

91   These are referred to as "auxiliary specialists" by Burling et al. in *The Give and Take in Hospitals*, 81. Relations between radiologists and hospital technology pose special problems for a university department. Radiology is stretched administratively and economically across two institutions, hospital and university—comparable in this respect with Anesthesiology, Pathology, and perhaps Pharmacy.

92   On interprofessional performances, and specialists' knowledge of each others' work, see Freidson, *Profession of Medicine:* "the visibility of performance granted by the division of labor is not holistic but fragmentary" (146).

93   The historical situation of University Hospital in the late 1990s—on the cusp of the film-to-digital transition—with screen images still playing mostly supplemental roles as yet—is discussed further in chapter 4, "Curating."

94   There are variations on the film handoff. I occasionally ferried films as participant-observer. Attendings sometimes went to the scanner room to pick up newly printed films from techs—when there were no compelling reading-room tasks, when they were rushed, or when there were other reasons to visit the scanner.

95   See chapter 2, "Cutting," and chapter 4, "Curating," for more on MagicView and PACS.

96   The magic of the MagicView is not about clever substitutions or sham causalities—notwithstanding radiologists' chuckles at animistic innuendo. See chapter 4,

"Curating" on magic and the function of contiguity. On magical thinking around the scanner apparatus itself in the 1980s, see Barley, "The Social Construction of a Machine."

97 On distributed cognition, see Hutchins, *Cognition in the Wild*.

98 Visits by radiologists to the scanner control room were much less frequent in 1996–97 than they were in Barley's observations of CT scanning in the 1980s. See "The Professional, the Semi-Professional, and the Machines," esp. 273–87.

99 Hanging films, though still a feature of radiological work at many community hospitals, is now a bygone art at University Hospital. The PACS system has, in a few short years, made film handling obsolete.

100 "Technique" in the lexicon of plain-film radiography comprises tech decisions regarding beam-energy, time, collimation/filters, etc. On development of protocols in early CT installations in the 1980s, see Barley, "The Professional, the Semi-Professional, and the Machine," 274–83.

101 For examples of CT protocols, see Castillo, *Neuroradiology Companion*, 3–7. For a nuanced discussion of protocols in general as negotiated and transformative instruments, see Berg, "Order(s) and Disorder(s): Of Protocols and Medical Practices."

102 This diagnosis is rare enough that even senior team members have never seen a case. On "indication criteria" as tools of distributing embodiments with interventions, see Mol, *The Body Multiple*.

103 Standardization of radiographic reporting was advocated around 1900, and systematically elaborated in 1922 by Preston Hickey (who introduced the term "interpretation" to describe the work of radiologists). He advocated comparability with pathology reporting. Gagliardi, "The Evolution of the X-ray Report."

Current standards for radiologic reporting are summarized in "ACR Standard for Communication: Diagnostic Radiology," *American College of Radiology Standards*.

For suggested wording of a report on a normal contrast-enhanced brain CT, see Castillo, *Neuroradiology Companion*, 33.

104 Another kind of formula is that of categories of pathology—e.g., for a CT of a brain: "no bleeding, infarction, or mass effect"; "no extra-axial fluid collections" . . .

105 Experience of interruption is not peculiar to radiologists. One Neurology resident spending an elective month in the Neuroimaging reading room, interrupted by a phone call about a patient's medication, comments: "See, you get bothered more by doctors, we get bothered more by patients." I have not sought to quantify interruptions of the viewbox workday. For careful timecoding and analysis of radiologists' work (in CT and in a "Main Department" at two hospitals), see Barley, "On Technology, Time, and Social Order." Barley documented eight interruptions per hour in a typical workday in a Main Department, and two to three interruptions per hour in CT. The hospitals where his fieldwork was sited had radiological routines markedly different from those I investigated—a function of those hospitals' nonteaching status, and of historical changes in radiology staff relations around CT technology.

106 On development of typed radiology reports, see Howell, *Technology in the Hospital*, 44–51.

107 See, for instance, Derrida's response to Lévi-Strauss's "The Writing Lesson" in *Of Grammatology*, 101–40.

108 On the field guide as observational prosthesis, see Law and Lynch, "Lists, Field Guides, and the Descriptive Organization of Seeing."

109 On choice of images for publication, see Dumit, *Minding Images*, esp. on "Visual Semiotics," 173ff.

110 This is one correlate of the observation by Law and Whittaker, in "On the Art of Representation," that a scientific document is assembled from heterogeneous parts. Such heterogeneity is related to connections between semiotic and political representation. Their analytic terms—suppression, scaling, stylization, labeling—are techniques of achieving "discrimination" for persuasion of viewers, readers, and polities.

111 The classic study of this subject is Max Weber's "Science as a Vocation," which considers (with Nietzsche) "the demon who holds the fibers of his [the scientist's] very life" (156).

2. Cutting

1 "Phantoms" are test objects of standard density with which CT scanners are calibrated at least monthly. They specify one way that evidential status of CT images is predicated on ghostly mimesis of flesh. Specimens placed on scanner beds over the years: pig's head (by CT's inventors), mummies, frozen cadavers, anesthetized cougars, coffins, barnacle-encrusted swords, unidentified persons. See Melcher et al., "Non-Invasive Computed Tomography and Three-Dimensional Reconstruction of the Dentition of a 2,800-Year-Old Mummy"; Notman et al., "Modern Imaging and Endoscopic Biopsy Techniques in Egyptian Mummies"; Mazansky, "CT in the Study of Antiquities."

2 This is true for CT scanners through the third generation. In fourth-generation scanners, there is a fixed ring of detectors: the locus of the *active* detectors moves.

3 "Here your countenance brightened up, and, perceiving your lips move, I could not doubt that you murmured the word 'stereotomy,' a term very affectedly applied to this species of pavement"("The Murders in the Rue Morgue," in *Collected Works*, 2:536). An early draft: "You continued the same inaudible murmur, with a knit brow, as is the \<habit> custom of a man tasking his memory, until I considered that you sought the Greek derivation of the word stereotomy" (536).

    The conventional radiological notion that best corresponds to "stereotomy" is that of "virtual" cutting. See Derrida, "The Sans of the Pure Cut," in *The Truth in Painting*, for playful but rigorous philosophical reflection on the status of the "*coupure pur*" which liberates Kantian beauty.

4 Mayer, Poland, 1914: his idea was to smudge rib shadows; he does not mention sectional images. See Webb, *From the Watching of Shadows*, 83. Webb's is the definitive history of tomography. It is great-man and machine history, full of portraits and

diagrams—occasionally depicting a schematized patient, never a tech—but it also addresses careers and contexts, and sometimes craft practices. This distinguishes it from most historical treatments of tomography, which focus almost exclusively on proper names and engineering principles.

5 Ibid., 81–82.

6 Bocage, "Methods of, and Apparatus for, Radiography on a Moving Plate" (1921). This volume also reproduces works of the tomographic pioneers Vallebona and Ziedses des Plantes (below).

7 World War I galvanized tomographic innovations; World War II prompted development of cheaper and portable machines. Webb, *From the Watching of Shadows*, 114–17.

8 Tomography was a technique of engineering, a draftsman's more than a tinkerer's project. However, nonexperts in local machine shops, with personally owned tools, were responsible for many developments. See the history by Ziedses des Plantes cited by Webb (107).

9 On black boxes as consummated automatism, see Latour, *Science in Action*, 130–31. On black-boxing as "reified theory," see Latour and Woolgar, *Laboratory Life*, 150, 242ff.

10 Vallebona's prototype involved movement of the object (a skull) rather than the x-ray tube and film, which remained stationary.

11 Kieffer, "The Laminagraph and Its Variations." Its namer, Sherwood Moore, noted that a planigraph is an instrument for changing the scale of drawings, hence the term has prior meaning in the printing industry; stratigraphy has "geological connotations"; "tomograph" is a trade name. He suggests: "'Body section roentgenography' (adopted from Andrews) appears to me to be the most suitable general term to be applied to this method of examination, regardless of the type of apparatus used. Possibly 'sectional roentgenography' or 'analytical roentgenography' might even be preferable." Moore, "Body Section Roentgenography with the Laminagraph," 522. Webb discusses terminologies in *From the Watching of Shadows*, 54n1, 103–104. The International College of Radiology's adoption of "tomography" in 1962 was a compromise between Vallebona (stratigraphy) and Ziedses des Plantes (planigraphy).

12 Webb, *From the Watching of Shadows*, 72.

13 Cartwright and Goldfarb, "Radiography, Cinematography and the Decline of the Lens."

14 Theoretical correspondences between the work of Lumière and Bocage were earlier summarized in Duhamel and Roques, "Louis Lumière et les origines photographiques de la tomographie."

15 Webb, *From the Watching of Shadows*, 73–76.

16 Ibid., 117–22.

17 Ibid., 125–36.

18 Most of these workers were aware of each other's work. One exception was an analogue CT scanner in Kiev in 1957, unknown to many European workers for years.

19 Cummings to Watson, 1925, quoted in Cartwright and Goldfarb, "Radiography, Cinematography, and the Decline of the Lens."

20 Benjamin, *Charles Baudelaire*, 132.

21 The first energy source used was a gamma emitter, not a Crooke's tube.

22 Hounsfield's first apparatus did not collect transmission data in slices but in long wedges—with source and detector making one full traverse along the specimen before the specimen was rotated one degree. Thus the lathe was not spinning the specimen fast or continuously.

23 Reiser, *Medicine and the Reign of Technology*.

24 Cormack was a mathematician. See their Nobel lectures: Cormack, "Early Two-Dimensional Reconstruction (CT Scanning) and Recent Topics Stemming from It"; Hounsfield, "Computed Medical Imaging." On the early history of CT, see Webb, *From the Watching of Shadows*; Webb, "Historical Experiments Predating Commercially Available Computed Tomography"; Susskind, "The Invention of Computed Tomography."

25 Baker, "The Diffusion of High Technology Medical Innovations"; Banta, "The Diffusion of the Computed Tomography (CT) Scanner in the United States"; Trajtenberg, *Economic Analysis of Product Innovation*, esp. chap. 2.

26 Baker ("The Diffusion of High Technology Medical Innovations," 160) compares CT purchases in the mid-1970s with CT articles in the medical literature: critical evaluations were few until 1976. Medical conventions and collegial experience were main sources of information for early adopters of CT. See Harper, "CT Scanners."

27 Baker, "Historical Vignette."

28 Susskind, "The Invention of Computed Tomography," 65–67.

29 Baker, "The Diffusion of High Technology Medical Innovation," 161.

30 For one critique of the diffusion model of technological change (as distinct from a translational model), see Latour, *Science in Action*, 132ff.

31 In 1979, there were more CT scanners in Los Angeles County than in the UK, and more in California than in all Europe (Latour, *Science in Action*, 156). In 1996, America had 27 scanners per million citizens; Japan had 69.7; Australia and Germany had 16 and 18, respectively; Canada had 8, the UK 6.3. Anderson, *Multinational Comparisons of Health Care*.

32 Spurgeon and Burton, "Screen Tests." They report that the U.S. Army is spending a million dollars to scan 4,000 troops and has requested funding for full-body scans for all troops. One provider of preventive scans has developed software to enable patients to take a "virtual tour" of their bodies.

33 Goffman, *Asylums*, 340–50.

34 Larger institutions may have more. The Head/Body division is common. At University Hospital there are (in 1997) three diagnostic scanners. The third diagnostic scanner is in an ambulatory clinic (ACC). Both Head and Body scanners have associated reading rooms; the ACC scanner does not. CT scans of pediatric patients are read separately, in a dedicated Pediatric Reading Room. In this book, the ACC scanner and Pediatric Reading Room are mostly disregarded. My account also omits two

scanners dedicated to radiation therapy of cancer, largely outside the purview of *diagnostic* CT. This is, however, a distinction with fuzzy edges. Diagnoses are made and revised in Radiation Oncology scanners—just as lesion targeting is discussed in diagnostic CT. Moreover, "Rad Onc" techs occasionally use diagnostic CT scanners when their scanners are broken. For an analysis of cross-disciplinary discussion of CT images among members of a Rad Onc (gamma knife) team, see Simon, "Images and Image."

35 In 1967, the British DHSS was more enthused about procuring an image of the brain than screening for body tumors. Hounsfield was referred to neuroradiologist Jamie Ambrose, who applied the EMI scanner to clinical studies of the brain at Atkinson Morley Hospital in 1971.

36 On the Neuro side, the Service is called Neuroradiology. Neuro studies include a high proportion of MRIs and MR angiograms—so the reading room is not so modality-centered. I frequently use the term "Neuroimaging" to describe that reading room.

37 Advantages of dividing radiology by organ system or by technique have been much debated. See, for example, Potts, "The Division of Radiology"; Bergeron, "Leaving Academia and Bringing Organ-System Radiology to a Community Hospital."

38 Douglas, *Purity and Danger*, esp. chap. 2.

39 The alcove contains vital sign monitors, oxygen/suction/electricity fixtures, chairs.

40 These are dimensions of the Neuro CT scanner room (not including the control room). The Body CT scanner room is somewhat smaller.

41 Liminality is a threshold state in rites of transition (van Gennep, Turner). Van Gennep, in *Rites of Passage*, suggests the limen is often marked with something manmade, whereupon are conducted rites of purification (signaling separation from prior contexts); upon passage, the rite is often consummated by incorporation—e.g., a shared meal. In CT, there is incorporation of contrast material prior to traversing the scanner—though arguably receipt of a diagnosis, somewhere on the other side, constitutes a more significant mode of incorporation. The plane of section is actually crossed and recrossed by the patient.

42 Technically, this phase of "slicing" is understood as data acquisition. Each slice of raw data is "reconstructed" to produce a slice of density distributions. These slices are stacked into a volume. This volume can be resliced—"reformatted"—at finer thicknesses and along different planes from that of data acquisition. So this plane of acquisition may differ from that of a displayed image-slice.

43 Confinement anxieties are more acute in MRI machines, which are bulkier, with deeper tunnels. Open-magnet MRIs have kept the market open to claustrophobics. For a literary portrayal of scanner claustrophobia, see Dewar, "Body Scan."

44 The table is so narrow that some patients feel reassured by the Velcro straps that fold across their chest and hips.

45 Air exchange is noisy in CT. Background rumble impaired recording of shop-floor conversations in the control room, even more so in the scanner room.

46 Partial sample of memos on walls in Body CT Control: repair requests; tube change log; instructions for Needlestick Hotline; memo about a patient transport project;

notice of an epidural catheter study; a parking plan; Radiation Oncology Guidelines; composite of current residents' and fellows' photographs; memo about dictations; Attending Schedule; managed care plan ID cards; annual calendar marked with tech vacations and birthdays; bootup codes for workstations; Siemens support numbers; continuing education schedule from the Society of Radiologic Technologists. Directly opposite the control console is a large whiteboard bearing several scrawled notes.

47 Goffman, "Role Distance": "The system or pattern only borrows a part of the individual" (86). In "Surgery as an Activity System," Goffman notes: "Surgery requires acts unbecoming a surgeon." In representing work regimes, I have tried to note participants' discretions and indiscretions, criticisms and complaints. Throughout this book, when I cite offhand, humorous, or surprising remarks, it is to take advantage of the light of critique they can shed on defining rhythms and textures of the shop floor. On CT tech roles and role relations, see Barley, "The Professional, the Semi-Professional, and the Machines," 233–430.

48 Interactions between techs and radiologists are described in Barley, "Technology as an Occasion for Structuring." Barley observed tech/radiologist exchanges in two Massachusetts hospitals that received new body scanners in 1982, and he documented different structures of collaboration over time—one rigidly hierarchical, the other more cooperative. In general, Barley found radiologists more involved in affairs at the scanner console—including decisions about patient position, contrast injection speed, etc. These practices were then brand new; in my field site fourteen years later they are more routinized. Barley also made an interesting typological classification of conversational elements in workplace social order: "unsought validation; anticipatory question; preference stating; clandestine teaching; role reversal; blaming the technologist; direction giving; countermand; usurping the controls; direction seeking; unexpected criticism; accusatory question; technical consultation; mutual execution" (101). In another paper, Barley reports the time radiologists spent in the scanner room at two community hospitals as 26 percent and 68 percent of their day, respectively. Barley, "On Technology, Time, and Social Order."

49 For related ironies regarding "life" in functional brain imaging, see Cohn, "Increasing Resolution, Intensifying Ambiguity." The injunction not to move is not specific to radiology: stillness, or strict limitation of specimen movement, is requisite for much physical examination, and surgery too.

50 On historical associations of nurses with technologies, see Sandelowski, *Devices and Desires*, esp. 9–15.

51 In April 1997, the American Nurses Association officially recognized radiology nursing as a specialty—though the American Radiological Nurses Association had been in existence for years.

52 On tech relations with patients in the 1980s, see Barley, "The Professional, the Semi-Professional, and the Machine," 369–75.

53 On protocols, see Timmermans and Berg, "Standardization in Action."

54 *American College of Radiology Standards* (1996). CT protocols include volume, route, and timing of contrast administration; collimation (slice thickness); slice spacing (table increment); kVp and mAs for small, medium, and large patients; superior and inferior extent of area to be examined; level and window settings of permanent images (9).

55 This list is not exhaustive—and it does not include special protocols adapted to ongoing research projects ("Stan's Epilepsy Protocol," for instance).

56 On CT tech pragmatics in the 1980s, see Barley, "The Social Construction of a Machine."

57 This procedure is presented as one continuous event, but it is actually an amalgam of several.

58 On low-osmolar contrast use as a case study in resource allocation, see Eddy, "Broadening the Responsibilities of Practitioners." See also Berg's discussion of selective-use contrast protocols in *Rationalizing Medical Work*, 103ff.

59 Attending and fellow, jokingly mimicking rhetorics of formal dictation as they plan a perinephric biopsy, in reading room of another University Hospital, 1991.

60 Kleinfield, *A Machine Called Indomitable*.

61 About this conjuring in the situation of the Visible Human Project (chapter 4, "Curating"), Csordas uses the archaic term "shade"—"spirit in the netherworld"—in "Computerized Cadavers," 180.

3. Diagnosing

1 Poe, "The Murders in the Rue Morgue," in *Collected Works*, 2:528. Emphasis in original.

2 Ginzburg, "Clues," 102–3, 105.

3 For instance, Ginzburg's conjectural knowing relates to the "wily intelligence" the Greeks called *mētis*—explored by Detienne and Vernant, *Cunning Intelligence in Greek Culture and Society:* "attitudes and intellectual behavior which combine flair, wisdom, forethought, subtlety of mind, deception, resourcefulness, vigilance, opportunism, various skills, and experience acquired over the years" (3). On habitus, see note 21 in my introduction. On thought styles—characteristic of specific "thought collectives," often bound to technical and literary genres—as determinative of modes of perception, see Fleck, *Genesis and Development of a Scientific Fact.* "Every thought style contains vestiges of the historical, evolutionary development of various elements from another style" (100).

4 Benjamin, *Charles Baudelaire*, 48.

5 Ibid., 132.

6 Ibid., 36.

7 Benjamin (ibid., 38) cites Simmel's *Sociology*. On Poe's detective tales as critical supplement to urban tactics of the flâneur, see Brand, "Reconstructing the Flâneur." My book does not undertake close analysis of Poe and detection—about which there is

a substantial literature, including Irwin, *The Mystery to a Solution*; Rosenheim, *The Cryptographic Imagination*. See Harrowitz, "The Body of the Detective Model," addressed below.

8 Benjamin, *Charles Baudelaire*, 43, 40, respectively.

9 Borges, "The Detective Story," 16.

10 Poe, "The Man of the Crowd," in *Collected Works*, 2:515.

11 On the CPC as theater of clinical reasoning, see Eddy and Clanton, "The Art of Diagnosis"; Hunter, *Doctors' Stories*, 107–10. On history of the CPC in America (courtesy of Richard Cabot, who perhaps enhanced the ritual's "competitive" flavor), see Crenner, "Diagnosis and Authority in the Early Twentiety-Century Medical Practice of Richard C. Cabot."

12 Ross, "From a Purely Radiological Point of View," 42.

13 Sochurek and Miller, "Medicine's New Vision," 3–41; Kevles, *Naked to the Bone*, 2.

14 For exemplary usage, see Timmermans and Berg, *The Gold Standard*—though their analysis concerns standards in general, and making knowledge through controlled trials.

15 For literary-historical analysis of the gold standard, see Michaels, *The Gold Standard and the Logic of Naturalism*. On relations between symbolic and monetary exchange, pivoting on Adam Smith's social theory, see Herbert, "Desire, Wealth, and Value," in *Culture and Anomie*, 74–149. For analysis of the epistemological and theological status of the gold standard, see Taylor, "Christianity and the Capitalism of Spirit."

16 On "The Gold-Bug" and history of the gold standard, see Shell, "The Gold Bug."

17 On writing and death, see Foucault, *Death and the Labyrinth*. On writing and death in Poe, see Kennedy, *Poe, Death, and the Life of Writing*; Rosenheim, *The Cryptographic Imagination*.

18 Remarks trimmed from dialogue to monologue: deletions marked with ellipses.

19 On cases and case presentations in medicine more generally, see especially Hunter, *Doctors' Stories*; Atkinson, *Medical Talk and Medical Work*, chap. 5; Epstein, *Altered Conditions*, chap. 3.

20 In differential diagnosis as process, the making of an inclusive list of possible diseases is only the first step, prefatory to comparison, judgment, selection.

21 Poe, "The Murders in the Rue Morgue" in *Collected Works*, 2:548.

22 Ibid., 529.

23 Ibid., 528.

24 Saunders, "Insection and Decryption." For another version of divided faculties in Poe's sociology of detection, see Lacan's "Seminar on 'The Purloined Letter'" and its commentaries in Muller and Richardson, *The Purloined Poe*. Lacan maps two superimposed triads of intersubjective gazes: "The first is a glance that sees nothing: the King and the police. The second, a glance which sees that the first sees nothing and deludes itself as to the secrecy of what it hides: the Queen, then the Minister. The third sees that the first two glances leave what should be hidden exposed to whomever would seize it: the Minister, and finally Dupin" (32).

25 "Abduction is the process of forming an explanatory hypothesis. It is the only logical operation which introduces any new idea" ("Lectures on Pragmatism," 1903).
The Peircean dossier on abduction remains quite fragmented. "Abduction and Induction" (1901–8) and "Deduction, Induction, and Hypothesis" (1878) are starting
points. Bergman and Paavola's collated entry on abduction in their online "Commens Dictionary of Peirce's Terms" is helpful. For exemplary appropriations of
Peirce in semiotics and literary theory, see Eco and Sebeok, eds., *The Sign of Three*,
esp. Ginzburg, "Morelli, Freud, and Sherlock Holmes" (version of essay cited in note
2 above), and Harrowitz, "The Body of the Detective Model." Peirce derived his term
from a "doubtful" retranslation of Aristotle's *Prior Analytics* chapter 25 ("A Letter to
Calderoni," cited in Bergman and Paavola). Other terms used to convey this logical mode have sometimes been formulated as special cases of deduction or induction—e.g., "hypothetico-deductive." See the discussion of this term in Campbell,
"The Diagnosing Mind," drawing on analogies with mapping, and in Berg, *Rationalizing Medical Work*, 27–28.

26 Induction may not lead by itself to proof, however. Hume's critique: it is like superstition. For discussion of induction with respect to disease etiologies, see Sutter,
"Assigning Causation in Disease."

27 I have resisted referring to this, here and in later discussions of judgment, as "in-
tuition"—to avoid psychological mystifications that sometimes attend use of this
term.

28 See selections in Cassiday, *Roots of Detection*.

29 On Poe's scientific rhetoric and bricolage, see Tresch, "'The Potent Magic of Verisimilitude'"; Tresch, "Extra! Extra! Poe Invents Science Fiction!"

30 Cuvier, *Histoire des progres des sciences naturelles*, 310.

31 Coleman, *Georges Cuvier*; Foucault, *The Order of Things*, 263–79.

32 Poe, "Preface and Introduction to 'The Conchologist's First Book.'"

33 These sixteenth- and seventeenth-century precursors to late-modern museums emphasized aesthetics of wonder and marvel. Some are cataloged in Kenseth, ed., *The
Age of the Marvelous*. On early modern "curiosity," see Benedict, *Curiosity*.

34 Huxley, "On the Method of Zadig." Huxley referred to these procedures of discern-
ing causes from effects and then predicting further effects, as a form of "divination."
"Zadig" is a famous story by Voltaire which exemplifies some of these procedures.

35 Bennett, *The Birth of the Museum*, 178. A fine study of the establishment of one such
"deep-time," albeit in the strata of scientific society meetings rather than museums,
is Rudwick, *The Great Devonian Controversy*.

36 This foreshortened account conflates two epistemic stages which both Foucault and
Bennett are at pains to distinguish. Taxonomy, the table of comparisons and its
gridwork of relationships, belonged to the (early modern) era of natural history.
"Biology" was distinctive of later modernity: from Cuvier's knitting up of the
paleo- with new notions of organism, it derived a new relation to time. See Bennett,
The Birth of the Museum, 95–98; Foucault, *The Order of Things*, 268.

37 Joseph Leidy, a vertebrate paleontologist, also versed in pathology—like Cuvier—is wearing the tophat. Poe is seated. When I encountered this photo in a conference presentation by the AFIP historian James Connor, "Tissues by Sunlight," I was enchanted with his suggestion that the youth pictured could be J. J. Woodward, who later would become pathologist and curator at the Army Medical Museum (see chapter 6, "Exposition"). Though this identification proved erroneous, the photo still knits up several historical points with admirable cogency. It is discussed at length in McFarland and Bennett, "The Image of Edgar Allan Poe." Benjamin would have been delighted by their suggestion that this may be America's earliest naturally lit daguerreotype of an indoor scene with human figures.

38 On phrenology, see Baltrusaitis, "Animal Physiognomy." For explicit associations between craniology and the work of Cuvier, and detective fiction, see Messac, *Le Detectif novel et l'influence de la pensée scientifique*. Taylor, *Hiding*, reflects philosophically on how phrenology made bodily surface "a sign to be deciphered" and explicitly addresses Poe and detective tales. For entailments of craniological thought in various race theories, see Stocking, *Victorian Anthropology*, esp. 65–70. Cerebral localization is an important domain in history of neurosciences. Dumit discusses it in relation to PET scanning in *Picturing Personhood*, 23–24.

39 On game theory in science, see Latour and Woolgar, *Laboratory Life*, 244–52.

40 These reflections unfolded in dialogue; my own brief interjections are removed to simplify.

41 Weber, *The Protestant Ethic and the Spirit of Capitalism*, 182. See also "Science as a Vocation."

42 See Peirce on "perceptual judgments . . . as an extreme case of abductive inference," in "Perceptual Judgments," 304. See also Polanyi on "aesthetic recognition" as a component of taxonomic effort, *Personal Knowledge*, 351.

43 Case from a different university hospital, 1991.

44 The fellow implies that calcium was deposited gradually, over time, in small areas of sequential tissue death. Thorotrast is an older contrast material.

45 Poe, "The Purloined Letter," *Collected Works*, vol. 2.

46 Sekula, "The Body and the Archive," 27.

47 Ibid., 27.

4. Curating

1 On curating criminals, see Sekula, "The Body and the Archive." On curating death, see Hacking's work on actuarial technologies, mortality bills, and the invention of *statistik* in *The Taming of Chance*; see also Porter, "Life Insurance, Medical Testing, and the Management of Mortality."

2 Anderson, "Census, Map, Museum." On archives at public academic hospitals in the 1990s—comprising records of patient care, plus administration, education, research, and health promotion—see Krizack, "Documentation Planning and Case Study."

3 On *locality* of archival formations within medical institutions, see Berg, *Rationalizing Medical Work*, 79–102.

4 Anderson, "Census, Map, Museum"; Bennett, *The Birth of the Museum*; Mitchell, *Colonising Egypt*; Richards, *Imperial Archive*.

5 Hospitals were among the disciplinary institutions whose architectural and operational sightlines emulated Bentham's Panopticon, designed for optimized surveillance of prison inmates (Foucault, *Discipline and Punish*, 195–217). Peripherally arranged cells opened to a central surveillance tower, so that inmates could be watched without being able to see their watchers. Constant surveillance, in service of reform—Foucault: "panopticism"—remains a constitutive ideal of institutional discipline. For innovations in European hospital architecture, see Foucault et al., *Les Machines à guérir*.

6 Prophetic charisma: see Weber, *Ancient Judaism*, 267–95. The Weberian complex of charisma includes charisma of lineage, of office (ordained through training), of magic (vested in artisans), and also of special objects, like fetishes. Weber, *The Sociology of Religion*, 2–3 and passim. What I am calling curatorial charisma derives both from objects, through care and handling, and from the institution, through special training.

7 Other customers of Starpoint, in 1997, included Blue Cross/Blue Shield, Northern Telecom, and Nationsbank. Starpoint has installations in several U.S. cities.

8 Bennett, *The Birth of the Museum*, 59–88.

9 At University Hospital, mammographic films are retained ten years: changes over longer periods are held to be important in breast cancer screening. Films are also retained longer for pediatric patients.

10 Expiration of the utility of photographs with time is discussed in Tagg, *The Burden of Representation*, in regard to early institutional contexts of photography—orphanages and police culprit galleries; children and recidivant criminals change appearance especially rapidly.

11 Poe referred to the time-dependent buoyancy of the submerged corpse in "The Mystery of Marie Roget," *The Complete Tales and Poems*, 183. An average CT study—six sheets of standard film (14 x 17 inches) in a folder—weighs just under a pound. A more extended CT or MR study can weigh two pounds. Film folders containing additional studies—various plain x-rays, several CTs—can weigh over five pounds. Thus a several-inch stack of folders could become a two-arm load.

12 See Latour on "immutable mobiles"—artifacts capable of undergoing "translation without corruption"—in "Visualization and Cognition."

13 At the Jardin des Plantes, curators were professors and lectured. Peale's Museum in Philadelphia was housed with the American Philosophical Society. Museums in other cities, American and European, often included lecture halls. See chapter 6, "Exposition," on pedagogical projects of the Army Medical Museum. Scientific advance and edification of the public were not sole goals; museums competed with other forms of public amusement, and in the mid-nineteenth century some exhibited curiosities and freaks alongside ordered taxonomic displays. See Harris, "The American

Museum," in *Humbug*, on how Barnum's museum management exploited sensationalism, exaggeration, hoaxes, and scientific controversy to win public attention.

14 On the label/caption, see Kirshenblatt-Gimblett, "Objects of Ethnography," 393ff: "The priority of objects over texts in museum settings was reversed during the second half of the nineteenth century." She cites G. Brown Goode, director of the U.S. National Museum (circa 1891), whose definition of a well-arranged museum is quoted in Bennett, *The Birth of the Museum*, 42: "a collection of instructive labels illustrated by well-selected specimens." Bennett elaborates on museological functions of labels in "Pedagogic Objects."

15 Contiguity in exhibition is discussed in chapter 5, "Testifying and Teaching."
      Drayton, in "Collection and Comparison in the Sciences," emphasizes the epistemic priority of exotic specimens and remote collectors in comparative projects and criticizes Latour's emphasis on "the centre's myth of its organizing efficiency."

16 Peale opened America's first natural historical museum in 1786. Nineteenth-century taxidermy involved toxic chemicals—e.g., arsenic. Glass cases in museums contained a potentially dangerous, albeit dead, Nature: subject/object divide. See Lloyd and Bendersky, "Of Peales and Poison."

17 See Haraway, "Teddy Bear Patriarchy," in *Primate Visions*, 26–58, for critique of taxidermic craft in realist aesthetics in the American Museum of Natural History in the early twentieth century. On glass, see Bennett, *The Birth of the Museum*, 101: England's Crystal Palace (1851) exhibited its own interior along with industrial apparatus. Progress was linked to visibility of internal mechanisms—and the self-monitoring of the museum-going public. In the same era, the glass of Parisian shopping arcades conditioned sensibilities of consumers and their removal from objects of desire. Friedberg, *Window Shopping*, esp. 60–68, and Benjamin, *The Arcades Project*.

18 On this "reordering of things" in the civilizing process, the transition from royal collections to public galleries, see Bennett, *The Birth of the Museum*, 33–40.

19 This elides transitions that art museums underwent, similar to those described with natural history museums: from collections of singular items to collections organized by type—period, school, etc.; see Bann, "Poetics of the Museum." On the extent to which art museums have, in producing typologies of "masterpieces," failed to live up to pedagogical agendas—have been "machines for the suppression of history"—see Burgin, "The End of Art Theory."

20 On historical changes in the "masterpiece"—once the work by a medieval apprentice which demonstrated competence, later a Romantic fetish of original handwork—and its entanglement with creativity, genius, etc., see Burgin, ibid., 153–54.

21 Early cultic images traded on resemblance (e.g., in portraiture) or denotation (e.g., in allegory or history) in depicting subjects or events worthy of memorialization. Conventions of resemblance or of denotation, of course, vary historically.

22 Aura—Walter Benjamin's term, developed in several essays, including "The Work of Art in the Age of Mechanical Reproduction"—has to do with associations that cluster around an object of perception, augmented by cumulative attentions, including

handling, especially the handling of the maker. Benjamin's treatment of technical versus auratic artistic production has been criticized by, among others, his friend Adorno (see, e.g., *Aesthetic Theory*, 49).

23 Ideals of art exhibition are not necessarily congruent with responses of museumgoers. Some recognize the art museum as a domain structured not so much around the vitality of the original as around death. Adorno compared Proust's and Valéry's views on this. Valéry found the museum a place of barbaric "reification and dilapidation" faithless to "the discipline of the artist's work." Proust found the art museum a productive memory-theater of spectatorial intuitions—albeit one predicated on the afterlives of decaying artifacts. See Adorno, "Valéry Proust Museum."

24 Mechanical image reproduction began with fifteenth-century woodcuts, then copper engravings and etchings. A second watershed in image reproduction (and publication) around 1800 included techniques of aquatint, wood engravings, steel engravings, and lithographs. Lithography eliminated the middleman of the engraver. These issues are discussed in relation to a form of sectional representation (read: slices) in Rudwick, "The Emergence of a Visual Language for Geological Science."

25 On digital reproduction, see Mitchell, *The Reconfigured Eye*.

26 Marxian accounts of the triumph of exchange value over use value include Debord, *The Society of the Spectacle*: "The image has become the final form of commodify reification." Also Baudrillard, *L'Echange symbolique et la mort*: "All the great humanist criteria of value, all the values of a civilization of moral, aesthetic, and practical judgment, efface themselves in our system of images and signs. All becomes indecidable, this is the characteristic effect of the domination of code, which rests throughout on the principle of neutralization and indifference. That is the generalized bordello of capitalism . . . a bordello of substitution and commutation" (21; my translation).

27 Benjamin, "The Work of Art in the Age of Mechanical Reproduction," in *Illuminations*, 225.

28 On motivated signs, see Todorov, *Theories of the Symbol*, esp. "Imitation and Motivation," 129–46. The question of motivation arises for signs which are not "imitative" or "natural."

29 A series of essays by William Pietz is invaluable in sorting out critical implications of fetishism in different historical settings. He suggests that "fetish" exemplifies four themes: "1) . . . untranscended materiality . . . as the locus of religious activity or psychic investment; 2) the radical historicality of the fetish's origin: arising in a singular event fixing together otherwise heterogenous elements . . . ; 3) the dependence of the fetish for its meaning and value on a particular order of social relations, which it in turn reinforces; and 4) the active relation of the fetish object to the living body of an individual . . . outside the affected person's will." "The Problem of the Fetish, II," 23.

Marx's use of the term critiqued the effacement of, alienation from, labor dimensions of value in the "reification" or "fetishism" of the commodity. See *Capital*, part 1, chap. 1, "Commodities."

30 Tagg, *The Burden of Representation*, 34–59.

31 The violence of the shutter, and the copy machine, forefront mechanical aspects of these aesthetics of alienation. From Benjamin's "A Small History of Photography": "And the difference between the copy . . . and the picture is unmistakable. Uniqueness and duration are as intimately conjoined in the latter as are transience and reproducibility in the former. The stripping bare of the object, the destruction of the aura, is the mark of a perception whose sense of the sameness of things has grown to the point where even the singular, the unique, is divested of its uniqueness—by means of its reproduction" (250). On aesthetics of shock, see Shiff, "Handling Shocks."

32 The *simulacrum*, a Platonic term, is for Baudrillard that which images refer to in the era of the "hyperreal." See *Simulations*.

33 The Visible Woman was/is composed of 5,100 transverse slices, at 1/3 mm increments; the Visible Man is approximately 1,800 slices, at 1 mm increments. CT datasets were both produced at 1 mm increments.

34 The main sponsor for the Visible Humans is the National Library of Medicine. On cultural implications of this project—development of an (inter)national gold standard from bodies of a criminal man and an anonymous woman—see Cartwright, "A Cultural Anatomy of the Visible Human Project"; Csordas, "Computerized Cadavers"; Waldby, *The Visible Human Project*.

35 Anderson, "Census, Map, Museum," 170–78. See Latour, "Visualization and Cognition," on maps and imperial agendas.

36 The genre was inaugurated (without the term "atlas") by his contemporary Ortelius (*Theatrum Orbis Terrarum*, 1570).

37 On relations of maps to pictures, see Alpers, *The Art of Describing*, 119–68.

38 There are two couriers on duty during the day: one in the File Basement, the other serving Film Assembly. Sally is the latter. The File Basement Courier is a step below in the hierarchy, covering more regular routes, less engaged with divergent tugs.

39 See Richards, *The Imperial Archive*, esp. "Archive and Utopia," for an account of Montgomerie's surveying of Tibet in 1862, using Indian monks and "body-based units of measure" (17–20).

40 MagicView in the 1990s was a product in a division known as SIENET: "networking and image management solutions." In 2004 Siemens still provides training for MagicView, but this family of tools (and trademark) has largely been superseded.

41 A few catchy PACS names from 1990s: "Impax," by Agfa; "Synapse," by Fuji; "StatView," by Image Datacorp; "*Quick*LINX," by DuPont.

42 From Siemens proprietary web page of 2000 describing benefits of filmless network: http://www.med.siemens.com/med/shs/en/prod_2/prod_223_nf.htm.

43 Other PACS companies also make use of rhetorical and pictorial references to magic or supernatural power in their advertisements. An AccuSoft PACS ad (1998) depicts the Sistine Chapel hands of God and Adam as x-rays.

44 Hume's definition of causality is "constant conjunction" of succession and contiguity, famously exemplified in repeated observation of billiard balls. *A Treatise of Human Nature*, book 1, section 14, 114–73.

Associations between magic and viewing are ancient. Reformation suspicion of Papist idolatry, which gave rise to textual methods of hermeneutics, was one early modern symptom. On imputed affiliations of images with trickery, concealment, and quackery in the early Enlightenment, see Stafford, "Conjuring."

45 "Contiguity" is a Humean term: one of the fundamental forms of association configuring human understanding—another of which is resemblance. Contiguity is definitional in Frazer's distinction of "contagious" magic (action at a distance after two objects have lost contact) from "homeopathic" magic (like produces like—based on resemblance). Frazer, *The Golden Bough*, 12–13.

46 Institutions also make radiographic images and transmit them electronically to distant locations for review and interpretation. This is "teleradiology"—radiology over distance. The first telephonic transmission of x-ray images was in 1947, between physicians in Pennsylvania. University Hospital participated in teleradiology projects and services throughout the 1990s. For history of teletechnologies and computers in medicine, see Reiser, *Medicine and the Reign of Technology*, 196–226. For general, somewhat technocratic overviews of telemedicine, see Bashshur et al., eds., *Telemedicine*. For critical cultural analysis of telemedicine, see Braitberg, "Liberators, Innovators, and Experts." Most applications of telemedicine manage visual information in "store-and-forward" routines—well suited to radiology, pathology, and dermatology. For a brief history of PACS within a complex ethnographic case study, see Kelty, *Scale and Convention*.

47 And this dynamism extends into the images themselves: there is, they say, more "dynamic range" in digital information captured by the scanner than film can preserve.

48 Departmental "Computed Body Tomography" guide.

49 Benjamin, *Arcades Project*, 205. See also "Unpacking My Library," in *Illuminations*, 59–68.

50 Here is the first-order classificatory arrangement of one of the online teaching files at University Hospital: Chest; CV / Interventional; GastroIntestinal; Genito-Urinary; Mammography; Musculoskeletal; Neuroradiology; Nuclear Medicine; Pediatric; Ultrasound. Note the entanglement of modalities and body systems in this listing. There is no category for CT as such: CT scans are found under each of the body system headings.

51 This is not universal with CTs and MRs, however. A plain x-ray displays patient information in one corner, where it can be marked through readily or concealed with a label. On CTs and MRIs, identifying information accompanies each slice—unless the tech removed it in copying. In the 1990s a few CT and MR films in teaching files retain patient identifiers (remedied since stricter HIPAA legislation). Most show the date of the study, some the patient's age.

52 Radiology Information System—a generic term. (Each letter is pronounced separately: R-I-S.) The RTAS is part of the local version.

53 Reiser, *Medicine and the Reign of Technology*, 205–9. See also Timmermans and Berg, *The Gold Standard*, 34–40.

54 Howell, *Technology in the Hospital*, 30–68. This history comprises careful comparison of records from selected Pennsylvania and New York hospitals.

55 In 1895, Taylor used the stopwatch to analyze actions of steel workers. His managerial strategies were particularly successful in the railroad industry. Portions of his analysis are excerpted in "Scientific Management."

56 Howell, *Technology in the Hospital*, 33–40. See also Timmermans and Berg, *The Gold Standard*, chap. 2.

57 Ibid., 42.

58 Ibid., 42–45.

59 Reiser, *Medicine and the Reign of Technology*, 207; Howell, *Technology in the Hospital*, 52.

60 Howell, *Technology in the Hospital*, 45–56.

61 This time frame—of which the largest component is the wait for the radiologist's review of the transcript—is often shorter at smaller hospitals with few radiologists: one to two working days.

62 Of twenty-four chapters in the revised text (two volumes), colleagues from University Hospital authored or coauthored eleven.

63 On joint authorship and economics of responsibility and credit, see Biagioli, "Aporias of Scientific Authorship." On the translation of research findings from journal science to "vademecum" (handbook/textbook) form, see Fleck, *Genesis and Development of a Scientific Fact*, 117–24.

64 On history of research trials, see Matthews, *Quantification and the Quest for Medical Certainty*.

65 Statistical treatments of large numbers—the techniques that underwrite "evidence" in evidence-based medicine—tend to be taught in medical schools as a set of value-neutral techniques without any historical reference. The kinds of framing offered by Ian Hacking—to wit, that *statistik* meant first and foremost "facts about the state" (*The Taming of Chance*)—or by Benedict Anderson, who writes about the colonial genealogy of the census—would be a salutary corrective. Kundel, in "The Origin of Health Services Research in Radiology," addresses technology assessment and associated statistical methods. He begins with 1940s TB screening. He cites a VA study of different radiographic modalities that showed observer variability (five readers) exceeded variation among techniques. Remarking how analysis in this and related studies treated false negative and false positive readings as independent, Kundel calls statistical decision theory in the 1950s a "new way of thinking about error," in which false negatives and false positives are treated as dependent and varying with decision criteria. He suggests that a particular radiology resident, Lusted, played an important role by applying a mode of analysis then being used to evaluate automated cytological analysis of sputum smears (also TB screening).

66 Kundel, "The Origin of Health Services Research in Radiology," suggests that the earliest applications of decision analysis in radiology were in 1969 and 1974, evaluating new screen systems. The first major study using ROC analysis came in 1979, comparing CT to radionuclide scanning for brain tumor evaluation.

67 Kundel, ibid., reviews other measures of imaging technologies, including diagnostic yield (problematic because it does not value negative report or differentiate among kinds of value in diagnosis). In 1971, in an efficacy study committee for the American College of Radiology, the criterion used to measure the impact of radiological reading on clinical management was the difference in pre-test and post-test probability of illness for a clinical decisionmaker.

68 "The trail of progress can be followed in the scientific literature." Chew, "*AJR: The 50 Most Frequently Cited Papers in the Last 50 Years*," 227. Nine of the fifty papers Chew discusses (a third of those published in the 1970s–80s) reference CT in their titles. Two pertain to body CT, two to draining abscesses, one to needle biopsy; one is a cross-modality study of hepatic neoplasms; one is a prospective comparison of CT and conventional tomography with pathologic correlation. Another, older paper involving pathologic correlation is cited as exemplar of the long-term value of such studies—a still-cited work on pulmonary infarcts, from which a chest x-ray sign ("Hampton's Hump") was derived. In 1982, Hounsfield's original description of CT in *British Journal of Radiology* (1973) was the "second most frequently cited article ever published in a radiologic journal" (232). Chew plots an exponential rise in number of early CT papers compared to CT scanners themselves. He also displays bibliometric careers of two early studies of body CT, cited extensively over three years following publication (but no longer cited after ten years). On the yoking of bibliometric methods to scientist curricula vitae and the mapping of citation networks, see Latour and Woolgar, *Laboratory Life*, 187–233; Latour, *Science in Action*, esp. 33–50. On bibliometrics and politics of persuasion, see Law and Whittaker, "On the Art of Representation."

5. Testifying and Teaching

Material derived from this chapter is published in *Skilled Visions*, edited by Cristina Grasseni (Berghahn Press, 2006).

1 Connotations of testimony in this chapter relate to forensic and interrogative contexts of radiological discourse. I do not unpack testimony as a category of religious rhetoric—though some terms of my analysis might lend themselves to such an enterprise.

2 See Burling et al., *The Give and Take in Hospitals*, 81–82.

3 Anatomy theaters are no longer primary sites of autopsic dissection. See Liebow, "The Autopsy Room as a Hall of Learning"; Fox, "The Autopsy: Its Place in the Attitude-Learning of Second-Year Medical Students," in *Essays in Medical Sociology*, 51–77, on teaching hospital autopsy settings in the 1950s—"as immaculate, well-organized and brightly lit as a modern laboratory or surgical amphitheater" (53).

4 Or older. Teaching was a radiological mission long before most medical schools had departments of radiology. In 1924, the roentgenologist commented in the annual report of the Brigham Hospital: "Situated as we are, where medicine, surgery, and

pathology meet, we command a view ideal for teaching. The students are taught not Roentgenology, not Medicine, Surgery, or Pathology, but are taught to observe the records and changes left by disease, to correlate these with, or subordinate them to, the findings by other methods, and then to apply their individual judgment in each case." Not until 1944 did Harvard Medical School establish a department of roentgenology. Holman and Edwards, "A History of Radiology at the Peter Bent Brigham Hospital."

5 This formulation—that a particular ritual form (Radiology Conference) is heir to another (the CPC)—is overstated. Both rituals, and the rooms where they have been conducted, have coexisted and exerted interdependent influences on diagnostic teaching in American hospitals, throughout most of the twentieth century and into the twenty-first. Historical studies of sites of clinical teaching in the early twentieth century have tended to emphasize ward or bedside, or tensions between laboratory and bedside—e.g., Rosenberg, "The Ward as Classroom," in *The Care of Strangers*, 190–211, and Thompson and Goldin, *The Hospital*.

6 "Theater of proof" is a key term from Latour's *The Pasteurization of France*, describing Pasteur's canny experimental demonstrations: "Nobody really knew what an epidemic was; to acquire such knowledge required a difficult statistical knowledge and long experience. But the differential death that struck a crowd of chickens in the laboratory was something that could be seen 'as in broad daylight'" (85). See also Crawford, "Imaging the Human Body," on theatrics of ocular demonstration in three historical instances of body imaging—culminating with an ultrasound system. Another crucial reference is Shapin and Schaffer's *Leviathan and the Air-Pump*, on early modern experimental philosophy, and the critiques of Hobbes, for whom Boyle's peer-witnessed experiments in the Royal Society's assembly hall constituted a theater of mystification and "priestcraft." The "artificial forum" of the CPC as a device of exhibiting clinical reasoning is discussed by Eddy and Clanton, "The Art of Diagnosis."

7 Morbidity and Mortality ("M&M") conferences are constituted to review difficult cases—especially those resulting in unexpected deaths or serious complications—within a particular clinical jurisdiction. Their orientation toward decisions and errors affiliates them strongly with *ethoi* of quality improvement and risk management—which may account for their persistence in nonteaching settings. See Orlander, Barber, and Fincke, "The Morbidity and Mortality Conference." For exemplary discussion of these conferences in departments of Surgery, where they have a robust tradition, see Bosk, *Forgive and Remember*.

8 Once radiology was a distinctively media-heavy discipline. But all hospital specialties have become more visual and media-intensive, and along with them University pedagogy in general, well beyond health affairs. In this era of multimedia and Internetworking, teleconferencing and telemedicine, this conference room is at best middle-tech.

9 "Case" as labeled, cross-referenced film stack is different from "case" as medical literary/textual genre—about which there is a substantial analytic literature. On case

history and case presentation, see in particular Hunter, "The Representation of the Patient," in *Doctors' Stories*, 51–68.

10 Workday roughly 7 to 7, or 8 to 5, depending on who is asked, Monday through Friday. See chapter 4, "Curating," on hospital scheduling and rhythms of production.

11 See chapter 1, "Reading and Writing," for eating in viewbox logistics; see below on "eating the film" as the stake in a bet.

12 CT cases occupy a larger wedge of conference activity than might appear here. CT images figure often and prominently under many headings—especially Neuro, Chest, GI, GU, Peds. Moreover, MRI and Nuclear Medicine—as well as Ultrasound—are "sectional" modalities and thus directly relevant to many issues raised in this book.

13 See Myers, "Fictions for Facts," on authority in the scientific dialogue and the role of the "simpleton."

14 On a distant but relevant historical form of inquest, see Burney, *Bodies of Evidence*.

15 Another rationale for this confrontational style of teaching: the American Board of Radiology still subjects candidates to oral examination. For a brief history of graduate education in radiology, see Krabbenhoft, "Certification and Education." For analysis and variations of resident teaching modalities, see Felson et al., "Viewbox Seminar"; Miller and Andrew, "View Box Exercises for Teaching Problem Solving in Radiology"; Imray et al., "A Modification of the Radiology Teaching Conference."

16 This case is a montage from several hot-seat conferences. Consent for audio recording was obtained for only one conference.

17 Attending to resident, interrupting a long perusal at the viewbox: "Has the shot clock run out yet? You're going to have to go for the three-pointer on this one."

18 Early hospital radiographers included physicists, engineers, and photographers as well as clinicians. Physician-radiographers' strategies for distancing themselves from photographer-radiographers included, in the early 1900s, taking possession of radiographs (patients purchase an interpretation, not a photograph—1913) and not showing patients photographs (would one show a patient a microscopic slide of urinary sediment?—1907). Howell, *Technology in the Hospital*, 159–60. The American Roentgen Ray Society had a physicist as its president in 1902, but in 1905, it eliminated "questionable practitioners" and established the goal of making radiology a clinical specialty. Kevles, *Naked to the Bone*, 85. Early English and Dutch radiological societies are considered in Pasveer, "Knowledge of Shadows."

19 Term from the judicial opinion in *Smith v. Grant* (Denver, 1896). Withers, "The Story of the First Roentgen Evidence."

20 Howell, *Technology in the Hospital*, 106. Kevles, *Naked to the Bone*, 30, discusses the inaugural case of *Cunnings v. Holder* (Court of Queen's Bench, Montreal, February 1896).

21 A phrase recalled by the plaintiff's attorney in *Smith v. Grant*—citing judges' sustaining of objections to x-rays in prior cases. See Withers, "The Story of the First Roentgen Evidence."

22 Ibid., 100. On the importance of radiographs of hands in securing popular belief in the power of x-rays in the early 1900s, see Cartwright, *Screening the Body*, 115ff; Howell, *Technology in the Hospital*, 138.

23 The Supreme Court of Tennessee declared in 1897 that if maps, diagrams, and photographs were admissible in court, so were x-rays (assuming credible expert testimony). Reiser, *Medicine and the Reign of Technology*, 66; Kevles, *Naked to the Bone*, 93.

24 Technique and safety were the early tasks of radiological societies on both sides of the Atlantic. In the later 1920s, radiological societies addressed economic relations of radiologists and hospitals. Hospitals sought to employ radiologists as salaried technical personnel; but radiologists sought to be independent experts and to control fees appropriate to complexities of image interpretation. Howell, *Technology in the Hospital*, 126–27. See Krabbenhoft, "Certification and Education," on the beginnings of American organized radiology, culminating in the uniting in 1939 of the American Roentgen Ray Society and the Radiological Society of North America under the umbrella of the American College of Radiology. See also Kevles, *Naked to the Bone*, 84–85.

25 In considering early medicolegal x-rays that depart from bony anatomy, Kevles wonders "what, if anything, had really been 'seen' by doctors and jurors alike in the early years and if everyone had colluded in an emperor's new clothes scenario." *Naked to the Bone*, 96.

26 Ibid., 94.

27 Garland, "The Interpretation of X-Rays in Court Hearings," *American Journal of Medical Jurisprudence* 1 (1938), as quoted in Kevles, *Naked to the Bone*, 94–95.

28 On radiologic testimony in forensic settings, see Dumit, *Minding Images*, 195–219; Collins, "Origins of Medico-Legal and Forensic Roentgenology," in Bruwer, ed., *Classic Descriptions of Diagnostic Roentgenology*, 1578–1604. One interesting episode involved a proposal in the late 1890s for x-ray supplementation of Bertillon's anthropometric methods of criminal identification (see chapter 3, "Diagnosis").

29 When radiologists and trainees interpret images, they know they may be held accountable for their mistakes or may hold others to account. Pressures of conference testimony are in some measure informed by the prospect of summons to court and requirements of legal judgment. For a less facile comparison of courtroom and biomedical domains, see Latour, "Scientific Objects and Legal Objectivity."

30 With "exudative and . . . organizing" the attending shifts to physiology—and then abruptly shifts again to a different imaging modality. As if, on a turret of microscopic objectives, he has rotated two new lenses into place, one after another, within seconds. The first shows a physiologic process—inflammatory goo "organizing"; the second shows ultrasound attributes of structure—surfaces and gas/fluid interfaces.

31 To say these terms are descriptive is to miss the sense in which they are also already evaluative, interpretive. Descriptions are not "pure." Recall fallacies of photographic naturalism (chapter 1, "Reading and Writing").

32 Miller described a weekly ceremonial in a U.S. radiology department, circa 1978, to confer a "White Cane Award"—for "outstanding ability in the art of being unable

to see the lesion on the roentgenogram." " [The nominee] is given one last chance to reread the films and discover the lesion that he has missed. Early in the game he does not see it even after this warning. The lesion is then pointed out to him . . . The small embarrassment caused by one's name being on the plaque for his fellow resident to see sensitizes one to his mistake. He generally does not repeat it." Miller noted that everyone on the staff (residents, attendings, chiefs) had had the award at some time. See "Tunnel Vision Award," 97–98.

33 Radiological literature also has a "trailing edge." Radiology (a century-old discipline) has little use for cases stretching much more than thirty years into the past. This is a particularly acute form of the presentism which typifies much biomedical practice. Some diagnostic traditions maintain very different relations to distant pasts—for instance those of traditional Chinese doctors, as discussed in Farquhar, *Knowing Practice*.

34 For discussion of image:image comparison in securing the evidential status of x-rays in the early twentieth century, see Pasveer, "Knowledge of Shadows," 374–76.

35 On contiguity as one of three kinds of association between ideas in the imagination (along with resemblance and causality), see Hume, *A Treatise of Human Nature*, 11.
      I begin with contiguity in a material sense. But images continue to interact with one another off the viewbox, in recollections and imaginations of radiologists. For instance, the same case is seen differently in different clusters or series. See Egglin and Feinstein, "Context Bias."

36 See Bosk on "putting on the hair shirt," *Forgive and Remember*, 138ff.

37 In another conference, discussing residents' experiences of a qualification exam, the attending jokes: "One of the first things we teach you is how to do CT reconstructions from raw data." Clearly preposterous: this tedious function is black-boxed into the CT computer.

38 Radiologists are not alone in these tissue-procurement responsibilities. Clinicians perform biopsies and draw fluid from bodily spaces. Venipuncturists procure materials for the pathologist's lenses. But this is an expanding responsibility for radiologists.
      For sociological description of interventional radiology practices, see Barley, *The Professional, the Semiprofessional, and the Machines*, 197–220; Lammer, "Horizontal Cuts and Vertical Penetration."

39 "Interventional" and "Vascular" are interchangeable in this discussion: both refer to a service which performs biopsies and plumbing procedures with needles, catheters, filters, and contrast.

40 An oversimplification: whole-body screening CT is a growing entrepreneurial enterprise—notwithstanding a statement of the American College of Radiology (2002) disputing its cost-effectiveness.

41 On history of correlating imaging with symptoms and physical signs, see Pasveer, "Knowledge of Shadows," 371–74.

42 See chapter 2, "Cutting," on spiral CT: scanners now collect information continuously during the steady movement of the table, rather than incrementally. Faster

scans produce better images of the chest, reducing blurring due to breathing. In 1996 and 1997 the diagnostic implications of this improvement were still new. In the ensuing decade it became common to use CT to diagnose PE—indeed CT is becoming a new "gold standard." Quiroz et al., "Clinical Validity of a Negative Computed Tomography Scan in Patients with Suspected Pulmonary Embolism."

43 In a different university hospital (1991): a CT radiologist cites the importance of "good in-house cytology"—allowing small samples, small needles, and fewer complications. He also praises a particular pathologist who has "skill and courage" to "make a diagnosis from cells"—and asserts one can "almost go anywhere in the body with impunity with a number 22 needle."

44 "Fascinoma" is slang playing on a medical lexeme signifying "tumor"—as in carcinoma. It can mean any interesting lesion. (Perhaps it can also connote the uncontrolled growth of fascination itself.)

45 See Virilio, "Public Image," *The Vision Machine*, 33–45. He offers a genealogy of detective interest seeded in the French Revolution, in which Poe is only part of an arc from Balzac to Baudelaire, thence to Flaubert and the aesthetics of analytic realism. Virilio suggests that the "coup de theatre" of this aesthetic has become a casualty of the "hyperreal" condition of bodies speaking for themselves: "Now that the court arena has become first a movie-projection room, then a video chamber, legal representatives of all stripes have lost all hope of creating within it, with the means at their disposal, a *reality-effect* capable of captivating the jury and the audience" (44).

6. Exposition

1 Ninety-five percent of U.S. radiology residents attend the course—in their second or third year. Residents from Canada, Spain, Switzerland, the Netherlands, and other nations also attend.

2 The department also teaches one-week courses annually, in each of its six primary sections, in Washington, D.C., and weekend courses elsewhere in the United States. Internationally, it offers courses in Mexico, Brazil, Austria, Germany, France, Netherlands, Spain, Portugal, Saudi Arabia, and Taiwan. Its enrollees come from 355 institutions—200 U.S. residency programs.

3 These began with the Crystal Palace exhibition in 1851. The first American exposition, in New York in 1853, of "Machinery," was directed by P. T. Barnum. The Paris Exposition of 1867 featured archaeological and ethnological materials. Hinsley, "The World as Marketplace," 345: "Virtually all subsequent fairs embodied these two aspects: displays of industrial achievement and promise for the regional or national metropolis, and exhibits of primitive 'others' collected from peripheral territories or colonies."

The Army Medical Museum exhibited photos of microscopic sections prepared by J. J. Woodward at the Centennial Exposition in Philadelphia in 1876 (figures 12 and 35a); some of these were also among the eighty-two specimens exhibited at the Columbian Exhibition in Chicago in 1893 (figure 35b) (Henry, *The Armed Forces Institute of Pathology*, 98–99).

4   Other sponsors include a manufacturer of CT scanners and several makers of contrast.

5   Express goals of the course are to enable participants to: "apply the principles of radiologic-pathologic correlation to the interpretation of radiologic studies; apply an understanding of the clinical and pathologic implications of the radiologic appearances of imaging interpretation; refine differential diagnoses in various organ systems based on specific imaging features." Http://radpath.org/index.cfm?pid=course&sub=description (accessed 2000).

6   Dr. Rosado de Christenson is one of my few interlocutors not assigned a pseudonym in this book. She graciously permitted me to quote her. Her comments are not official representations of views of the AFIP or the Department of Defense.

7   Henry, *The Armed Forces Institute of Pathology*, 11, 250.

8   Wagner, Greenspan, et al., "From the Director," http://www.afip.org (last accessed 2003). The AFIP is referred to here as "the People's Institute."

9   Henry, *The Armed Forces Institute of Pathology*, 11–15. The first curator of the Museum was John Hill Brinton.

10 Ibid., 55. From *Lippincott Magazine* (1871).

11 Ibid., 56. Quoted from memoirs of Brinton (1912).

12 Ibid.

13 Ibid., 62, citing Berenger-Feraud from the *Gazette des hôpitaux civils et militaires* (1870).

14 Ibid., 65.

15 Billings summarized purposes of the Museum's work in the *Medical News* of Philadelphia (1886): "1. To illustrate the effects, both immediate and remote, of wounds and of the diseases that prevailed in the Army. 2. To illustrate the work of the Army Medical Department; models of transportation of sick and wounded, and of hospitals; medical supplies; instruments; etc. 3. To illustrate human anatomy and pathology of both sexes and of all ages. 4. To illustrate the morphological basis of ethnological classification, more especially of the native races of America, including anthropometry, and craniology. 5. To illustrate the latest methods and apparatus for biological investigations and the various methods of preparing and mounting specimens." Henry, *The Armed Forces Institute of Pathology*, 84. After his military career, Billings went on to become the first director of the New York Public Library (ibid., 103).

16 Ibid., 101.

17 Ibid., 164–66.

18 Ibid., 160–63, 179–80.

19 Ibid, 207–25. According to Major Callender, curator of the Museum (1930), the Registry's purpose was to "collect data and specimens from patients, especially those with tumors, with a view to accumulating a sufficient number of instances of each

disease to determine its characteristic course, the criteria for diagnosis, and to evaluate methods of treatment" (209).

20   Henry, *The Armed Forces Institute of Pathology*, 235.

21   Ibid., 205, quoting Callender, the curator, in memorandum of 1922.

22   The first Institute, the Army Institute of Pathology, was initially subordinate to the Army Medical Museum (1944). Then these relations were inverted, with the Institute supervising work of the Museum (1946). Then the Institute was expanded as the Armed Forces Institute of Pathology, and the Museum renamed the Medical Museum of the AFIP (1949). In 1974 the Museum was renamed the Armed Forces Medical Museum, and in 1989 the National Museum of Health and Medicine.

23   See also Connor and Rhode, "Shooting Soldiers."

24   The images are scanned at 300dpi and stored in TIF format. A full 14 x 17 inch film is converted into an 18Mb file. Dr. Rosado de Christenson indicated that this change from filmic to digital archiving was a difficult decision—only partly related to dwindling supplies of Log-E film. Difficulties in the transition included technical issues; standard PACS (Picture Archiving and Communications Systems) architectures were not immediately useful for the AFIP's purposes.

25   The Repository holds 2.7 million cases, according to Dr. Rosado de Christenson.

26   http://radpath.afip.org/radpath/select.html (accessed 2000). These case preparation instructions suggest it may be easier to "start with a case that already has the gross photographs." (Instructions have subsequently been revised.) Requirement of gross photography is waived, for one of two cases a resident submits, for only these reasons: "1. specimen removed in small fragments; 2. needle biopsy (no surgical resection and no autopsy); 3. pathognomonic radiologic appearance (AND no surgery, no autopsy)" (*pathognomonic* = characteristic or indicative of disease). Requirement of pathologic material (minimally, histologic slides) is likewise only waived, for one case, if "the diagnosis can be made without pathologic material (e.g., congenital heart disease, some CNS conditions)."

27   Ibid.

28   Departmental policy prohibits Rad-Path faculty/staff from writing up residents' materials as case reports—though the AFIP does retain rights to use gross and radiologic images in publications and educational materials. The AFIP does not copyright material in its archives. Since 2001, *Radiographics* has regularly published "Best Cases from the AFIP": http://www.rsna.org/Education/archive/bestcases.cfm (accessed 2008).

29   Dr. Rosado de Christenson credits Dr. Jeffrey Galvin with the Department of Radiologic Pathology's commitment to point-of-care service, which has culminated most recently in the "Ask AFIP" web portal: www1.askafip.org (2007).

30   http:/radpath.afip.org/radpath/organ.html (accessed 2000).

31   Interview with Dr. John E. Madewell by Charles Stuart Kennedy, AFIP Oral History Program, October 20, 1994.

32   Madewell interview, 12.

33   There are two Technical Exhibit areas, one each in the North and East buildings.

34 Voxel is a unit of volume in a volumetric (3D) matrix: a little box. The term is meant to correspond to a pixel in a 2D image array. A voxel can be designated by x/y/z coordinates and assigned various kinds of qualities or values: density, intensity, name. Some of these can be displayed pictorially (in grayscale or color) and some textually, as when a particular voxel is indexed with the cursor. The term "voxel" was invented in 1973 by the team that developed the Dynamic Spatial Reconstructor, or DSR Scanner, at the Mayo Clinic. Personal communication, Eric Hoffman (University of Iowa), RSNA 1996.

35 From the "Guide to Technical Exhibitors," *On Display* (1996), 159: "VOXGRAMS simultaneously evoke all four physical depth cues—focus, parallax, convergence, and stereo, in the same fashion as any real 3D object."

36 *On Display* (1996) lists twenty-one vendors for CT equipment.

37 In 1975, GE launched its CT development effort, focusing on "fan-beam" technology. By 1978, it had captured 60 percent of the CT market, and led thereafter. There is limited discussion of GE's early development process in Lynn et al., "Marketing and Discontinuous Innovation."

38 Renato is a pseudonym.

39 In most hospitals, cardiologists perform and review coronary cine-angiograms—whereas nearly all other angiography is performed by radiologists.

40 In 1999, radiologists accounted for roughly half of 24,645 healthcare professional registrants. Another 6,000 were radiology and hospital staff and executives; 2,700 were residents and students; 1,700 were information technology, commercial R&D, and consultant attendees; 315 were "other physicians." http://www.rsna.org/rsna2000/advanceregistration/distribution.html (accessed 2000).

41 One example of this interconnection, apart from corporate sponsorship of individual research projects, is the contribution of corporate sponsors to the RSNA Research and Education Fund. "RSNA 1996 Scientific Program," *Radiology* Supplement 201 (P) (November 1996), 9.

42 Quoted from *Radiographics* advertisement in RSNA, *On Display* (1996), 83.

Impression

1 This aspect of CT may not be so atypical in rituals of cutting. Ancient Greek sacrifice, for instance, focused less on the edge of the blade and the sacrificial animal than on distribution of portions to the social body. Detienne, "Culinary Practices and the Spirit of Sacrifice."

2 *Capital*, part 1, chap. 1, "Commodities."

3 Pietz, "The Problem of the Fetish, II," 23.

4 Benedict, *Curiosity*, 2–3, 9.

CLINICAL TERMS AND JARGON

amiodarone: cardiac rhythm drug, potentially toxic to lungs

anasarca: generalized *edema*, fluid in interstitial spaces of flesh

autonomic neuropathy: problem with nerve networks that support basic functions like blood vessel tone and gut motility

barium enema: x-ray of lower colon lined with contrast material

BCNU: cancer chemotherapy drug

bitemporal: anatomically, on both sides of the brain or skull; visually, involving outer halves of visual fields of both eyes

bronchoscope: (as verb) to insert a brochoscope into a person's airway to look down its branches, suck out secretions, sometimes biopsy adjacent tissue

call case: case seen initially by the on-call resident

central lines: intravenous lines in central veins (close to the heart)—typically subclavian or internal jugular veins—kept meticulously clean

centrum semiovale: anatomic region of brain

Chance fracture: transverse fracture of a vertebral body

contrast or contrast medium: any substance—ingested or injected—used to enhance visibility of structures or fluid in imaging studies of the body

creatinine: a laboratory blood test of renal function

desmoplastic response: fibrous proliferation

dissection of aorta: spontaneous splitting of the aortic wall by blood under pressure

DL_{CO}: Pulmonary Function Lab analysis of expired breath, from a subject who has breathed a mixture including helium and carbon monoxide (CO)—held to measure how efficiently gas is transferred from airway to capillary blood

Echinococcus: cyst-forming parasite, uncommon in the United States

foley: catheter for draining urinary bladder

ground-glass opacity: homogenous, fine-grained opacification of lung areas on CT, held to correlate with inflammation or edema

hemangioma: vascular tumor, usually purplish

hemianopsia: loss of sight in half of an eye's visual field

hemosiderosis: accumulation of heme pigments (blood breakdown products) in tissues

heparin: blood thinner: for treatment of clotting, usually given intravenously, thus requires hospitalization (and sometimes affects platelet counts)

herpes zoster: "shingles," painful rash from recurrent chicken pox in an adult, usually along a strip of skin

histology: study of microscopic structure, usually in thin slices of stained tissue (in pathology departments, histopathology is distinguished from cytopathology and from surgical pathology)

Hounsfield number: calculated density (named for one of CT's inventors)

hypercoagulability: tendency for flowing blood to clot too readily

infarction: tissue death

intercostal block: anesthetic injection near nerve between ribs to block pain sensation

low molecular weight heparin: newer blood thinner (in 1997) administered by subcutaneous injection—potentially at home—thus an attractive alternative to intravenous heparin

MAI: Mycobacterium Avium Intracellulare, a TB-like organism

mediastinum: central chest area, including heart and central vessels

mesentery: supportive tissue of bowel, including blood supply and fat

metastases: secondary cancer lesions arising remotely (sometimes in a different organ) from a primary cancer

MRA: magnetic resonance angiogram

nephrostomy: procedure to drain urinary outflow from an obstructed kidney

orthostatic hypotension: drop in blood pressure with postural shift to upright position

paraneoplastic syndrome: cluster of systemic effects associated with a neoplasm (cancer)

percutaneous: through the skin

peritoneal irritation: sign of something in normally empty peritoneal space (surrounding abdominal viscera)—blood, fluid, air, pus, or bowel leakage

platelets: blood particles involved in forming clots

PML: Progressive Multifocal Leukoencephalopathy, disease of demyelinating brain lesions, usually at the gray-white junction, associated with JC viral infections, often in immunocompromised patients

Pneumocystis: organism which can infect lungs and impair gas exchange

pulmonary embolus: clot that forms in a vein, breaks loose, and travels through the venous system, through the right heart, to lodge in a pulmonary artery

pulmonary thrombus: clot that forms within a pulmonary artery

pyonephrosis: pus in or around kidney

saccadic: jerky—usually referring to eye movements

saddle thrombus: blood clot lodged in the split of the main pulmonary artery, impeding blood flow to lungs

sed rate: sedimentation rate of red blood cells in a tube—rapid in less viscous serum, suggesting inflammatory condition

sickle-thal: combination of abnormal hemoglobins—a sickle cell anemia and a thalassemia

spectroscopy: MR spectroscopy, a refinement of MR technique—mostly a research tool, in 1997—that may identify molecular constituents or metabolites specific to tissue types

T1, T2: MRI imaging sequences

Thorotrast: older, radioactive contrast material noted for very slow clearance from the body

thymoma: overgrowth of the thymus, a gland in the upper chest

TIPS: Transjugular Intrahepatic Portosystemic Shunt, a procedure (through a catheter in a neck vein) to create a shunt to modify liver blood flow

walkie-talkie: slang for "outpatient"

BIBLIOGRAPHY

Ackerknecht, Erwin H. "Cuvier and Medicine." *Gesnerus* 45 (1988): 313–15.
———. *Medicine at the Paris Hospital, 1794–1848*. Baltimore: Johns Hopkins University Press, 1967.
Adorno, Theodor W. "Valéry Proust Museum." In *Prisms*, translated by Samuel Weber and Sherry Weber, 175–85. Cambridge, Mass.: MIT Press, 1981.
Agamben, Giorgio. *Homo Sacer: Sovereign Power and Bare Life*. Translated by Daniel Heller-Roazen. Stanford, Calif.: Stanford University Press, 1998.
Alpers, Svetlana. *The Art of Describing: Dutch Art in the Seventeenth Century*. Chicago: University of Chicago Press, 1983.
———. "The Museum as a Way of Seeing." In Karp and Lavine, eds., *Exhibiting Cultures*, 25–32.
Amann, Klaus, and Karin Knorr Cetina. "Thinking through Talk: An Ethnographic Study of a Molecular Biology Laboratory." *Knowledge and Society* 8 (1989): 3–26.
American College of Radiology. "ACR Standard for Communication: Diagnostic Radiology." *American College of Radiology Standards*. Reston, Va.: American College of Radiology, 1996.
American Hospital Association. *CT Scanners: A Technical Study*. Chicago: American Hospital Association, 1977.
Anderson, Alan, and Alexander Tsiaras. "The Body Made Vivid: Safer Scans Improve Diagnosis." *Life*, May 1984, 126–34.
Anderson, Benedict. "Census, Map, Museum." In *Imagined Communities: Reflections on the Origin and Spread of Nationalism*, 163–85. New York: Verso, 1991.
Anderson, Gerard. *Multinational Comparisons of Health Care: Expenditure, Coverage, and Outcomes*. New York: The Commonwealth Fund, 1998.

Ariès, Philippe. *Western Attitudes toward Death from the Middle Ages to the Present.* Translated by Patricia Ranum. Baltimore: Johns Hopkins University Press, 1974.

Atkinson, Paul. *Medical Talk and Medical Work: The Liturgy of the Clinic.* London: Sage, 1995.

Bachelard, Gaston. *The Poetics of Space.* Translated by Maria Jolas. Boston: Beacon Press, 1969.

Baker, Hillier L., Jr. "Historical Vignette: Introduction of Computer Tomography in North America." *American Journal of Neuro-Radiology* 14 (March/April 1993): 283–87.

Baker, Stephen R. "The Diffusion of High Technology Medical Innovations: The Computed Tomography Scanner Example." *Social Science and Medicine* 13 (1977): 155–62.

———. "PACS and Radiology Practice: Enjoy the Benefits but Acknowledge the Threats." *American Journal of Roentgenology* 173 (1999): 1173–74.

Baltrusaitis, Jurgis. "Animal Physiognomy." In *Aberrations,* 1–58. Cambridge, Mass.: MIT Press, 1989.

Bann, Steven. "Poetics of the Museum: Lenoir and Du Sommerard." In *The Clothing of Clio: A Study of the Representation of History in Nineteenth-Century Britain and France,* 77–92. Cambridge: Cambridge University Press, 1984.

———. "'Views of the Past': Reflections on the Treatment of Historical Objects and Museums of History." In *The Inventions of History: Essays on the Representation of the Past,* 122–47. Manchester: Manchester University Press, 1990.

Banta, H. David. "The Diffusion of the Computed Tomography (CT) Scanner in the United States." *International Journal of Health Services* 10, no. 2 (1980): 251–69.

Barbour, John. "Imaging Wizardry Helping Doctors: Science Making Body Transparent." *Herald Sun,* July 18, 1993, G1.

Barker, Francis. *The Tremulous Private Body: Essays on Subjection.* New York: Methuen, 1984.

Barley, Stephen R. "On Technology, Time, and Social Order: Technically Induced Change in the Temporal Organization of Radiological Work." In *Making Time: Ethnographies of High-Technology Organizations,* edited by Frank A. Dubinskas, 123–69. Philadelphia: Temple University Press, 1988.

———. "The Professional, the Semi-Professional, and the Machine: The Social Ramifications of Computer Based Imaging in Radiology." Ph.D. diss., Massachusetts Institute of Technology, 1984.

———. "The Social Construction of a Machine: Ritual, Superstition, Magical Thinking and Other Pragmatic Responses to Running a CT Scanner." In *Biomedicine Examined,* edited by Margaret Lock and Deborah Gordon, 497–540. Dordrecht: Kluwer Academic Publishers, 1988.

———. "Technology as an Occasion for Structuring: Observations on CT Scanners and the Social Role of Radiology Departments." *Administrative Science Quarterly* 31 (1986): 78–108.

Barnes, Barry, and David Edge, eds. *Science in Context: Readings in the Sociology of Science.* Cambridge, Mass.: MIT Press, 1982.

Barret-Kriegel, Blandine. "L'Hôpital comme équipement." In *Les Machines à guérir: Aux origines de l'hôpital moderne,* edited by Michel Foucault et al., 19–30. Belgium: Solédi, 1979.

Barrett, James, et al. "Unobtrusive Evaluation of Mammographers' Eye Movements during Diagnosis of Mammograms." *Radiographics* 16 (1996): 167–72.

Barthes, Roland. *Camera Lucida: Reflections on Photography.* Translated by Richard Howard. New York: Hill and Wang, 1981.

———. "The Rhetoric of the Image." In *Image/Music/Text,* translated by Stephen Heath, 32–51. New York: Hill and Wang, 1977.

Bashshur, Rashid, Jay Sanders, and Gary Shannon, eds. *Telemedicine: Theory and Practice.* Springfield, Ill.: C. C. Thomas, 1997.

Baudrillard, Jean. *L'Echange symbolique et la mort.* Paris: Gallimard, 1976.

———. *Simulations.* New York: Semiotexte, 1983.

———. "Symbolic Exchange and Death." In *Selected Writings,* edited by Mark Poster, 119–48. Stanford, Calif.: Stanford University Press, 1988.

Beard, David, R. Eugene Johnston, Osamu Toki, and Claire Wilcox. "A Study of Radiologists Viewing Multiple Computed Tomography Examinations Using an Eyetracking Device." *Journal of Digital Imaging* 3, no. 4 (November 1990): 230–37.

Beaulieu, Anne. "Images Are Not the (Only) Truth: Brain Mapping, Visual Knowledge, and Iconoclasm." *Science, Technology, and Human Values* 27, no. 1 (winter 2002): 53–86.

Benedict, Barbara M. *Curiosity: A Cultural History of Early Modern Inquiry.* Chicago: University of Chicago Press, 2001.

Benjamin, Walter. *The Arcades Project.* Edited by Rolf Tiedemann; translated by Howard Eiland and Kevin McLaughlin. Cambridge, Mass.: Belknap Press, 1999.

———. *Charles Baudelaire: A Lyric Poet in the Era of High Capitalism.* Translated by Harry Zohn. London: New Left Books, 1973.

———. "Doctrine of the Similar." Translated by Knut Tarnowski. *New German Critique* 17 (spring 1979): 65–69.

———. *Illuminations.* Edited by Hannah Arendt. Translated by Harry Zohn. New York: Schocken Books, 1969.

———. "A Small History of Photography." In *One Way Street and Other Writings,* translated by Edmund Jephcott and Kingsley Shorter, 240–57. London: New Left Books, 1979 (1931).

Bennett, Tony. *The Birth of the Museum.* New York: Routledge, 1995.

———. "Pedagogic Objects, Clean Eyes, and Popular Instruction: On Sensory Regimes and Museum Didactics." *Configurations* 6, no. 3 (1998): 345–71.

Bentley, Thomas Leslie James. "The Evolution of Radiography." *Radiography* 12 (1946): 14–33.

Berbaum, Kevin D., E. A. Franken Jr., D. D. Dorfman, R. T. Caldwell, and E. A. Krupin-ski. "Role of Faulty Visual Search in the Satisfaction of Search Effect in Chest Radiography." *Academic Radiology* 5, no. 1 (January 1998): 9–19.

Berg, Marc. "Order(s) and Disorder(s): Of Protocols and Medical Practices." In Berg and Mol, eds., *Differences in Medicine,* 226–46.

———. *Rationalizing Medical Work: Decision Support Techniques and Medical Practices.* Cambridge, Mass.: MIT Press, 1997.

Berg, Marc, and Geoffrey Bowker. "The Multiple Bodies of the Medical Record: Towards a Sociology of an Artifact." *Sociological Quarterly* 38, no. 3 (1997): 513–37.

Berg, Marc, and Annemarie Mol, eds. *Differences in Medicine: Unraveling Practices, Techniques, and Bodies.* Durham, N.C.: Duke University Press, 1998.

Berger, John. *Ways of Seeing.* Harmondsworth, Middlesex, England: Penguin, 1972.

Bergeron, R. Thomas. "Leaving Academia and Bringing Organ-System Radiology to a Community Hospital." *Clear Images,* July 1987, 20–23.

Bergman, Mats, and Sami Paavola, eds. "The Commens Dictionary of Peirce's Terms." http://www.helsinki.fi/science/commens/dictionary.html.

Bergvall, Ulf, Torgny Greitz, and Ladislau Steiner. "Computer Tomography in Post-Mortem Examination of the Brain and Other Specimens." *Acta Radiologica (Stockholm): Supplementum* 346 (1975): 39–44.

Bertillon, Alphonse. *Identification anthropométrique, instructions signalétiques.* Melun: Imprimerie Administrative, 1893.

Besançon, Alain. *The Forbidden Image: An Intellectual History of Iconoclasm.* Translated by Jane Marie Todd. Chicago: University of Chicago Press, 2000.

Biagioli, Mario. "Aporias of Scientific Authorship." In Biagioli, ed., *The Science Studies Reader,* 12–30.

———, ed. *The Science Studies Reader.* New York: Routledge, 1999.

Blume, Stuart. *Insight and Industry: On the Dynamics of Technological Change in Medicine.* Cambridge, Mass.: MIT Press, 1992.

Bocage, André-Edmund-Marie. "Methods of, and Apparatus for, Radiography on a Moving Plate." In Bruwer, ed., *Classic Descriptions of Roentgenology,* 1414–17.

Bohm, Christian, Torgny Greitz, and Lars Eriksson. "A Computerized Adjustable Brain Atlas." *European Journal of Nuclear Medicine* 15, no. 11 (1989): 687–89.

Bone, Roger C. "The ARDS Lung: New Insights from Computed Tomography." *Journal of the American Medical Association* 269, no. 16 (April 28, 1993): 2134–35.

Bonitzer, Pascal. "Hitchcockian Suspense." In *Everything You Always Wanted to Know About Lacan (but Were Afraid to Ask Hitchcock),* edited by Slavoj Žižek, 15–30. New York: Verso, 1992.

Borck, Cornelius. "Writing Brains: Tracing the Psyche with the Graphical Method." *History of Psychology* 8, no. 1 (February 2005): 79–94.

Borges, Jorge Luis. "The Detective Story." Translated by Alberto Manguel. *Descant* 51 (winter 1985–86): 15–24.

Bosk, Charles L. *Forgive and Remember: Managing Medical Failure.* Chicago: University of Chicago Press, 1979.

Bouillion, Victor. "War and Medicinema: The X-Ray and Irradiation in Various Theatres of Operations: A Selected 100-Year Chronology." In *Incorporations,* edited by Jonathan Crary and Sanford Kwinter, 253. New York: Zone, 1992.

Bourdieu, Pierre. *The Logic of Practice.* Translated by Richard Nice. Stanford, Calif: Stanford University Press, 1990.

———. "The Specificity of the Scientific Field and the Social Conditions of the Progress of Reason." In Biagoli, ed., *The Science Studies Reader,* 31–50.

Bowker, Geof. "Pictures from the Subsoil, 1939." In *Picturing Power: Visual Depiction and Social Relations,* edited by Gordon Fyfe and John Law, 221–54. London: Routledge, 1988.

Bracegirdle, Brian. *A History of Microtechnique: The Evolution of the Microtome and the Development of Tissue Preparation.* Lincolnwood, Ill.: Science Heritage, 1986.

Braitberg, Victor. "Liberators, Innovators, and Experts: Struggles for the Telemedical Future in the Shadow of Neoliberal Reform." Ph.D. diss., University of North Carolina, 2002.

Brand, Dana. "Reconstructing the 'Flâneur': Poe's Invention of the Detective Story." *Genre* 18 (spring 1985): 36–56.

Brecher, Ruth, and Edward Brecher. *The Rays: A History of Radiology in the United States and Canada.* Baltimore: Williams and Wilkins, 1969.

Brodwin, Paul E., ed. *Biotechnology and Culture: Bodies, Anxieties, Ethics.* Bloomington: Indiana University Press, 2000.

Bronzino, Joseph D., Vincent H. Smith, and Maurice L. Wade. *Medical Technology and Society: An Interdisciplinary Perspective.* Cambridge, Mass.: MIT Press, 1990.

Brown, Richard Harvey. "Logics of Discovery as Narratives of Conversion: Rhetorics of Invention in Ethnography, Philosophy, and Astronomy." *Philosophy and Rhetoric* 27, no. 1 (1994): 1–34.

Bruno, Giuliana. "Spectatorial Embodiments: Anatomies of the Visible and the Female Bodyscape." *Camera Obscura* 28 (1992): 239–61.

Bruwer, André J., ed. *Classic Descriptions of Roentgenology,* vol. 2, translated by James D. Bricker. Springfield, Mass.: Charles C. Thomas, 1964.

Bucher, Rue. "Pathology, a Study of Social Movements within a Profession." In *Medical Men and Their Work,* 113–27. Chicago: Atherton Press, 1972.

Buck-Morss, Susan. *The Dialectics of Seeing: Walter Benjamin and the Arcades Project.* Cambridge, Mass.: MIT Press, 1989.

Bull, James. "The History of Computed Tomography." In *Radiology of the Skull and Brain: Technical Aspects of Computed Tomography,* vol. 5, edited by Thomas Newton and D. Gordon Potts, 3835–52. St. Louis: Mosby, 1981.

Burack, Robert C., et al. "The Challenging Case Conference: An Integrated Approach to Resident Education and Support." *Journal of General Internal Medicine* 6 (July/August 1991): 355–59.

Burgener, Francis A., and Martti Kormano. *Differential Diagnosis in Conventional Radiology.* New York: Thieme-Stratton, 1985.

Burgin, Victor. "The End of Art Theory." In *The End of Art Theory: Criticism and Postmodernity*, 140–215. Atlantic Highlands, N.J.: Humanities Press International, 1986.

Burling, Temple, Edith Lentz, and Robert Wilson. *The Give and Take in Hospitals: A Study of Human Organization in Hospitals*. New York: G. P. Putnam's Sons, 1956.

Burney, Ian A. *Bodies of Evidence: Medicine and the Politics of the English Inquest, 1830–1926*. Baltimore: Johns Hopkins University Press, 2000.

Burton, Elizabeth, Dana Troxclair, and William Newman III. "Autopsy Diagnoses of Malignant Neoplasms: How Often are Clinical Diagnoses Incorrect?" *Journal of the American Medical Association* 280, no. 14 (1998): 1245–48.

Butler, Judith. "Bodies That Matter." In *Bodies That Matter: On the Discursive Limits of "Sex,"* 27–56. New York: Routledge, 1993.

Bynum, William F., and Roy Porter, eds. *Companion Encyclopedia of the History of Medicine*. 2 vols. London: Routledge, 1993.

Bynum, William F., Stephen Lock, and Roy Porter, eds. *Medical Journals and Medical Knowledge: Historical Essays*. London: Routledge, 1992.

Campbell, E. J. Morgan. "The Diagnosing Mind." *Lancet* 1 (1987): 849–51.

Canguilhem, Georges. "Machine and Organism." Translated by Mark Cohen, and Randall Cherry. In *Incorporations*, edited by Jonathan Crary and Sanford Kwinter, 45–69. New York: Zone, 1992.

———. *On the Normal and the Pathological*. Edited by Robert S. Cohen, translated by Carolyn R. Fawcett. Dordrecht: D. Reidel, 1978.

Carter, Daniel, Debra Wiedmeyer, Piero Antuono, and Khang-cheng Ho. "Correlation of Computed Tomography and Postmortem Findings of a Diffuse Astrocytoma: A Case Report." *Computerized Medical Imaging and Graphics* 13, no. 6 (November/December 1989): 491–94.

Cartwright, Lisa. "A Cultural Anatomy of the Visible Human Project." In *The Visible Woman*, edited by Paula Treichler, Constance Penley, and Lisa Cartwright. New York: New York University Press, 1998.

———. "'Experiments of Destruction': Cinematic Inscriptions of Physiology." *Representations* 40 (fall 1992): 129–52.

———. *Screening the Body: Tracing Medicine's Visual Culture*. Minneapolis: University of Minnesota Press, 1995.

———. "Women, X-Rays, and the Public Culture of Prophylactic Imaging." *Camera Obscura* 29 (May 1992): 18–54.

Cartwright, Lisa, and Brian Goldfarb. "Radiography, Cinematography and the Decline of the Lens." In *Incorporations*, edited by Jonathan Crary and Sanford Kwinter, 190–201. New York: Zone, 1992.

Casey, Edward S. "The Place of Space in *The Birth of the Clinic*." *Journal of Medicine and Philosophy* 12, no. 4 (November 1987): 351–56.

Cassiday, Bruce, ed. *Roots of Detection: The Art of Deduction before Sherlock Holmes*. New York: Frederick Ungar, 1983.

Castillo, Mauricio. *Neuroradiology Companion: Methods, Guidelines, and Imaging Fundamentals.* Philadelphia: J. B. Lippincott, 1995.

Chew, Felix S. "*AJR:* The 50 Most Frequently Cited Papers in the Last 50 Years." *American Journal of Roentgenology* 150 (February 1989): 227–33.

Cicourel, Aron. "The Integration of Distributed Knowledge in Collaborative Medical Diagnosis." In *Intellectual Teamwork: Social and Intellectual Foundations of Cooperative Work,* edited by Jolene Galegher, Robert E. Kraut, and Carmen Egido, 221–42. Hillsdale, N.J.: Lawrence Erlbaum Associates, 1990.

Clifford, James. "Power and Dialogue in Ethnography: Marcel Griaule's Initiation." In *The Predicament of Culture: Twentieth-Century Ethnography, Literature, and Art,* 55–90. Cambridge, Mass.: Harvard University Press, 1988.

Coffman, Jeffrey A. "Computed Tomography." In *Brain Imaging: Applications in Psychiatry,* edited by Nancy C. Andreasen, 1–66. Washington, D.C.: American Psychiatric Press, 1989.

Cohn, Simon. "Increasing Resolution, Intensifying Ambiguity: An Ethnographic Account of Seeing Life in Brain Scans." *Economy and Society* 33, no. 1 (February 2004): 52–76.

———. "Seeing and Drawing: The Role of Play in Medical Imaging." In *Skilled Visions: Between Apprenticeship and Standards,* edited by Christina Grasseni, 91–105. Oxford: Berghahn Books, 2007.

Coleman, William. *Georges Cuvier: Zoologist: A Study in the History of Evolution Theory.* Cambridge, Mass.: Harvard University Press, 1964.

Collins, Vincent. "Origins of Medico-Legal and Forensic Roentgenology." In Bruwer, ed., *Classic Descriptions of Roentgenology,* 1578–1704.

Comaroff, Jean. "Medicine: Symbol and Ideology." In *The Problem of Medical Knowledge: Examining the Social Construction of Medicine,* edited by Peter Wright and Andrew Treacher, 49–68. Edinburgh: Edinburgh University Press, 1982.

Connor, James. "Tissues by Sunlight: J. J. Woodward, Biomedical Science, and the Convergent Technology of Photomicrography in Victorian America." Presentation to Society for the Social History of Medicine, Manchester UK, June 25, 2003.

Connor, James, and Michael Rhode. "Shooting Soldiers: Civil War Medical Images, Memory, and Identity in America." *Invisible Culture* 5 (winter 2003). http://www.rochester.edu/in_visible_culture/.

"Consultants or Pairs of Hands?" *Lancet* 707, no. 7958 (1976): 520–21.

Cormack, Allan M. "Early Two-Dimensional Reconstruction (CT Scanning) and Recent Topics Stemming from It." *Journal of Computer Assisted Tomography* 4, no. 5 (October 1980): 658–64.

Court of Appeals of Ohio, Jackson County. *Lambert v. Goodyear Tire & Rubber Co.* 79 Ohio App. 3d 15; 606 N.E.2d 983; 1992 Ohio App.

Crary, Jonathan. *Techniques of the Observer: On Vision and Modernity in the Nineteenth Century.* Cambridge, Mass.: MIT Press, 1992.

Crary, Jonathan, and Sanford Kwinter, eds. *Incorporations.* New York: Zone, 1992.

Crawford, T. Hugh. "Imaging the Human Body: Quasi Objects, Quasi Texts, and the Theater of Proof." *PMLA* 111, no. 1 (1996): 66–79.

Crease, Robert P. "Biomedicine in the Age of Imaging." *Science* 261, no. 5121 (1993): 554.

Crenner, Christopher. "Diagnosis and Authority in the Early Twentiety-Century Medical Practice of Richard C. Cabot." *Bulletin of the History of Medicine* 76 (2002): 30–55.

Csordas, Thomas. "Computerized Cadavers." In *Biotechnology and Culture: Bodies, Anxieties, Ethics,* edited by Paul E. Brodwin, 173–92. Bloomington: Indiana University Press, 2000.

Cunningham, Andrew, and Perry Williams, eds. *The Laboratory Revolution in Medicine.* Cambridge: Cambridge University Press, 1992.

Cuvier, Georges. *Histoire des progres des sciences naturelles depuis 1789 jusqu'à ce jour,* vol. 1. Paris: Baudouin Frères, N. Delangle, 1816.

Cuvier, Georges, and Alexandre Brongniart. *Essai sur la géographie minéralogique des environs de Paris.* Paris: Baudouin, 1811.

Daston, Lorraine, and Peter Galison. "The Image of Objectivity." *Representations* 40 (fall 1992): 81–128.

Debord, Guy. *The Society of the Spectacle.* Translated by Donald Nicholson-Smith. New York: Zone, 1995 (1967).

De Lauretis, Teresa. *The Cinematic Apparatus.* New York: St. Martin's Press, 1980.

Derrida, Jacques. *Of Grammatology.* Translated by Gayatri Spivak. Baltimore: Johns Hopkins University Press, 1976.

———. *The Truth in Painting.* Translated by Geoff Bennington and Ian McLeod. Chicago: University of Chicago Press, 1987.

Descartes, René. *Discourse on Method, Optics, Geometry, and Meteorology.* Translated by Paul Olscamp. Indianapolis, Ind.: Hackett, 2001.

Detienne, Marcel. "Culinary Practices and the Spirit of Sacrifice." In *The Cuisine of Sacrifice among the Greeks,* edited by Marcel Detienne and Jean-Pierre Vernant, translated by Paula Wissing, 1–20. Chicago: University of Chicago Press, 1989.

Detienne, Marcel, and Jean-Pierre Vernant. *Cunning Intelligence in Greek Culture and Society.* Translated by Janet Lloyd. Chicago: University of Chicago Press, 1974.

Dewar, Catharine. "Body Scan." *New England Journal of Medicine* 339, no. 19 (November 5, 1998): 1401.

Dijck, José van. "Bodies without Borders: The Endoscopic Gaze." *International Journal of Cultural Studies* 4, no. 2 (2001): 219–37.

———. *The Transparent Body: A Cultural Analysis of Medical Imaging.* Seattle: University of Washington Press, 2005.

Dodier, Nicolas. "Expert Medical Decisions in Occupational Medicine: A Sociological Analysis of Medical Judgment." *Sociology of Health and Illness* 16, no. 4 (1994): 489–514.

Doubilet, Peter, and Peter G. Herman. "Interpretation of Radiographs: Effect of Clinical History." *American Journal of Roentgenology* 137 (November 1981): 1055–58.

Douglas, Mary. *Purity and Danger: An Analysis of Concepts of Pollution and Taboo.* London: Routledge and Kegan Paul, 1978.

Drayton, Richard. "Collection and Comparison in the Sciences: A Seminar Manifesto," *Occasional Papers* 1 (Oxford: Museum of the History of Science, 1996).

Duden, Barbara. "A Repertory of Body History." In *Fragments for a History of the Human Body: Part Three,* vol. 3, edited by Michel Feher, Ramona Naddaff, and Nadia Tazi, translated and edited by Siri Hustvedt, 470–554. New York: Zone, 1989.

Duhamel, Joseph, and Jean-Claude Roques. "Louis Lumière et les origines photographiques de la tomographie." *La Presse Medicale* 67, no. 46 (1959): 1723–25.

Dumit, Joseph. "Minding Images: PET Scans and Personhood in Biomedical America." Ph.D. diss., University of California, Santa Cruz, 1995.

———. *Picturing Personhood: Brain Scans and Biomedical Identity.* Princeton, N.J.: Princeton University Press, 2004.

Duncan, Carol. "Art Museums and the Ritual of Citizenship." In Karp and Lavine, eds., *Exhibiting Cultures,* 88–103.

Eco, Umberto, and Thomas Sebeok, eds. *The Sign of Three: Dupin, Holmes, Peirce.* Bloomington: Indiana University Press, 1983.

Eddy, David M. "Broadening the Responsibilities of Practitioners: The Team Approach." *Journal of the American Medical Association* 269, no. 14 (April 14, 1993): 1849–55.

Eddy, David M., and Charles H. Clanton. "The Art of Diagnosis: Solving the Clinicopathological Exercise." *New England Journal of Medicine* 308, no. 21 (1982): 1263–68.

Egglin, Thomas K. P., and Alvan R. Feinstein. "Context Bias: A Problem in Diagnostic Radiology." *Journal of the American Medical Association* 276, no. 21 (1996): 1752–55.

Eisenberg, John M. "Sociological Influences on Decision-Making by Clinicians." *Annals of Internal Medicine* 90 (1979): 957–64.

Eisenberg, Ronald L. *Radiology: An Illustrated History.* St. Louis, Mo.: Mosby Year Book, 1992.

Eisenberg, Ronald L., and Philip C. Goodman. "Countdown to Discovery." *American Journal of Roentgenology* 156, no. 5 (May 1991): 944.

Elkins, James. *The Domain of Images.* Ithaca: Cornell University Press, 1999.

———. *Pictures of the Body: Pain and Metamorphosis.* Stanford, Calif.: Stanford University Press, 1999.

———. "On Visual Desperation and the Bodies of Protozoa." *Representations* 40 (fall 1992): 33–56.

———. "Why Are Our Pictures Puzzles? Some Thoughts on Writing Excessively." *New Literary History* 27, no. 2 (1996): 271.

Elks, Martha L. "Rituals and Roles in Medical Practice." *Perspectives in Biology and Medicine* 39, no. 4 (summer 1996): 601.

Ell, Stephen R. "Radiology and History." *Investigative Radiology* 23, no. 12 (1988): 956–58.

Elmore, Joann G., et al. "Variability in Radiologists' Interpretations of Mammograms." *New England Journal of Medicine* 331, no. 22 (December 1, 1994): 1493–99.

Engelhardt, H. Tristram, Jr. "Clinical Judgment." *Metamedicine* 2 (1981): 301–17.

Epstein, Julia. *Altered Conditions: Disease, Medicine, and Storytelling.* New York: Routledge, 1995.

Ettlin, Thierry, Hannes Staehelin, and Udo Kischka. "Computed Tomography, Electroencephalography and Clinical Features in the Differential Diagnosis of Senile Dementia: A Prospective Clinicopathologic Study." *Archives of Neurology* 46 (1989): 1217–20.

Evens, Ronald G. "History of the Mallincrodt Institute of Radiology." *American Journal of Roentgenology* 160 (1993): 1343–48.

———. "Röntgen Retrospective: One Hundred Years of a Revolutionary Technology." *Journal of the American Medical Association* 274, no. 11 (September 20, 1995): 912–16.

Evens, Ronald G., and R. Gilbert Jost. "Economic Analysis of Body Computed Tomography Units Including Utilization." *Radiology* 127 (1978): 151–57.

Evidence-Based Medicine Working Group. "Evidence-Based Medicine: A New Approach to Teaching the Practice of Medicine." *Journal of the American Medical Association* 268, no. 17 (1982): 2420–25.

Eycleshymer, Albert, and Daniel Schoemaker. *A Cross-Section Anatomy.* New York: Appleton-Century-Crofts, 1938.

Fabian, Johannes. *Time and the Other: How Anthropology Makes Its Object.* New York: Columbia University Press, 1983.

Farquhar, Judith. *Appetites: Food and Sex in Postsocialist China.* Durham, N.C.: Duke University Press, 2002.

———. *Knowing Practice: The Clinical Encounter of Chinese Medicine.* Boulder, Colo.: Westview Press, 1993.

———. "Multiplicity, Point of View, and Responsibility in Traditional Chinese Healing." Paper presented at Association for Asian Studies Annual Meeting. Washington, D.C., 1989.

Feher, Michel, Ramona Naddaf, and Nadia Tazi, eds. *Fragments for a History of the Human Body: Part Two.* New York: Zone, 1989.

Feinstein, Alvan R. *Clinical Judgment.* Baltimore: Williams and Wilkins, 1967.

Felker, Marcia Elliott. "Ideology and Order in the Operating Room." In *The Anthropology of Medicine: From Culture to Method,* edited by Lola Romanucci-Ross, Daniel E. Moerman, and Laurence R. Tancredi, 349–65. South Hadley, Mass.: J. F. Bergin; New York: Praeger, 1983.

Felson, Benjamin, et al. "Viewbox Seminar: A New Method for Teaching Roentgenology." *Radiology* 116 (July 1975): 75–78.

Fischer, Harry W., ed. *The Radiologist's First Reader.* Chelsea, Mich.: BookCrafters, 1988.

Fischman, Elliot. "Retracing Röntgen's Discovery." *Diagnostic Imaging* 48, no. 5 (1979): 294–303.

Fishman, Elliot K. "CT Imaging in the Year 2000: Looking Forward." In *Body CT Categorical Course Syllabus,* edited by Elliot K. Fishman and Michael P. Federle, 215–17. Leesburg, Va.: American Roentgen Ray Society, 1994.

Fleck, Ludwik. *Genesis and Development of a Scientific Fact.* Edited by Fred Bradley and Thaddeus J. Trenn. Translated by Thaddeus J. Trenn and Robert K. Merton. Chicago: University of Chicago Press, 1979.

Flodmark, Olof, et al. "Correlation between Computed Tomography and Autopsy in Premature and Full-Term Neonates That Have Suffered Perinatal Asphyxia." *Radiology* 137, no. 1 (October 1980): 93–103.

Foucault, Michel. *The Birth of the Clinic: An Archaeology of Medical Perception.* Translated by A. M. Smith. New York: Random House, 1975.

———. *Death and the Labyrinth: The World of Raymond Roussel.* Translated by Charles Ruas. Garden City, N.Y.: Doubleday, 1986.

———. *Discipline and Punish: The Birth of the Prison.* Translated by Alan Sheridan. New York: Random House, 1979.

———. *The Order of Things: An Archaeology of the Human Sciences.* New York: Random House, Vintage, 1970.

Foucault, Michel, Blandine Kriegel, et al. *Les machines à guérir: Aux origines de l hôpital moderne.* Belgium: Solédi, 1979.

Fox, Renée C. *Essays in Medical Sociology: Journeys into the Field.* New Brunswick, N.J.: Transaction Books, 1988.

———. *The Sociology of Medicine: A Participant Observer's View.* Foundations of Modern Sociology Series. Englewood Cliffs, N.J.: Prentice Hall, 1989.

Frazer, James. *The Golden Bough: A Study in Magic and Religion.* Abridged ed. New York: Macmillan, 1963.

Freidson, Eliot. *Profession of Medicine: A Study of the Sociology of Applied Knowledge.* 1970. Chicago: University of Chicago Press, 1988.

Friedberg, Anne. *Window Shopping: Cinema and the Postmodern.* Berkeley: University of California Press, 1993.

Fryback, Dennis G., and John R. Thornbury. "The Efficacy of Diagnostic Imaging." *Medical Decision Making* 11 (1991): 88–94.

Fyfe, Gordon, and John Law, eds. *Picturing Power: Visual Depiction and Social Relations.* London: Routledge, 1988.

Gagliardi, Raymond A. "The Evolution of the X-Ray Report." *American Journal of Roentgenology* 164 (1995): 501–2.

Gal, Anthony, and Philip Cagle. "The 100-Year Anniversary of the Description of the Frozen Section Procedure." *Journal of the American Medical Association* 294, no. 24 (2005): 3135–37.

Galdston, Iago. "Diagnosis in Historical Perspective." *Bulletin of the History of Medicine* 9 (1941): 367–84.

Gallagher, Catherine, and Thomas Laqueur, eds. *The Making of the Modern Body: Sexuality and Society in the Nineteenth Century.* Berkeley: University of California Press, 1987.

Gallop, Jane. *Thinking through the Body.* New York: Columbia University Press, 1988.

Gandelman, Claude. "The 'Scanning' of Pictures." *Communication and Cognition* 19, no. 1 (1986): 3–24.

Gawande, Atul. "Final Cut: Medical Arrogance and the Decline of the Autopsy." *New Yorker,* March 13, 2001, 94–99.

Geertz, Clifford. *The Interpretation of Cultures.* New York: Basic Books, 1973.

Gennep, Arnold van. *Rites of Passage.* Translated by Monika Vizedom and Gabrielle Caffee. Chicago: University of Chicago Press, 1960.

Gilman, Sander. *Disease and Representation: Images of Illness from Madness to AIDS.* Ithaca, N.Y.: Cornell University Press, 1988.

Ginzburg, Carlos. "Clues: Roots of an Evidential Paradigm." In *Clues, Myths, and the Historical Method,* translated by John and Anne Tedeschi, 96–125. Baltimore: Johns Hopkins University Press, 1989.

Gladwell, Malcom. "The Picture Problem: Mammography, Air Power, and the Limits of Looking." *New Yorker,* December 13, 2004, 74–81.

Goffman, Erving. *Asylums: Essays on the Social Situation of Mental Patients and Other Inmates.* Garden City, N.Y.: Doubleday Anchor, 1961.

———. "Role Distance." In *Encounters: Two Studies in the Sociology of Interaction,* 85–152. Indianapolis: Bobbs-Merrill, 1961.

Goldberg, Marshall. "Ten Rules for the Doctor Detective." *Postgraduate Medicine* 101, no. 2 (February 1997): 23–26.

Goldenberg, Maya. "On Evidence and Evidence-Based Medicine." *Social Science and Medicine* 62 (2006): 2621–32.

Goodman, Nelson. *Languages of Art: An Approach to a Theory of Symbols.* Indianapolis, Ind.: Hackett, 1976.

———. "Seven Strictures on Similarity." In *How Classification Works: Nelson Goodman among the Social Sciences,* edited by M. Douglas and D. Hull, 13–23. Edinburgh: Edinburgh University Press, 1992.

Goodwin, Charles. "Professional Vision." *American Anthropologist* 96, no. 3 (September 1994): 606–33.

———. "Seeing in Depth." *Social Studies of Science* 25, no. 2 (May 1995): 237–74.

Gordon, Deborah. "Clinical Science and Clinical Expertise." In Lock and Gordon, eds., *Biomedicine Examined,* 257–98.

Gordon, Gerald, and G. Lawrence Fisher. *The Diffusion of Medical Technology.* Cambridge, Mass.: Ballinger, 1975.

Granshaw, Lindsay, and Roy Porter, ed. *The Hospital in History.* London: Routledge, 1989.

Grasseni, Cristina, ed. *Skilled Visions: Between Apprenticeship and Standards.* Oxford: Berghahn, 2006.

Gritzer, Glenn. "Occupational Specialization in Medicine: Knowledge and Market Explanations." *Research in the Sociology of Health Care* 2 (1981): 251–83.

Grove, Allen W. "Röntgen's Ghosts: Photography, X-Rays, and the Victorian Imagination." *Literature and Medicine* 16, no. 2 (fall 1997): 141–73.

Hacking, Ian. "Do We See through a Microscope?" In *Images of Science: Essays on Realism and Empiricism,* edited by Paul M. Churchland and Clifford A. Hooker, 132–52. Chicago: University of Chicago Press, 1985.

———. *The Taming of Chance.* Cambridge: Cambridge University Press, 1990.

Haglund, William D., and Corinne L. Fligner. "Confirmation of Human Identification Using Computerized Tomography (CT)." *Journal of Forensic Sciences* 38, no. 3 (May 1993): 708–12.

Hankins, Thomas L. "Blood, Dirt, and Nomograms: A Particular History of Graphs." *Isis* 90 (1999): 50–80.

Haraway, Donna. "The Biopolitics of Postmodern Bodies: Determinations of Self in Immune System Discourse." In *Knowledge, Power, and Practice: The Anthropology of Medicine and Everyday Life,* edited by Shirley Lindenbaum and Margaret Lock, 364–410. Berkeley: University of California Press, 1993.

———. "A Manifesto for Cyborgs: Science, Technology, and Socialist Feminism in the 1980s." *Socialist Review* 80 (1985): 65–107.

———. *Primate Visions: Gender, Race, and Nature in the World of Modern Science.* New York: Routledge, 1989.

———. "When Man Is on the Menu." In *Incorporations,* edited by Jonathan Crary and Sanford Kwinter, 39–43. New York: Zone, 1992.

Harper, Clark. "CT Scanners: The Industry behind the Science." *Radiologic Technology* 51, no. 2 (September–October 1979): 199–202.

Harris, Neil. *Humbug: The Art of P. T. Barnum.* Chicago: University of Chicago Press, 1981.

Harrowitz, Nancy. "The Body of the Detective Model." In *The Sign of Three: Dupin, Holmes, Peirce,* edited by Umberto Eco and Thomas Sebeok, 179–97. Bloomington: Indiana University Press, 1983.

Hartouni, Valerie. "Fetal Exposures: Abortion Politics and the Optics of Allusion." *Camera Obscura* 28 (1992): 130–49.

Hay, John. "The Human Body as a Microcosmic Source of Macrocosmic Values in Calligraphy." In *Theories of the Arts in China,* edited by Susan Bush and Christian Murck, 75–102. Princeton. N.J.: Princeton University Press, 1983.

Hayles, N. Katherine. "The Materiality of Informatics." *Configurations* 1, no. 1 (winter 1993): 147–70.

———. "Virtual Bodies and Flickering Signifiers." *October* 66 (fall 1993): 69–91.

Heath, Christian. *Body Movement and Speech in Medical Interaction.* Cambridge: Cambridge University Press, 1986.

Heidegger, Martin. "The Age of the World Picture." In *The Question Concerning Technology and Other Essays,* translated by William Lovitt, 115–54. New York: Harper and Row, 1977.

———. "The Question Concerning Technology." In *Basic Writings,* 283–318. New York: Harper and Row, 1977.

Helman, Cecil. "The Radiological Eye." In *The Body of Frankenstein's Monster: Essays in Myth & Medicine,* 13–18. New York: W. W. Norton, 1991.

Hendee, William R. "Cross Sectional Medical Imaging: A History." *Radiographics* 9, no. 6 (November 1989): 1155–80.

Henderson, Linda Dalrymple. "X-Rays and the Quest for the Invisible Reality in the Art of Kupka, Duchamp, and the Cubists." *Art Journal* 47 (winter 1988): 323–40.

Henry, Robert S. *The Armed Forces Institute of Pathology: Its First Century, 1862–1962.* Washington, D.C.: Office of the Surgeon General, 1964.

Herbert, Christopher. *Culture and Anomie: Ethnographic Imagination in the Nineteenth Century.* Chicago: University of Chicago Press, 1991.

Higson, Gordon R. "The Beginning of CT Scanning—A Personal Recollection." *Bulletin of the British Institute of Radiology* 5, no. 1 (March 1979): 3–4.

Hillman, Bruce J., et al. "The Value of Consultation among Radiologists." *American Journal of Roentgenology* 127 (1976): 807–9.

Hinsley, Curtis. "The World as Marketplace: Commodification of the Exotic at the World's Columbian Exposition, 1893." In Karp and Lavine, eds., *Exhibiting Cultures,* 344–65.

Holman, B. Leonard, and Sally Edwards. "A History of Radiology at the Peter Bent Brigham Hospital." *American Journal of Roentgenology* 161 (1993): 437–441.

Holman, B. Leonard, et al. "Medical Impact of Unedited Preliminary Radiology Reports." *Radiology* 191 (1994): 519–21.

Holmes, Oliver Wendell. "On the Stereoscope and the Stereograph." *Atlantic Monthly,* June 1859, 738–48.

Hopwood, Nick. "'Giving Body' to Embryos: Modeling, Mechanism, and the Microtome in Late Nineteenth-Century Anatomy." *Isis* 90, no. 3 (September 1999): 462–96.

Hounsfield, Godfrey N. "Computed Medical Imaging." *Journal of Computer Assisted Tomography* 4, no. 5 (October 1980): 665–74.

———. "Computed Medical Imaging." Review. *Science* 210, no. 4465 (October 3, 1980): 22–28.

———. "Historical Notes on Computerized Axial Tomography." *Journal of the Canadian Association of Radiologists* 27 (1976): 135–42.

Howell, Joel. *Technology in the Hospital: Transforming Patient Care in the Early Twentieth Century.* Baltimore: Johns Hopkins University Press, 1995.

Hughes, Gordon. "Coming into Sight: Seeing Robert Delaunay's Structure of Vision." *October* 102 (fall 2002): 87–100.

Hume, David. *A Treatise of Human Nature.* Edited by L. Selby-Bigge. 2nd ed. Oxford: Clarendon Press, 1978.

Hunter, Kathryn Montgomery. *Doctors' Stories: The Narrative Structure of Medical Knowledge.* Princeton, N.J.: Princeton University Press, 1991.

———. "Narrative, Literature, and the Clinical Exercise of Practical Reason." *Literature and Medicine* 21, no. 3 (June 1, 1996): 303.

Hutchins, Edwin. *Cognition in the Wild.* Cambridge, Mass.: MIT Press, 1995.

Huxley, Thomas Henry. "On the Method of Zadig: Retrospective Prophecy as a Function of Science." In *Science and Culture, and Other Essays*, 128–48. New York: D. Appleton, 1884.

Imray, Thomas J., Lincoln L. Berland, and Guillermo F. Carrera. "A Modification of the Radiology Teaching Conference." *Journal of the Canadian Association of Radiologists* 30 (March 1979): 53.

Institute of Medicine. *Computed Tomographic Scanning: A Policy Statement.* Washington, D.C.: National Academy of Sciences, 1977.

Irwin, John T. *The Mystery to a Solution: Poe, Borges, and the Analytic Detective Story.* Baltimore: Johns Hopkins University Press, 1994.

Jacobs, Lawrence, William R. Kinkel, and Reid Heffner Jr. "Autopsy Correlations of Computerized Tomography: Experience with 6,000 CT Scans." *Neurology* 26, no. 12 (December 1976): 1111–18.

Jaffe, C. Carl. "Medical Imaging." *American Scientist* 70 (1982): 576–85.

———. "Medical Imaging, Vision, and Visual Psychophysics." *Medical Radiography and Photography* 60, no. 1 (1984): 3–48.

James, A. Everette, Jr., William J. Curran, Henry P. Pendergrass, and John E. Chapman. "Academic Radiology, Turf Conflict and Antitrust Laws." *Investigative Radiology* 25, no. 2 (1990): 200–202.

Janssen, Timotheus, and Frans Blommaert. "Image Quality Semantics." *Journal of Imaging Science and Technology* 41, no. 5 (1 September 1997): 555.

Janssens, V., et al. "Post-Mortem Limitations of Body Composition Analysis by Computed Tomography." *Ergonomics* 37, no. 1 (January 1994): 207.

Jardine, Nicholas. "The Laboratory Revolution in Medicine as Rhetorical and Aesthetic Accomplishment." In *The Laboratory Revolution in Medicine*, edited by Andrew Cunningham and Perry Williams, 304–23. Cambridge: Cambridge University Press, 1992.

Jay, Martin. *Downcast Eyes: The Denigration of Vision in Twentieth-Century French Thought.* Berkeley: University of California Press, 1993.

———. "Scopic Regimes of Modernity." In *Vision and Visuality*, edited by Hal Foster, 3–28. Seattle: The New Press: 1988.

Jonas, Hans. "The Nobility of Sight: A Study in the Phenomenology of the Senses." In *The Phenomenon of Life: Toward a Philosophical Biology.* Chicago, 1982.

Kant, Immanuel. "The Conflict of the Philosophy Faculty with the Faculty of Medicine." In *The Conflict of the Faculties: Der Streit der Fakultäten*, translated by Mary J. Gregor, 175–214. New York: Abaris Books, 1979.

———. *The Critique of Judgment.* Translated by Werner Pluhar. Indianapolis, Ind.: Hackett, 1987.

Karp, Ivan, and Steven D. Lavine, eds. *Exhibiting Cultures: The Poetics and Politics of Museum Display.* Washington, D.C.: Smithsonian Institution Press, 1991.

Kassirer, Jerome P. "Diagnostic Reasoning." *Annals of Internal Medicine* 110 (1989): 893.

Katz, Pearl. "Ritual in the Operating Room." *Ethnology* 20 (1981): 335–50.

Keating, Peter, and Alberto Cambrosio. "Biomedical Platforms." *Configurations* 8 (2000): 337–87.

Keats, Theodore. "Origins of Stereoscopy in Diagnostic Roentgenology." In Bruwer, ed., *Classic Descriptions in Diagnostic Roentgenology*, 983–1021.

Kelty, Chris. *Scale and Convention: Programmed Languages in a Regulated America*. Dissertation. Cambridge, Mass.: MIT Press, 2000.

Kennedy, J. Gerald. *Poe, Death, and the Life of Writing*. New Haven: Yale University Press, 1987.

Kenseth, Joy, ed. *The Age of the Marvelous*. Hanover, N.H.: Hood Museum of Art, 1991.

Kevles, Bettyann Holtzmann. *Naked to the Bone: Medical Imaging in the Twentieth Century*. New Brunswick, N.J.: Rutgers University Press, 1997.

Khan, Arfa, et al. "Solitary Pulmonary Nodules: Comparison of Classification with Standard, Thin-Section, and Reference Phantom CT." *Radiology* 179 (1991): 477–81.

Kidel, Mark, and Susan Rowe-Leete. "Mapping the Body." Translated and edited by Siri Hustvedt. In Feder et al., eds., *Fragments for a History of the Human Body*, 448–69.

Kieffer, Jean. "The Laminagraph and Its Variations." *American Journal of Roentgenology* 39, no. 4 (April 1938): 497–513.

King, Lester S. *Medical Thinking: A Historical Preface*. Princeton, N.J.: Princeton University Press, 1982.

———. "What Is a Diagnosis." *Journal of the American Medical Association* 202, no. 8 (1967): 714–17.

King, Lester S., and M. C. Meehan. "A History of Autopsy: A Review." *American Journal of Pathology* 73 (1973): 514–44.

Kirshenblatt-Gimblett, Barbara. "Objects of Ethnography." In Karp and Lavine, eds., *Exhibiting Cultures*, 386–443.

Kleinfield, Sonny. *A Machine Called Indomitable*. New York: Times Books, 1985.

Knorr-Cetina, Karin, and Michael Mulkay, editors. *Science Observed: Perspectives on the Social Study of Science*. New York: Sage, 1983.

Knowles, John H. "Radiology: A Case Study in Technology and Manpower." *New England Journal of Medicine* 280 (1969): 1271–78.

Kofman, Sarah. "L'oeil de boeuf: Descartes et l'après-coup idéologique." In *Camera obscura de l'idéologie*, 71–76. Paris: Galilée, 1976.

Krabbenhoft, Kenneth. "Certification and Education." In *A History of the Radiological Sciences: Diagnosis*, edited by Raymond A. Gagliardi and Bruce L. McClennan, 511–17. Reston, Va.: Radiology Centennial, 1996.

Krauss, Rosalind. "Grids." In *The Originality of the Avant-Garde and Other Modernist Myths*, 9–22. Cambridge, Mass.: MIT Press, 1986.

———. *The Optical Unconscious*. 1993. Cambridge, Mass.: MIT Press, 1994.

Krizack, "Documentation Planning and Case Study." In *Documentation Planning for the U.S. Health Care System*, edited by Joan Krizack, 207–35. Baltimore: Johns Hopkins University Press, 1994.

Kroker, Arthur, and Marilouise Kroker. "Theses on the Disappearing Body in the Hyper-Modern Condition." In *Body Invaders: Panic Sex in America,* 20–34. New York: St. Martin's Press, 1987.

Kundel, Harold L. "The Origin of Health Services Research in Radiology: A Review." *American Journal of Radiology* 166 (1996): 1255–57.

Kundel, Harold. L., and Paul. S. La Follette. "Visual Search Patterns and Experience with Radiological Images." *Radiology* 103 (1972): 523–28.

Lacan, Jacques. "Of the Gaze as *Objet Petit a.*" In *The Four Fundamental Aspects of Psychoanalysis,* edited by Jacques-Alain Miller, translated by Alan Sheridan, 65–119. Harmondsworth, Middlesex, England: Penguin, 1979 (1964).

———. "Seminar on 'The Purloined Letter.'" Translated by Jeffrey Mehlman. In Muller and Richardson, eds., *The Purloined Poe,* 28–54.

Lammer, Christina. "Horizontal Cuts and Vertical Penetration: The Flesh and Blood of Image Fabrication in the Operating Theaters of Interventional Radiology." *Cultural Studies* 16, no. 6 (2002): 833–47.

Landecker, Hannah. "Immortality, In Vitro: A History of the HeLa Cell Line." In Brodwin, ed., *Biotechnology and Culture,* 53–74.

Laqueur, Thomas. "Bodies, Details, and the Humanitarian Narrative." In *The New Cultural History,* edited by Lynn Hunt, 176–204. Berkeley: University of California Press.

Larkin, Gerald. "Medical Dominance and Control: Radiographers in the Division of Labor." *Sociological Review* 26, no. 4 (November 1978): 843–58.

———. "Radiographers." Chapter 3 in *Occupational Monopoly and Modern Medicine.* London: Tavistock Publications, 1983.

Latour, Bruno. "The Costly Ghastly Kitchen." In *The Laboratory Revolution in Medicine,* edited by Andrew Cunningham and Perry Williams, 295–303. Cambridge: Cambridge University Press, 1992.

———. *The Pasteurization of France.* Translated by Alan Sheridan and John Law. Cambridge, Mass.: Harvard University Press, 1988.

———. *Science in Action.* Cambridge, Mass.: Harvard University Press, 1987.

———. "Scientific Objects and Legal Objectivity," translated by Alain Pottage. In *Law, Anthropology and the Constitution of the Social: Making Persons and Things,* edited by Alain Pottage and Martha Mundy, 73–114. Cambridge: Cambridge University Press, 2004.

———. "Visualization and Cognition." *Knowledge and Society: Studies in the Sociology of Culture Past and Present* 6 (1986): 1–40.

———. *We Have Never Been Modern.* Translated by Catherine Porter. Cambridge: Harvard University Press, 1993.

Latour, Bruno, and Steve Woolgar. *Laboratory Life: The Construction of Scientific Facts.* 2nd ed. Princeton, N.J.: Princeton University Press, 1986.

Lave, Jean. "The Practice of Learning." In *Understanding Practice: Perspectives on Activity and Context,* edited by Seth Chaiklin and Jean Lave, 3–34. Cambridge: Cambridge University Press, 1993.

Law, John, and Michael Lynch. "Lists, Field Guides, and the Descriptive Organization of Seeing." In Lynch and Woolgar, eds., *Representation in Scientific Practice,* 267–99.

Law, John, and John Whittaker. "On the Art of Representation." In Fyfe and Law, eds., *Picturing Power,* 160–83.

Ledley, R. S., G. Di Chiro, A. J. Luessenhop, and H. L. Twigg. "Computerized Transaxial Tomography of the Human Body." *Science* 186 (October 1974): 207–12.

Lee, Joseph K. T., Stuart S. Sagel, and Robert J. Stanley, eds. *Computed Body Tomography with MRI Correlation.* 2nd ed. New York: Raven Press, 1989.

Lerner, Barron. "Fighting the War on Breast Cancer: Debates over Early Detection, 1945 to the Present." *Annals of Internal Medicine* 129, no. 1 (July 1, 1998): 74–78.

———. "The Perils of 'X-Ray Vision': How Radiation Images Have Historically Influenced Perception." *Perspectives in Biology and Medicine* 35, no. 3 (spring 1992): 382–97.

———. "Seeing What Is Not There: Mammography and Images of Breast Cancer." Cancer Workshop. University of North Carolina at Chapel Hill, April 2000.

Levitsky, David B., Mark S. Frank, Michael L. Richardson, and Robert J. Shneidman. "How Should Radiologists Reply When Patients Ask about Their Diagnoses? A Survey of Radiologists' and Clinicians' Preferences." *American Journal of Radiology* 161 (1993): 433–36.

Liebow, Averill A. "The Autopsy Room as a Hall of Learning." *American Journal of Medicine* 21 (October 1956): 485–86.

Lloyd, Phoebe, and Gordon Bendersky, "Of Peales and Poison," *MD* (1994): 27–29.

Lock, Margaret, and Deborah Gordon, eds. *Biomedicine Examined.* Dordrecht: Kluwer Academic Publishers, 1988.

Lock, Margaret, and Nancy Scheper-Hughes. "A Critical-Interpretive Approach in Medical Anthropology: Rituals and Routines of Discipline and Dissent." In *Medical Anthropology: Contemporary Theory and Method,* edited by T. M. Johnston and C. F. Sargent, 47–72. New York: Praeger.

Lukács, Georg. "Reification and the Consciousness of the Proletariat." In *History and Class Consciousness: Studies in Marxist Dialectics,* translated by Rodney Livingstone, 83–222. Cambridge, Mass.: MIT Press, 1971.

Lynch, Michael. *Art and Artifact in Laboratory Science: A Study of Shop Work and Shop Talk in a Research Laboratory.* London: Routledge and Kegan Paul, 1984.

———. "Discipline and the Material Form of Scientific Images: An Analysis of Scientific Visibility." *Social Studies of Science* 15 (1985): 37–66.

———. "Sacrifice and the Transformation of the Animal Body into a Scientific Object: Laboratory Culture and Ritual Practice in the Neurosciences." *Social Science of Medicine* 18 (1988): 265–89.

Lynch, Michael, and Samuel Edgerton. "Aesthetics and Digital Image Processing." In Fyfe and Law, eds., *Picturing Power,* 184–220.

Lynch, Michael, and Steve Woolgar. "Introduction: Sociological Orientations to Representational Practice in Science." *Human Studies: A Journal for Philosophy and the Social Sciences* 11, nos. 2–3 (April–July 1988): 99–116.

Lynch, Michael, and Steve Woolgar, eds. *Representation in Scientific Practice.* Cambridge, Mass.: MIT Press, 1990.

Lynn, Gary, Joseph Morone, and Albert Paulson. "Marketing and Discontinuous Innovation: The Probe and Learn Process." *California Management Review* 38 (1996): 8–37.

Maasilta, P., et al. "Correlations between Findings at Computed Tomography (CT) and at Thoracoscopy/Thoracotomy/Autopsy in Pleural Mesothelioma." *European Respiratory Journal* 4, no. 8 (September 1991): 952–54.

Madewell, John E. Interview by Charles Stuart Kennedy, AFIP Oral History Program, October 20, 1994.

Marwick, Charles. "Pathologists Request Autopsy Revival." *Journal of the American Medical Association* 273, no. 24 (June 28, 1995): 1889–91.

Marx, Karl. *Capital: A Critique of Political Economy,* Part I. Edited by Frederick Engels. Translated by Samuel Moore and Edward Aveling. New York: International Publishers, 1967 (1867).

Matthews, J. Rosser. *Quantification and the Quest for Medical Certainty.* Princeton, N.J.: Princeton University Press, 1995.

Maulitz, Russell C. *Morbid Appearances: The Anatomy of Pathology in the Early Nineteenth Century.* Cambridge: Cambridge University Press, 1987.

Mazansky, Cyril. "CT in the Study of Antiquities: Analysis of a Basket-Hilted Sword Relic from a 400-Year-Old Shipwreck." *Radiology* 186, no. 3 (1993): 55A–61A.

McFarland, Benjamin J., and Thomas P. Bennett. "The Image of Edgar Allan Poe: A Daguerreotype Linked to the Academy of Natural Sciences of Philadelphia." *Proceedings of the Academy of Natural Sciences of Philadelphia* 147 (1996): 1–32.

McKinlay, John B. "From 'Promising Report' to 'Standard Procedure': Seven Stages in the Career of a Medical Innovation." *Milbank Memorial Fund Quarterly* 59, no. 3 (1981): 374–411.

Melcher, Antony, Stephanie Holowa, and Peter Lewin. "Non-Invasive Computed Tomography and Three-Dimensional Reconstruction of the Dentition of a 2,800-Year-Old Mummy Exhibiting Extensive Dental Disease." *American Journal of Physical Anthropology* 103, no. 3 (July 1997): 329.

Meléndez, J. Carlos, and Ernest McCrank. "Anxiety-Related Reactions Associated with Magnetic Resonance Imaging Examinations." *Journal of the American Medical Association* 270, no. 6 (August 11, 1993): 745–47.

Merleau-Ponty, Maurice. *The Visible and the Invisible.* Edited by Claude Lefort. Translated by Alphonso Lingis. Northwestern University Studies in Phenomenology and Existential Philosophy. Evanston, Ill.: Northwestern University Press, 1968.

Messac, Régis. *Le Detectif novel et l'influence de la pensée scientifique.* Paris: H. Champion, 1929.

Michaels, Walter Benn. *The Gold Standard and the Logic of Naturalism: American Literature at the Turn of the Century.* Berkeley: University of California Press, 1987.

Miles, Margaret R. "Image." In *Critical Terms for Religious Studies,* 160–72. Chicago: University of Chicago Press, 1997.

————. *Image as Insight: Visual Understanding in Western Christianity and Secular Culture.* Boston: Beacon Press, 1985.

Miller, Roscoe. "Tunnel Vision Award." In Fischer, ed., *The Radiologist's First Reader,* 97–100.

Miller, Roscoe E., and Barbara J. Andrew. "View Box Exercises for Teaching Problem Solving in Radiology." *American Journal of Roentgenology* 128 (1977): 271–72.

Mirzoeff, Nicholas, ed. *The Visual Culture Reader.* New York: Routledge, 1998.

Mitchell, Timothy. *Colonising Egypt.* Berkeley: University of California Press, 1991.

————. "Orientalism and the Exhibitionary Order." In *Colonialism and Culture,* edited by Nicholas B. Dirks, 289–317. Ann Arbor: University of Michigan Press, 1992.

Mitchell, W. J. T. *Iconology: Image, Text, Ideology.* Chicago: University of Chicago Press, 1986.

Mitchell, William J. *The Reconfigured Eye: Visual Truth in the Post-Photographic Era.* Cambridge, Mass.: MIT Press, 1992.

Mol, Annemarie. *The Body Multiple: Ontology in Medical Practice.* Durham, N.C.: Duke University Press, 2002.

Montgomery, Kathryn. *How Doctors Think: Clinical Judgment and the Practice of Medicine.* Oxford: Oxford University Press, 2006.

Moore, Sherwood. "Body Section Roentgenography with the Laminagraph." *American Journal of Roentgenology and Radium Therapy* 39, no. 4 (April 1938): 514–22.

Morris, George H., and Ronald J. Chenail. *The Talk of the Clinic: Explorations in the Analysis of Medical and Therapeutic Discourse.* Hillsdale, N.J.: Lawrence Erlbaum Associates, 1995.

Muller, John P., and William J. Richardson, eds. *The Purloined Poe: Lacan, Derrida, and Psychoanalytic Reading.* Baltimore: Johns Hopkins University Press, 1988.

Myers, Greg. "Fictions for Facts: The Form and Authority of the Scientific Dialogue." *History of Science* 30, no. 3 (1992): 221–47.

Nicolson, Malcom. "The Art of Diagnosis: Medicine and the Five Senses." In Bynum and Porter, eds., *Companion Encyclopedia of the History of Medicine,* 2:826–51.

Nodine, Calvin F. "Eye Tracks Lead Sleuths to 'Invisible' Lesions." *Diagnostic Imaging* (October 1990): 149–55.

Notman, Derek, et al. "Modern Imaging and Endoscopic Biopsy Techniques in Egyptian Mummies." *American Journal of Roentgenology* 146 (January 1986): 93–96.

Office of Technology Assessment, U.S. Congress. *The Computed Tomography (CT or CAT) Scanner and Its Implications for Health Policy.* Washington, D.C.: Government Printing Office, 1976.

Oldendorf, William H. "Some Possible Applications of Computerized Tomography in Pathology." *Journal of Computer Assisted Tomography* 4, no. 2 (April 1980): 141–44.

Orlander, Jay D., Thomas W. Barber, and B. Graeme Fincke. "The Morbidity and Mortality Conference: The Delicate Nature of Learning from Error." *Academic Medicine* 77, no. 10 (October 2002): 1001–6.

Orphanidou, Charitini, et al. "Accuracy of Subcutaneous Fat Measurement: Comparison of Skinfold Calipers, Ultrasound, and Computed Tomography." *Journal of the American Dietetic Association* 94, no. 8 (August 1994): 855.

Pagden, Anthony. "The Autoptic Imagination." In *European Encounters with the New World: From Renaissance to Romanticism,* 51–88. New Haven, Conn.: Yale University Press, 1993.

Pagel, W. "The Speculative Basis of Modern Pathology." *Bulletin of the History of Medicine* 18 (1945): 1–43.

Pantin, Carl F. A. "The Recognition of Species." *Science Progress* 42 (1954): 587–98.

Parsons, Talcott. *The Social System.* Glencoe, Ill.: Free Press, 1951.

Pasveer, Bernike. "Knowledge of Shadows—the Introduction of X-Ray Images in Medicine." *Sociology of Health and Illness* 11, no. 4 (1989): 361–81.

Peirce, Charles S. "Abduction and Induction." In *The Philosopy of Peirce: Selected Writings,* edited by Justus Buchler, 150–56. London: Routledge and Kegan Paul, 1940.

———. "Deduction, Induction, and Hypothesis" and "[The Logic of Abduction]." In *Charles S. Peirce,* edited by Vincent Tomas, 125–43, 235–55. New York: Liberal Arts Press, 1957,

———. "Perceptual Judgments." In *The Philosopy of Peirce,* 302–5.

Peterson, Harold O. "First a Radiologist." *American Journal of Roentgenology, Radium Therapy and Nuclear Medicine* 92, no. 6 (1964): 1227–31.

Pickering, Andrew, ed. *Science as Practice and Culture.* Chicago: University of Chicago Press, 1992.

Pickering, Andrew. "The Mangle of Practice." In Biagioli, ed., *The Science Studies Reader,* 372–93.

Pietz, William. "The Problem of the Fetish, I." *Res* 9 (spring 1985): 5–17.

———. "The Problem of the Fetish, II: The Origin of the Fetish." *Res* 13 (spring 1987): 23–45.

———. "The Problem of the Fetish, IIIA: Bosman's Guinea and the Enlightenment Theory of Fetishism." *Res* 16 (autumn 1988): 105–123.

Poe, Edgar Allan. *Collected Works of Edgar Allan Poe,* edited by Thomas O. Mabbott. 3 vols. Cambridge, Mass.: Belknap Press, 1978.

———. *Complete Tales and Poems.* New York: Vintage, 1985.

———. "Preface and Introduction to 'The Conchologist's First Book.'" In *Complete Works,* edited by James Harrison, 14:96–99. New York: Thomas Crowell, 1902.

Polanyi, Michael. *Personal Knowledge: Towards a Post-Critical Philosophy.* Chicago: University of Chicago Press, 1962.

Pollock, Della. *Telling Bodies Performing Birth: Everyday Narratives of Childbirth.* New York: Columbia University Press, 1999.

Porter, Theodore M. "Life Insurance, Medical Testing, and the Management of Mortality." In *Biographies of Scientific Objects,* edited by Lorraine Daston, 226–46. Chicago: University of Chicago Press, 2000.

———. *Trust in Numbers: The Pursuit of Objectivity in Science and Public Life.* Princeton, N.J.: Princeton University Press, 1995.

Potts, D. Gordon. "The Division of Radiology." *Radiology* 140, no. 3 (September 1981): 839–40.

Pratt, Mary Louise. *Imperial Eyes: Travel Writing and Transculturation.* London: Routledge, 1992.

Quiroz, Rene et al. "Clinical Validity of a Negative Computed Tomography Scan in Patients with Suspected Pulmonary Embolism." *Journal of the American Medical Association* 293, no. 16 (April 27, 2005): 2012–17.

"Radiologists Ask: Who's Reading Medical Images?" *American Medical News,* August 1998, 42–44.

Rapp, Rayna. "Accounting for Amniocentesis." In *Knowledge, Power, and Practice: The Anthropology of Medicine and Everyday Life,* edited by Shirley Lindenbaum and Margaret Lock, 55–76. Berkeley: University of California Press, 1993.

Reed, Harvey R. "The X-Ray from a Medico-Legal Standpoint." *Journal of the American Medical Association* 30 (1898): 1016.

Reiner, Bruce, et al. "Impact of Filmless Radiology on Frequency of Clinician Consultations with Radiologists." *American Journal of Roentgenology* 173 (1999): 1169–72.

Reiser, Stanley Joel. *Medicine and the Reign of Technology.* Cambridge: Cambridge University Press, 1978.

———. "The Science of Diagnosis." In *Companion Encyclopedia of the History of Medicine,* edited by W. F. Bynum and Roy Porter, 2:826–51. London: Routledge, 1993.

Richards, Thomas. "Archive and Utopia." *Representations* 37 (winter 1992): 104.

———. *The Imperial Archive: Knowledge and the Fantasy of Empire.* London: Verso, 1993.

Roberts, K. B., and J. D. W. Tomlinson. *The Fabric of the Body: European Traditions of Anatomical Illustration.* New York: Oxford University Press, 1992.

Roepstorff, Andreas. "Transforming Subjects into Objectivity: An Ethnography of Knowledge in a Brain Imaging Laboratory. *FOLK, Journal of the Danish Ethnographic Society* 44 (2002): 145–70.

Rosai, Juan, ed. *Guiding the Surgeon's Hand: The History of American Surgical Pathology.* Washington, D.C.: American Registry of Pathology, AFIP, 1997.

Rosenberg, Charles E. *The Care of Strangers: The Rise of America's Hospital System.* New York: Basic Books, 1987.

Rosenberg, Charles E., and Janet Golden, eds. *Framing Disease: Studies in Cultural History.* New Brunswick, N.J.: Rutgers University Press, 1992.

Rosenheim, Shawn J. *The Cryptographic Imagination: Secret Writing from Poe to the Internet.* Baltimore: Johns Hopkins University Press, 1997.

Ross, Steven. "From a Purely Radiological Point of View." In Fischer, ed., *The Radiologist's First Reader,* 41–46.

Rössner, Stephan et al. "Adipose Tissue Determinations in Cadavers—a Comparison between Cross-Sectional Planimetry and Computed Tomography." *International Journal of Obesity* 14, no. 10 (October 1990): 893–902.

Rudwick, Martin J. S. "The Emergence of a Visual Language for Geological Science, 1760–1840." *History of Science* 14 (1976): 149–95.

———. *The Great Devonian Controversy: The Shaping of Scientific Knowledge among Gentlemanly Specialists.* Chicago: University of Chicago Press, 1985.

Sandelowski, Margarete. *Devices and Desires: Gender, Technology, and American Nursing.* Chapel Hill: University of North Carolina Press, 2000.

Saunders, Barry. "Insection and Decryption: Edgar Poe's "The Gold-Bug" and the Diagnostic Gaze." M.A. thesis, University of North Carolina at Chapel Hill, 1989.

Sawday, Jonathan. *The Body Emblazoned: Dissection and the Human Body in Renaissance Culture.* New York: Routledge, 1996.

Schaffer, Simon. "Experimenters' Techniques, Dyers' Hands, and the Electric Planetarium." *Isis* 88:3 (September 1997): 456–83.

Schaffner, Kenneth, ed. *Logic of Discovery and Diagnosis in Medicine.* Berkeley: University of California Press, 1985.

Scully, Robert E. "Ben Castleman—Champion of the CPC." *New England Journal of Medicine* 307, no. 6 (1982): 370–71.

Sekula, Allan. "The Body and the Archive." *October* 39 (winter 1986): 3–64.

Serres, Michel. *Hermes: Literature, Science, Philosophy.* Baltimore: Johns Hopkins University Press, 1982.

Shapin, Steven, and Simon Schaffer. *Leviathan and the Air-Pump: Hobbes, Boyle, and the Experimental Life.* Princeton, N.J.: Princeton University Press, 1985.

Sheiman, R. G., C. Fey, M. McNicholas, and V. Raptopoulos. "Possible Causes of Inconclusive Results on CT-Guided Thoracic and Abdominal Core Biopsies." *American Journal of Roentgenology* 170 (1998): 1603–7.

Shell, Marc. "The Gold Bug." *Genre* 13, no. 1 (spring 1980): 11–30.

Shiff, Richard. "Handling Shocks: On the Representation of Experience in Walter Benjamin's Analogies." *Oxford Art Journal* 15, no. 2 (1992): 88–103.

Simmel, Georg. "The Stranger." In *The Sociology of Georg Simmel,* translated by Kurt Wolff, 802–8. Glencoe, Ill.: Free Press, 1950.

Simon, Christian M. "Images and Image: Technology and the Social Politics of Revealing Disorder in a North American Hospital." *Medical Anthropology Quarterly* 13, no. 2 (1999): 141–62.

Singer, Charles. *A Short History of Anatomy and Physiology from the Greeks to Harvey.* New York: Dover, 1957.

Skolnick, Andrew A. "New Holographic Process Provides Noninvasive, 3-D Anatomic Views." *Journal of the American Medical Association* 271, no. 1 (January 5, 1994): 5–8.

Snyder, Joel, and Neil W. Allen. "Photography, Vision, and Representation." *Critical Inquiry* 2 (1975): 143–69.

Sochurek, Howard, and Peter Miller. "Medicine's New Vision." *National Geographic,* January 1987, 2–41.

Sontag, Susan. "AIDS and Its Metaphors." In *Illness as Metaphor and AIDS and Its Metaphors.* New York: Anchor Books, 1989.

Spurgeon, Devon, and Thomas M. Burton. "Screen Tests: For the Very Cautious, a Physical Exam Now Includes a CAT Scan." *Wall Street Journal*, March 23, 2000, A1.

Squire, Lucy Frank. *Fundamentals of Radiology*. Rev. ed. Cambridge, Mass.: Harvard University Press, 1964.

Stafford, Barbara. *Body Criticism: Imaging the Unseen in Enlightenment Art and Medicine*. Cambridge, Mass.: MIT Press, 1991.

———. "Conjuring: How the Virtuoso Romantic Learned from the Enlightened Charlatan." *Art Journal* (summer 1993): 22–30.

———. *Good Looking: Essays on the Virtue of Images*. Cambridge, Mass.: MIT Press, 1996.

———. "Voyeur or Observer? Enlightenment Thoughts on the Dilemmas of Display." *Configurations* 1, no. 1 (winter 1993): 95–128.

Star, Susan Leigh. "Epilogue: Work and Practice in Social Studies of Science, Medicine, and Technology." *Science, Technology, and Human Values* 20, no. 4 (autumn 1995): 501–7.

Starr, Paul. *The Social Transformation of American Medicine*. New York: Basic Books, 1982.

Stocking, George. *Victorian Anthropology*. New York: Free Press, 1987.

Sturken, Marita, and Lisa Cartwright. *Practices of Looking: An Introduction to Visual Culture*. Oxford: Oxford University Press, 2001.

Susskind, Charles. "The Invention of Computed Tomography." *History of Technology* 6 (1981): 39–80.

Sutter, Morley C. "Assigning Causation in Disease: Beyond Koch's Postulates." *Perspectives in Biology and Medicine* 39, no. 4 (summer 1996): 581.

Tagg, John. *The Burden of Representation: Essays on Photographies and Histories*. Minneapolis: University of Minnesota Press, 1993.

———. "The Currency of the Photograph." In *Thinking Photography*, edited by Victor Burgin, 110–41. London: Macmillan, 1982.

Taussig, Michael. "The Golden Bough: Magic and Mimesis." In *Mimesis and Alterity: A Particular History of the Senses*, 44–58. New York: Routledge, 1993.

———. "Reification and the Consciousness of the Patient." *Social Science and Medicine* 14 (1980): 3–13.

———. "Sympathetic Magic in a Post-Colonial Age." *Excursus* 5 (March 1992): 2–9.

———. "Tactility and Distraction." *Cultural Anthropology* 6, no. 2 (1991): 147.

Taylor, Frederick W. "Scientific Management." In *Critical Studies in Organization and Bureaucracy*, edited by Frank Fischer and Carmen Siriani, 44–54. Philadelphia: Temple University Press, 1984.

Taylor, Mark C. "Christianity and the Capitalism of Spirit." In *About Religion: Economies of Faith in Virtual Culture*, 1404–67. Chicago: University of Chicago Press, 1999.

———. *Hiding*. Chicago: University of Chicago Press, 1997.

Taylor, Mark C., and Esa Saarinen. *Imagologies: Media Philosophy.* New York: Routledge, 1994.

Thompson, John, and Grace Goldin. *The Hospital: A Social and Architectural History.* New Haven: Yale University Press, 1975.

Timmermans, Stefan, and Marc Berg. *The Gold Standard: The Challenge of Evidence-Based Medicine and Standardization in Health Care.* Philadelphia: Temple University Press, 2003.

———. "Standardization in Action: Achieving Local Universality through Medical Protocols." *Social Studies of Science* 27, no. 2 (1997): 273–305.

Todorov, Tzvetan. *Theories of the Symbol.* Translated by Catherine Porter. Ithaca, N.Y.: Cornell University Press, 1982.

Trajtenberg, Manuel. *Economic Analysis of Product Innovation: The Case of CT Scanners.* Cambridge, Mass.: Harvard University Press, 1990.

Traweek, Sharon. *Beamtimes and Lifetimes: The World of High-Energy Physics.* Cambridge, Mass.: Harvard University Press, 1988.

———. "Discovering Machines: Nature in the Age of Its Mechanical Reproduction." In *Making Time: Ethnographies of High-Technology Organizations,* edited by Frank A. Dubinskas, 39–91. Philadelphia: Temple University Press, 1988.

Tresch, John. "Extra! Extra! Poe Invents Science Fiction!" In *The Cambridge Companion to Edgar Allan Poe,* edited by Kevin Hayes, 113–32. Cambridge: Cambridge University Press, 2002.

———. "'The Potent Magic of Verisimilitude': Edgar Allan Poe within the Mechanical Age." *British Journal for the History of Science* 30 (1997): 275–90.

Turner, Victor. *The Ritual Process: Structure and Anti-Structure.* Chicago: Aldine, 1969.

Tyson, Ruel W., Jr. "A Definition of Religious Studies in the Context of Health Professional Education." *Proceedings of the Institute on Human Values in Medicine,* 31–39. Philadelphia: Society for Health and Human Values, 1974.

———. "Odysseus and the Cyclops." Unpublished essay.

Virilio, Paul. *The Vision Machine.* Bloomington: Indiana University Press, 1994.

Wacquant, Loic J. D. "Towards a Reflexive Sociology: A Workshop with Pierre Bourdieu." *Sociological Theory* 7, no. 1 (spring 1989): 26–63.

Wailoo, Keith. *Drawing Blood: Technology and Disease Identity in Twentieth-Century America.* Baltimore: Johns Hopkins University Press, 1997.

Waldby, Catherine. *The Visible Human Project: Informatic Bodies and Posthuman Medicine.* New York: Routledge, 2000.

Wartofsky, Marx. "Rules and Representation: The Virtues of Constancy and Fidelity Put in Perspective." In *Models: Representation and the Scientific Understanding,* 211–30. Dordrecht: D. Reidel, 1979.

Webb, Steve. *From the Watching of Shadows: The Origins of Radiological Tomography.* Bristol: A. Hilger, 1990.

———. "Historical Experiments Predating Commercially Available Computed Tomography." *British Journal of Radiology* 65, no. 777 (September 1992): 835–37.

Weber, Max. *Ancient Judaism.* Translated and edited by Hans Gerth and Don Martindale. New York: The Free Press, 1952.

———. *The Protestant Ethic and the Spirit of Capitalism.* Translated by Talcott Parsons. New York: Scribner's, 1958.

———. "Science as a Vocation." In *From Max Weber: Essays in Sociology.* Translated and edited by H. H. Gerth and C. Wright Mills, 129–56. London: Routledge and Kegan Paul, 1952.

———. *The Sociology of Religion.* Translated by Ephraim Fischoff. Boston: Beacon, 1964.

Weisz, George. "The Emergence of Medical Specialization in the Nineteenth Century." *Bulletin of the History of Medicine* 77 (2003): 536–75.

Wiener, Harry. *Findings in* CT: *A Clinical and Economic Analysis of Computed Tomography.* New York: Pfizer, 1979.

Withers, Stafford. "The Story of the First Roentgen Evidence." *Radiology* 17 (1931): 99–103.

Woolgar, Steve. *Science: The Very Idea.* New York: Tavistock Publications and Ellis Horwood, 1988.

Wright, Peter, and Andrew Treacher, eds. *The Problem of Medical Knowledge: Examining the Social Construction of Medicine.* Edinburgh: Edinburgh University Press, 1982.

"X Ray Photographs as Evidence." *Journal of the American Medical Association* 29 (1897): 870.

Young, Allan. "The Creation of Medical Knowledge: Some Problems of Interpretation." *Social Science and Medicine* 15B, no. 3 (July 1981): 379–86.

———. *The Harmony of Illusions.* Princeton, N.J.: Princeton University Press, 1995.

Yoxen, Edward. "Seeing with Sound: A Study of the Development of Medical Images." In *The Social Construction of Technological Systems: New Directions in the Sociology and History of Technology,* edited by Wiebe E. Bikjer, Thomas P. Hughes, and Trevor J. Pinch, 281–303. Cambridge, Mass.: MIT Press, 1987.

Zerubavel, Eviatar. *Patterns of Time in Hospital Life.* Chicago: University of Chicago Press, 1979.

ILLUSTRATIONS

9. Piet Mondrian, *Composition 1916*, 1916. Oil on canvas with wood strip at bottom edge, 119 x 75.1 cm (46 7/8 x 29 5/8 inches). Solomon R. Guggenheim Museum, New York. 49.1229. 39

10. Eadweard Muybridge, "Movements, Man, Walking, carrying 75-lb. stone on left shoulder." Plate 26 in *Animal Locomotion*. Philadelphia: University of Pennsylvania, 1896. Reproduction courtesy of Collections of the University of Pennsylvania Archives. 40

11. Slice of terrain. From Georges Cuvier and Alexandre Brongniart, *Essai sur la géographie minéralogique des environs de Paris, avec une carte géognostique, et des coupes de terrain*, figure 1. Paris: Baudouin, 1811. Reproduction courtesy of Cecil Schneer, Emeritus Professor of Geology, University of New Hampshire. 42

12. Joseph Woodward, Photomicrograph of tumor section, Philadelphia Exposition in 1876. Reproduction courtesy of National Museum of Health and Medicine, Armed Force Institute of Pathology. 43

13. Section from Albert Eycleshymer and Daniel Schoemaker, *A Cross-Section Anatomy*, p. 23 (drawn by Tom Jones). New York: Appleton-Century-Crofts, Inc., 1911. © 1938 Mary Elizabeth Eycleshymer. 45

14. Resident on the phone, Body CT reading room. Author photo. 63

15. Logbook, Body CT reading room. Author photo. 77

16. Gantry bore, patient, and tech: Neuro CT. Author photo. 94

17. Conventional tomography: plane with *b* remains in focus while *a* is blurred. Reprinted by permission of the publisher from Robert A. Novelline, *Squire's Fundamentals of Radiology*, 5th ed., p. 25. Cambridge, Mass.: Harvard University Press, copyright © 1964 the Commonwealth Fund; copyright © 1975, 1982, 1988, 1997 President and Fellows of Harvard College. 96

18. Metal shavings from a lathe. Author photo. 98

19. Tube and detector arrangements in four generations of gantry design. From Joseph Bronzino, Vincent Smith, and Maurice Wade, *Medical Technology and Society: An Interdisciplinary Perspective*, fig. 7.13, p. 433. Cambridge, Mass.: MIT Press, 1990. Reproduced with permission of MIT Press. 106

20. Tech positioning patient, Neuro CT. Author photo. 112

21. Scan in progress, Neuro CT. Author photo. 118

22. Michael Mahoney and Marc Kahn, "A Medical Mystery," *New England Journal of Medicine* 339 (September 1998): 745. Copyright © 2006 Massachusetts Medical Society, all rights reserved. 135

23. Edgar Poe with Joseph Leidy and assistant in Academy of Natural Sciences, Phila-delphia, ca. 1842. Academy of Natural Sciences, Coll. 49. Reproduction courtesy of The Academy of Natural Sciences, Ewell Sale Stewart Library. 147

24. **a and b.** Resemblance versus identity. From Alphonse Bertillon, *Identification anthropométrique, Instructions Signalétiques,* plates 59 and 60. Melun: Imprimerie Administrative, 1893. 153

25. Francis Galton, Photo composites: criminal and other types, 1907. From *Inquiries into Human Faculty and Its Development,* plate facing p. 8. London: J. M. Dent and Sons Ltd., 1928. 155

26. **a and b.** AFIP tables: rogues' galleries for diagnostic identification. "AFIP Subject Reviews: Range of Disease (Thymoma)" and "AFIP Subject Reviews: Differential Diagnosis (Thymoma)," at http://www.radpath.org/unknowns/differential/body.htm and http://www.radpath.org/reviews/subpages/range.htm (accessed November 2000). Courtesy of Melissa Rosado de Christenson and the Department of Radiologic Pathology, Armed Forces Institute of Pathology. 156

27. Satellite surveillance technology assists breast cancer detection. From National Information Display Lab project brochure for RSNA conference, 1996. Courtesy of the Sarnoff Corporation. 169

28. Clerk filing films, Basement File Room. Author photo. 171

29. Radiology Scheduling. Author photo. 185

30. Body CT Suite floorplan. Courtesy of University Hospital. 189

31. Anatomy theater as museum. Davidge Hall, Anatomical Hall. Courtesy of the Medical Alumni Association of the University of Maryland, Inc. 201

32. Divisions of expertise at the back viewbox, Body CT reading room. Author photo. 229

33. Columbian Exposition stereoscope card, 1893. Library of Congress. 277

34. Army Medical Museum Great Hall, 1890s. Reproduction courtesy of National Museum of Health and Medicine, Armed Forces Institute of Pathology. 279

35. **a and b.** Army Medical Museum exhibits at (a) the Philadelphia Exposition, 1876; and (b) the Columbian Exposition, 1893. Reproductions courtesy of National Museum of Health and Medicine, Armed Forces Institute of Pathology. 281

36. PIXY, Tissue-Equivalent Teaching Phantom, RSNA Technical Exhibits, 1996. Author photo. 289

37. Scientific Exhibits, RSNA Meeting, 1996. Author photo. 291

INDEX

abduction, 143–45, 148, 150, 157, 231, 305,
327 n. 25, 328 n. 42
actuarial status, 215, 225, 296, 328 n. 1
advertisement, 170, 293–96, 304, 332 n. 43
aesthetics: broadly-defined, 10, 310 n. 22;
classification and, 328 n. 42; cultural
potency of medical images and, 12; cutting and, 41, 301; of detection, 142–43,
157, 340 n. 45; grids and, 37–38, 303, 316
n. 60; of juxtaposition, 221, 292, 301,
303; museums and, 165, 296, 327 n. 33;
nonradiological images and, 167, 313
n. 23, 314 nn. 37–38, 317 n. 81; in radiological work, 30, 33, 48–49, 72, 91,
148–49, 179; urban crowds and, 131
affects: in case conferences, 142–43; in
diagnostic intrigues, 273, 304; at the
museum of pathology, 278, 280, 282; in
viewing images, 134, 304
AFIP (Armed Forces Institute of Pathology): history of, 342 n. 22; hybrid
missions of, 11, 278, 287, 296–97, 341
n. 8; pathologic tissue archives of, 280,
285; rad-path archives of, 270, 276, 282,
285, 287, 341 n. 28; as sponsor of course
in radiologic pathology, 275–76; visits
to, 278, 282
age of patients, 140, 215, 218, 225, 260
alienation, 129, 132, 149–50, 167–68, 331
n. 29, 332 n. 31
ambiguity, 19, 27, 31, 35, 208, 219, 244, 290,
299
analysis: of cases, through early accounting methods, 193; in detection, 130–31,
142–43, 199, 340 n. 45; of gesture, by the
camera, 22; of medical decisions, 197,
334 nn. 65–66; performance of, 134, 143;
of the radiological gaze, 22; as representational mode, 305; as skill, elided
in technology assessment, 290; by the
tomographic apparatus, 95, 321 n. 11
anthropology, 8, 147, 152, 288, 294, 309
n. 16
anthropometry, 152, 161, 164, 278, 280,
332 n. 39, 338 n. 28, 341 n. 15. *See also*
craniology

anxiety: cultural importance of medical images and, 12, 15, 100, 304; in diagnostic gaze, 35–36, 90; of patients, 36, 105, 127, 323 n. 43; about radiological performance, 127, 205, 238, 253, 277, 304; supervisory, 127, 304

apprenticeship, 15, 142–43, 151, 157, 273, 330 n. 20

archives, 160; for culprit identification, 152, 154, 328 n. 1; medical, 280, 286, 328 n. 2, 329 n. 3; in radiology, 159–60, 197–98. *See also* collection

Armed Forces Institute of Pathology. *See* AFIP

Army Medical Museum, 279, 341 n. 15, 341 n. 19; at the Expo, 43, 278, 280–81, 316 n. 72, 316 n. 75, 340 n. 3; as National Museum of Health and Medicine, 11, 164, 278, 280, 282, 296–97, 316 n. 75, 342 n. 22; pathologist J. J. Woodward and, 42–43, 280–81, 316 n. 72, 328 n. 37

artifacts in radiographs, 16, 24, 31–33, 90, 315 n. 48

artwork, 330 nn. 20–22; collection and display of, 165, 303, 312 n. 18, 330 n. 14, 330 nn. 18–19, 331 n. 23; Cubist, 26–28; dissective traditions in, 41; geological illustration and, 39; modernist grids in, 36–37, 91, 316 n. 60; northern European landscapes, 26; radiographs as, 49, 167, 169, 219, 303; on workspace walls, 184

assembly line, 58, 75, 79, 151, 172, 182, 193, 273, 303

atlas, 44, 84, 160, 169–70, 195, 277, 332 n. 36

attendings in radiology/CT, 7, 15, 19, 53, 62–63, 78–79, 203

attention: attracting, 304; in biopsies, 125; to chart, 113; in conferences, 204, 207; of curator, 166; economy of, 2, 60, 139; particularity of, 93, 207; personal knowledge and, 15, 17; regimes and rituals of, 2–3, 128, 149; serial, 287,

292–93; summons to, 76; at viewbox, 14, 33–35, 70, 80, 90, 151, 211, 231, 238, 250, 301; waiting and, 2, 26

attire, 50, 94, 108, 125–26, 258, 293, 288–9, 294

aura, 165–66, 169, 330–31 n. 22, 332 n. 1

authority: of cadaver/death, 139, 157, 259, 274, 280–82, 285, 300–302; of experience, 5, 222, 311; of medical imaging, 273–74, 285; of pathologist, 138–39, 274; of physician, 3, 303; of radiologist, 51, 66, 75, 212, 218, 244, 299, 318 n. 89; scientific, 160, 296, 303, 337 n. 13; of texts, 82, 311 n. 11. *See also* charisma

authorship, 75, 196, 234, 334 nn. 62–63

autopsis: of CT radiologists, 15–17, 19–20, 24, 90, 205, 216, 232, 245, 249, 272; early modernity and, 311–12 n. 11; historical changes in perception and, 286; library work vs., 159; patients' viewing their own images and, 5–6; public participation in the coroner's inquest and, 311 n. 10; visitors to CT viewbox and, 7, 55, 318 n. 89

autopsy, postmortem: Army Medical Museum and, 278–79; CT's subsitution for, 111, 138, 300, 308 n. 4, 335 n. 3; cutting in, 94; decline of, 273, 286

belief: in collegial testimony, 82, 197, 225, 255, 273; popular, 3, 338 n. 22; private psychology and, 8, 309 n. 14

Benjamin, Walter: on aura, 165–66, 330 n. 22, 332 n. 31; on collecting, 179; on cult value, 166; on detection, 131–32, 137–38, 305; on gesture and perception, 1, 22–23, 300, 313 n. 24; on history, 6, 9; on montage, 6, 308 n. 7; on photography, 22–23, 98, 132, 167, 330 n. 31; on shock, 98, 100, 129, 131–32, 137–38, 167; on urban experience, 6, 36, 131–32, 305, 325 n. 7

Bertillon, Alphonse, 152–54, 280, 338 n. 28

bet (wager), 148, 151, 227, 239–40, 250, 257, 269, 271, 282

biology, 146, 157, 164, 327 n. 36

biopsy, 51, 78, 83–84, 198, 226–27, 253, 260–62, 339 n. 38; complication of, 258–59, 265–66; CT-guided, 111, 125–28, 187, 190, 265–68, 293, 301, 325 n. 59; practices of, 125–28, 262–64, 267–68, 290, 340 n. 43; results of, 227, 297

black box, 95, 99, 110, 321 n. 9

bodies: cultural studies of, 8, 310 n. 17; curating of, 182–90, 198; of detectives/examiners, 16, 18, 143, 145, 203, 332 n. 39; film and, 163–66, 174, 329 n. 11; hospital management of, 11, 90, 182–84, 186–90, 304; images and, 11, 14, 26–27, 38, 41–44, 50, 91, 128, 148, 152, 159, 163, 168–70, 316 n. 61; institutions and, 186–87, 190, 198, 288–89, 328 n. 1; practices and, 15, 91, 130, 297, 304; texts and, 90, 152, 159, 187, 190, 192–93, 198

bureaucracy, 37, 160–62, 193, 303

cadaver/corpse: associated with biopsy, 128, 190; in contemporary medicine, 3, 12, 273, 280, 300; in CPC, 134, 157, 200; in detective story, 133, 137, 159, 329 n. 11; emulation of, 111, 190; in nineteenth-century medicine, 2, 200; untranscended by imaging, 131, 260, 300–301, 325 n. 61

camera obscura, 24–25, 301, 313 n. 28, 314 n. 31, 317 n. 78

cancer, 340 n. 44; in AFIP collections, 341 n. 19; biopsy and, 227, 290; CT in radiation oncology and, 323 n. 34; found in CT, 250–52, 260; imaging workups and, 70; known prior to CT scan, 78, 83–84, 86, 232, 238, 242–43, 254–57, 261, 290; mammography and, 312 n. 16, 329 n. 9; postmortem autopsy and, 292, 308 n. 4; suspected on CT, 63–64, 249, 266, 290

careers: of images, 175, 302; of lesions, 108, 163; of nurses, 108, 111; of physicians, 108; of radiologists, 49, 240, 304, 337 n. 18; of scientists, 335 n. 68; of techs, 108–9, 324 n. 46

carnival, 290, 292, 298

cases: building/preparation of, 151, 177, 282–84, 302, 305, 342 n. 26; closure of, 88–89, 141–42, 206–7, 219, 228, 272; enclosing specimens, 160, 165, 282, 330 n. 16; firsthand involvement in, 222, 302; interesting, 179, 269–70, 272, 282, 284, 302; in Peircean abduction, 144–45; presentation of, 52, 60, 134, 139–42, 202–3, 207–8, 216, 224, 271–72, 326 n. 19, 336 n. 9, 337 n. 12, 337 n. 15; as property, 80, 178–79, 181–82, 224; publication of, 196, 284, 342 n. 28; in radiology, 9, 139–41, 157, 202, 272, 302, 336 n. 9; as series, 9, 87–88, 140, 142, 144, 218, 272, 302, 339 n. 35; for teaching file, 177, 181, 269, 285

causality, 176, 318 n. 96, 332 n. 44; chain of, in diagnostic process, 52; of disease, 2, 170, 216, 246, 257–58, 272, 286, 327 n. 26, 327 n. 34

center and periphery, 25, 137, 160, 169, 195–97, 329 n. 5, 330 n. 15, 340 n. 3

certainty, 144, 150, 208, 227, 334 n. 64; vision and, 24–25, 238, 245, 256, 313 n. 27

charisma, 222; of curator, 160, 162, 165–66, 179, 181–82, 192, 198, 329 n. 6; of priest or prophet, 162, 329 n. 6

chart, hospital, 193–94, 334 n. 54

cinematography, 13, 22–23, 37–38, 40, 96–98, 314 n. 39

citation, 84, 197, 217–18, 335 n. 68

citizen, 132, 152, 282, 296

classification, 12, 131–32, 160; aesthetics and, 328 n. 42; in assigning CT protocols, 68; of diseases or findings, 2, 84, 133, 141–42, 150, 154–57, 165, 169, 198, 211, 220, 286, 297, 307 n. 3; ethnological/racial, 147, 152, 155, 341 n. 15; history

classification (*cont.*)

of, 146, 152–55, 193, 280, 327 n. 36; of
pure/impure, 106; of teaching file cases,
179, 333 n. 50

clerks: in hospital, 56, 186, 230; in radiology, 56, 69, 109, 119, 161–63, 170–2,
184–86, 188, 198, 302

clients, 7, 53, 55, 170, 199, 224, 271, 295–97;
criticism or suspicion of, 66–68, 163,
230; satisfaction of, 67–68, 70, 75, 193,
294

Clinical-Pathological Conference. *See*
CPC

clinicians. *See* correlations: rad-clin;
viewbox: visitors to

clues, 111, 142, 146, 226, 229, 238; in teaching conferences, 217, 219, 272

collection, 146, 197, 302, 330 n. 15; of
books, 195; history of, 157, 160, 165,
330 n. 18; of pathological specimens,
278–80, 296, 341 n. 19; possession and,
137, 177–79; of radiological images, 11,
160–63, 172, 177–79, 181–82, 269, 286

comfort, discomfort: of patients, 93, 103,
110, 120–21, 127, 182; of professionals, 18,
34–35, 51, 127, 205–6, 212

commodity: critique of, 167–68, 303, 331
n. 29; displayed behind glass, 6, 132, 330
n. 17; image/interpretation as, 26, 129,
168, 198, 300, 331 n. 26; rise of specialties and, 318 n. 90; at RSNA meeting,
288–89; scientism as, 91, 297

comparative anatomy, 26, 41, 145–47, 157,
159, 165, 278–80

comparison, 131; classification and, 12, 327
n. 36, 330 n. 15; in differential diagnosis, 211, 326 n. 20; of images, 21, 33, 81,
141–42, 154–55, 157, 182, 196, 220–21,
245–46, 253, 284, 301, 339 n. 34; of instant images with prior images, 63, 65,
74, 86, 172, 242, 266; with the normal,
33; of specimens, 165, 198. *See also* comparative anatomy; correlations

competition, 122, 174, 196, 208, 253, 294,
302; collegial, 54, 143, 228, 255, 318 n. 89;
in CPC, 5, 134, 326 n. 11

computed tomography. *See* CT

conferences in radiology, 199; attendance
at, 64, 75, 200, 203, 271; case selection
for, 58, 203–4, 220, 261–62, 284, 337
n. 12; formats of, 203–5, 228, 271, 301;
hot seat, 205–7, 210, 217, 271, 302, 337
nn. 15–16

confidence: in ability of trainees, 53; as
bravado, 127; naming and, 35, 206;
rhetoric of, 206, 243; seeing and, 24–26,
29, 35, 90; as trick, 218, 289, 299

confidentiality, 8, 14–15, 299, 333 n. 51

confinement in scanner, 105, 110–11, 129,
190, 323 n. 43

conjecture, hypothesis: abduction as,
143–44, 150, 327 n. 25; confirming, 83,
133–34, 243–44; in detection, 143, 147,
157, 216; in diagnostic reasoning, 52,
133–34, 155–57; ethos of, 146–48, 157,
280, 301, 325 n. 3; in geological sections,
40, 301; in radiology conferences, 142;
in relation to rules, 12; at viewbox,
230–31, 239, 243–44, 253, 266. *See also*
abduction

consent, 8, 63–64, 125, 294, 337 n. 16

consistency, 176, 179, 220, 244–49, 254,
289–90; as radiological term, 32, 35, 74,
206, 226, 233, 242, 244

consultation: of AFIP pathologists, 279,
283; as extra radiological service following x-ray study, 109; by pathologist
of radiologist, 227; by patient of clinician, 257–58; of radiologist by clinician,
50–55, 136, 302, 318 n. 89; training for,
68, 238

contradiction, 10, 29, 206–7, 210, 228, 302

contrast: agency of, 120, 122–23; economics of, 86, 120–23, 297, 325 n. 58; flow of,
64, 104, 120, 188, 190, 223; intravenous,
68, 86, 105–7, 109, 111–12, 115, 117, 119–21,

185, 209, 218, 255, 325 n. 58; oral, 34, 64,
103–4, 107, 114–15, 119–20, 125, 188–89,
205, 242, 323 n. 41; patient reactions
to, 64, 111, 120–21, 125, 185; rectal, 103,
232–36, 243
copy (vs. original), 30, 166–68, 176, 178, 331
n. 25, 332 n. 31
corporations, 296–98, 343 n. 41
correction, 75, 79, 210, 249, 270
correlations, 273; clin-path, 2, 8, 127,
133–34, 136–37, 139, 148, 313 n. 29;
among images, 172, 174, 210, 220, 224,
232, 245–54, 266, 272, 301, 339 n. 34; of
images and chart, 216; of images with
prior reports, 232, 250, 254; of images
with texts, 81–83, 221, 272; among
imaging modalities, 246, 249, 261, 301,
334 n. 65, 338 n. 30; rad-clin, 8, 10–11,
51–52, 54–55, 60, 78, 127, 136–37, 139–40,
149–51, 188, 198, 200–201, 219, 221,
224–26, 231, 243–44, 253–59, 270, 272,
296, 302, 305, 339 n. 41; rad-path, 8, 78,
126–29, 140, 164, 198, 200, 224, 226–28,
259–68, 270–71, 277, 283–86, 289, 292,
297, 299, 301–2, 342 n. 26
courtroom and radiology, 8, 207–8,
216–17, 271, 301, 337 nn. 20–21, 338 nn.
25–29
CPC (Clinical-Pathological Conference),
133, 136–41, 157, 216–17, 280, 302, 336
n. 6; history of, 134, 146, 200, 273, 326
n. 11, 336 n. 5
craft practice, 108; abduction as, 145, 305;
apprenticeship and, 15; curatorial, 198;
of film courier, 174; film handling as,
65; history of radiological innovation
and, 95–99, 128–29, 321 n. 8; image production, 165–66, 224; mystifications of,
14, 336 n. 6; radiological reading as, 7,
49, 90, 149, 305; referenced in radiological terminology, 317 n. 85; sectioning as,
41–42; tacit dimensions of, 7; valued in
display of objects, 330 n. 20

craniology, 27, 146–47, 280, 328 n. 38, 341
n. 15
credibility, 84, 148, 197, 199, 256
credit, 129, 148, 151, 177, 231, 245, 292; for
publication, 197, 284, 334 n. 63
crime, criminals: curating, 152–55, 157, 159,
167, 219, 280, 328 n. 1, 329 n. 5, 329 n. 10,
338 n. 28; detection and, 132–33, 142–43,
145, 151–52, 157, 216, 305; at viewbox, 217,
305; Visible Human and, 332 n. 34
crowd, crowding, 36, 172, 290, 296; detection and, 131–32, 152, 157, 305
CT (computed tomography): conventional x-rays vs., 18, 27, 48, 286;
economics of, 99–100, 103, 123, 170, 183,
287, 294–96; history of, 97–101, 106,
109–10, 138, 294, 301, 322 nn. 21–22, 322
nn. 24–26, 323 n. 35, 343 n. 34; neuro vs.
body, 54, 57, 101–2, 182, 317 n. 86, 322–23
nn. 34–35; research on, 22, 87–89, 293,
335 n. 68; scanners, 100–101, 105–6, 118,
293–97, 320 n. 2, 322 n. 31, 322–23 n. 34;
theory of, 97, 100, 115, 128, 322 n. 24
Cubism, 12, 26–28, 272, 314 n. 38
culprit, 132, 142–43, 152, 155, 157, 159,
161–62, 280, 294
cultic: image, 166, 330 n. 21; object, 162,
165, 198; routine, 8, 12
Cummings, E. E., 97–98
curator, 11; charisma of, 160, 162, 165–66,
179, 181–82, 192, 198, 329 n. 6
curiosity: of clerks, 161, 186; history of,
145–46, 157, 282, 327 n. 33; as psychological matter, 10, 148–49, 157; social
conditions of, 10, 125, 141, 143, 149,
151, 157, 219–20, 229, 253, 268, 273, 304;
within suite of affects, 143, 273
cutting, 1, 12, 40–42, 94, 129, 157, 168, 320
n. 3, 343 n. 1; "cutting edge" and, 84,
218, 293, 299; by spinning, 93, 97–99,
300–302, 322 n. 22. *See also* dissection
Cuvier, Georges, 41, 145–47, 159, 280, 316
n. 67, 327 n. 36, 328 n. 38

206, 220–21, 289–90, 292, 294, 303, 336
n. 6. *See also* exhibition; juxtaposition

dissection, 38, 41–42, 93–94, 145–46, 157,
313 n. 27; anatomy theater and, 2, 335
n. 3

distribution: of archives, 197–98; of CT
scanners, 322 n. 31; of diseases, 307 n.
3; of ignorance, 214; inequitable, of CT
scans, 100, 304; of practices and bodies,
149, 300–301, 305, 308 n. 6, 319 n. 97,
319 n. 102, 343 n. 1; of RSNA meeting
registrants, 290; of sites connected with
CT suite, 174–76

divination, 130–31, 303, 327 n. 34

division of labor, 7–8, 12, 296, 303, 318
n. 92; in conferences, 200–201; curatorial, 198; in detective story, 143, 305;
in history of hospital, 55, 318 n. 90;
in Radiology, 54–55, 240; in reading
room, 62–63, 101–3, 151, 228–29, 252–53;
between techs and radiologists, 109. *See
also* specialties

doc-at-viewbox, 3–4, 274, 296, 300, 303–4

eating and drinking, 282; in conferences,
202–3, 205; by patients, 51–52, 323
n. 41; at reading room viewbox, 22,
62, 65, 78–79, 91–92; in scanner room,
108, 269–71; in scheduling, 184. *See also*
contrast: oral

efficiency: of archives, 305, 330 n. 15; of
courier, 175; diagnostic, 123; of hospital,
193, 303; in radiological reporting,
193; in scanner operations, 113, 182; of
viewbox work, 22, 79, 303

embarrassment, chagrin, 79, 143, 205, 253,
338–39 n. 32, 339 n. 36

emergencies: in CT, 64–65, 111, 124–25; in
radiology, 183; scans for, 13, 56–57, 101,
111, 175, 186, 248

empire, 25, 160, 169, 173, 297, 305, 314 n. 30,
332 n. 39

enemy, 20, 35, 37. *See also* culprit

error, 176, 299; avoided, 134; cautionary
tale of, 33; clerical, 230; in conference,
142, 210–12, 216, 228; courtroom and,
208, 338 n. 29; in decision theory, 7,
334 n. 65; in detective intrigue, 143,
157, 270–71; exposed by postmortem
autopsy, 308 n. 4; in film review, 140,
148, 206, 210–12, 226, 252–53, 270–71;
in learning radiology, 277, 339 n. 32; of
manual practice, 79; message on computer, 119; reviewed in M&M conference, 336 n. 7, 339 n. 36

etiquette and manners, 14, 52–53, 92,
112–13, 207, 298

evidence: images as, 308 n. 5; medical
images as, 5, 136–37, 207–8, 216, 227,
232, 264, 270, 303, 320 n. 1, 339 n. 34;
producing and framing, 9, 91, 131, 137,
227, 270–73, 305, 308 n. 5, 311 n. 10. *See
also* courtroom; gold standard

evidence-based medicine, 12, 196–97, 308
n. 5, 311 n. 27, 334 nn. 64–65

example, exemplification, 7; of abduction/
judgment, 91, 145, 273; of morbid type,
48, 154, 164, 178–80, 285; of taxonomic
type, 165

exchange: affective, 134; of air in scanner
room, 104, 323 n. 45; archival, 160–62,
171; collegial, 68; of diagnostic images
for extra-diagnostic goods, 129, 151,
198, 271, 304; in diagnostic practices,
12, 51, 125, 131, 136–37, 169, 197, 221; gold
standard and, 137, 326
n. 15; involving patient, 180; value, 129,
137, 151, 182, 198, 222, 224, 271, 304, 331
n. 26; visual, 36

excitement, enthusiasm: of discovery, 140,
151, 298; of learning, 80, 205; of radiological vocation, 49, 148, 284; about
research question, 86, 90, 261

exhibition, 303; by AFIP, 284, 296; by
Army Medical Museum, 279; of art-
works, 330 n. 19, 331 n. 23; of CT

exhibition (*cont.*)

scans, 91, 162, 167; history of, 24–25, 160, 164–66, 313–14 n. 30, 330 n. 14, 330 nn. 16–18, 340 n. 3; learning judgment and, 134, 304; at RSNA, 288–94, 297–99; value, 166–69, 181, 198, 222, 287, 294–95

experience, 26, 131–32, 138, 325 n. 3; cumulative training and, 46, 53, 61, 70, 127, 141, 150, 287, 301, 312 n. 22; of patient, 5–6, 93, 104–5, 111–12, 125–29, 180, 190, 255–59, 303; professional, 12, 46, 139, 181, 212, 228, 255, 271, 311 n. 5

expertise: medical, 6, 10, 300, 303, 311 n. 10; nursing, 111; radiological, 54–55, 65, 91, 140–41, 203, 208, 303–5, 318 n. 89, 338 n. 24

exposition, 11, 136, 274–77, 279–80, 287, 292–99, 316 n. 72, 340–41 n. 3

facts, 70, 141, 217, 297; abduction and, 144–45; construction of, 308 n. 5, 315 n. 49

faculty (academic), 54–55, 102–3, 177, 203, 275–76. *See also* attendings

fellows in radiology/CT, 7, 30, 53, 55, 63–66, 70, 83, 203, 317 n. 86

fetish, fetishism, 3, 10, 132, 167, 198, 300, 303–5, 329 n. 6, 331 n. 29

field guide, 81–83, 157, 285, 302–3, 320 n. 108

film, x-ray: clerk, 161–63, 170–72, 198; courier for, 56–57, 91, 160–62, 171–76, 198, 332 n. 38; custodianship, 60–61, 124, 163, 176; disappearance/obsolescence of, 7, 162, 175–76, 309 n. 10, 318 n. 93; economics of, 163, 176; file room, 161–62, 170–72; folders for, 170–73, 202; handling of, 56, 61, 65–66, 91, 140, 160, 164, 166, 170–74, 189, 202, 238, 273, 318 n. 94; hanging of, 55, 65–66, 171, 265, 319 n. 99; lost, 74, 162–63, 171–73, 175–76, 178, 244–45, 273; management of, 57, 161, 170, 171–75, 301, 329 n. 9; material-

ity of, 164, 178, 202, 270–71, 329 n. 11; recycling of, 163, 166; in teaching files, 176–79; warehouse for, 162–63, 171, 174. *See also* images

filmless radiology, 7, 161, 175–76, 180–82, 198, 309 nn. 9–10, 342 n. 24. *See also* PACS

findings: diagnostic/pathognomic, 150; good pickups, 30, 71, 246–49; incidental, 150–51, 253, 256–57; missed, 30, 252–53, 269–70, 286–87, 338–39 n. 32; obvious, 27, 29, 148, 150, 207, 223, 273, 286, 301; recited in case presentation, 70, 206; resemblance among, 150, 154; unusual, 151, 219–20, 260, 263; on x-rays, 30–31, 50–51, 70, 90, 140–42. *See also* description

fixity: of body on scanner table, 93, 103, 110, 189–90, 324 n. 49; of CT scanner, 100; of gaze, 21, 312 n. 22; of radiograph, 14, 137, 165, 301; of tissue/specimen, 2, 10, 128, 137, 165, 280, 297, 307 n. 3

flâneur, 132, 297, 325 n. 7

footwork, 93, 126, 132, 191, 292; courier for film and, 56–57, 91, 160–62, 171–76, 198, 332 n. 38; traveling to images, 53, 58; at viewbox, 18, 55

Foucault, Michel, 307 nn. 2–3, 311 n. 6, 313 n. 29, 326 n. 17, 327 n. 36, 329 n. 5

friendship, 5, 33–35, 37, 53, 90, 143, 312 n. 19

fun, 87, 148–49, 188, 207, 273

furniture, 2, 275, 305; of bureaucracy, 160; in conference room, 201–2, 206; in film management, 170; in reading room, 18, 50, 55; in scanner room, 105–7, 323 n. 39; in scheduling, 184; in waiting room, 189

game, 116–17, 145, 294, 328 n. 39; diagnosis as, 136, 139, 148, 215, 254, 337 n. 17. *See also* bet (wager)

gaze, 25; conjoint/intersubjective, 19–20, 29, 302, 326 n. 24, 326; diagnostic, 11,

35, 49, 55, 90, 211, 300, 313; of image looking back, 13, 35–36; movement of, 21–24, 27, 58, 90, 312 n. 17; touch and, 1, 18; tracking of, 22–23, 312–13 n. 22. *See also* vision

GE (General Electric), 110, 116, 293–95, 298, 343 n. 37

gender, 6, 50, 108, 111, 185, 216, 288–89, 294

generality vs. particularity, 129, 133, 141, 144–45, 150, 186, 220

gestures, 93; in conference, 205, 211–12, 258, 294; in CT biopsy, 126–27; of detective, 145; gaze and, 312 n. 17; motion of x-ray beam and, 94–99, 128–29; performance and, 312 n. 19; regulated, in scientific management, 193; at RSNA meeting demonstration, 294; shock and, 132; styled practices of physical exam and, 15; of techs, 104, 112, 119; unconscious aspects of, 22; at viewbox, 18, 55. *See also* handwork

gift, 32, 49, 67, 83, 122, 240, 277, 284, 288

glass, 51, 317 n. 85; between control room and scanner, 107, 110; in exhibitions, 160, 277, 330 nn. 16–17; in museum, 165, 282, 296

gold standard, 168, 195, 326 nn. 14–16; death and, 137–38, 157; radiographs as, 137–39, 256, 340 n. 42; tissue as, 2, 137, 226, 270

grids: of CT display matrix, 3, 37–38, 168, 303; modernist, 37–38, 91, 303, 316 n. 60; perspectivist, 24; taxonomic, 160, 327 n. 36

guest-host relations: in CT reading room, 50, 52–55, 92, 151; in CT scanner room, 104–5, 268; in ethnography, 8, 14, 309 n. 11

habitus, 1, 10, 24, 53, 68, 131, 143, 205, 310 n. 21

handwork, 129; in biopsy, 125–27; in classical tomography, 96; in confer-

ence presentation, 203; in CT scanning, 107–8, 111–12, 115–19; of detective, 145; in film management, 172–74; in scheduling, 185–86; valued by art curators, 330 n. 20

haunting, 10–12, 94, 131, 168, 200, 222, 253, 301, 316 n. 58; Visible Human Project and, 325 n. 61

hiding, 36, 301; detection and, 132, 142, 169; Reformation suspicion of images and, 333 n. 44; withholding of clues and, 216–17, 228, 254, 272

hierarchy, 7, 84, 160, 172, 174–75, 197–98, 253, 296; of evidence, 308 n. 5; of radiologists and techs, 324 n. 48; in sociality of intrigue, 151, 157, 273, 305; in teaching, 52–55, 61–63, 91–92, 142–43, 201, 271

histology, histopathology, 224, 226–27, 264–66, 278, 280, 290; at AFIP, 283, 286, 342 n. 26; in CPC, 139; history of, 42, 316 n. 72

history, philosophical, 6, 9–10, 94, 100, 130, 301, 305

holography, 47, 288, 297, 317, 343 n. 35

horror, 5, 168, 217, 278, 282

hospital: academic, 7–8, 17, 55, 103, 181–82, 196, 201, 285; archives of, 328–29 nn. 2–3; departmentalization of, 318 nn. 90–91, 336 n. 4; economics of, 122, 166, 193; history of, 3, 55, 134, 148, 193, 280, 309 n. 12, 335–36 nn. 3–5; layout of, 183–84, 188–90, 194–95, 197, 200–202, 300, 307 n. 3, 329 n. 5

Hounsfield, Godfrey, 98–99, 322 n. 22, 323 n. 35

humanism, 91, 309 n. 14

humor and laughter, 92; about apparatus, 79, 318 n. 96, 339 n. 37; about collegial relations, 17, 67–68, 108, 177, 218, 245, 254, 262–63, 268–69; about contradiction or error, 52, 212, 227, 230, 282; about invasive procedure, 128, 245, 325 n. 59; about patients, 180, 225; about

humor and laughter (*cont.*)
trainees, 61, 79–80, 206, 212, 225, 240;
about uncertainty, 17, 35, 52, 73, 82, 218,
222, 241, 254, 262
hybridity, 297, 301, 320 n. 110

identification: of culprits, 153–54, 167, 280,
338 n. 28; of patients, 180, 282–83, 333
n. 51; of personnel, 108, 162, 174, 191,
278, 282, 294, 296, 298, 324 n. 46. *See
also* registration
images: curating, 165–67; making, 11, 49,
115–19, 123–24, 165–67, 177, 224, 319
n. 100, 331 n. 24; moral engagement
with, 12, 331 n. 26; printing, 123–24, 177,
189, 195, 269–70; quality of, 33, 38, 48,
124, 176, 178, 202, 232, 283; texts and,
198. *See also* correlations
Imatron, 294–96
impression, 12, 300; in radiological report,
30, 32, 70, 243, 248, 251, 315 n. 45
improvisation, 68, 114, 149, 187
inconsistency, 229, 249, 253, 255, 259, 283,
290, 299; in rad-path correlations, 228,
263–65
index, indexicality, 20, 105, 116, 128, 161,
166, 243, 265; clinical-pathological
gaze and, 313 n. 29; indication for an
imaging study and, 140, 151, 186, 188,
224, 229–32, 261, 319 n. 102. *See also*
pointing; targeting
inequity, 100, 287, 304
innovation, 95, 97, 100, 193, 293–94, 321
n. 8
interest: of artists in dissected depths,
41; in collecting, 177–79; in a confusing dictation, 70; of Cubists in x-rays,
26–28; culprit pursuit and, 161–62;
detective ethos and, 157, 340 n. 45; in
display of acumen, 134, 238; in exemplary cases, 179–80; at the Expo, 289,
298; in findings, 83, 148, 252, 269–70,
340 n. 44; in logistics of workup, mu-

seum displays and, 165, 278, 282, 296;
the *punctum* and, 36; region of, marked
on image, 75–76; in research questions,
85–87, 203, 218; suite of, 125, 272–73
internationalism, 100, 167, 170, 276, 294,
296–98, 303, 340 n. 2
Internet, 107, 168, 181, 191, 195, 284–85, 296,
336 n. 8, 342 n. 29
interpretation, 130, 134, 159, 165, 173, 198,
290, 304; in CT, 12, 27, 70, 91, 149, 190,
194, 265; radiological, 27, 35, 166, 194,
207–8, 222, 253, 286, 315 n. 45; as section
of x-ray report, 315 n. 45
interrogation, 61–62, 74, 204, 207–8, 215,
228, 271–73, 302
interruption, 6–7, 300, 305; of conference
presentation, 206, 208; death and, 10; of
dictation, 71–74; of gaze, 21–22; of radiologists by techs, 55; of reading, 49–50,
149, 151, 228, 271, 319 n. 105; shock and,
132; by telephone, 62, 132, 185; by writing, 80, 300–301
interventional radiology, 54, 108, 128, 236,
238–41, 243–45, 266, 339 nn. 38–39
intrigue, 10–11, 50, 145, 312 n. 19; in conference, 216–17, 228; diagnostic, 125, 131,
133–35, 142, 148–49, 151, 157–58, 198, 229,
253, 259, 268, 272–73, 296, 302, 304–5;
logistics of, 151; value of, 12, 151, 304
irritation and annoyance, 70; in affective economy of intrigue, 143; about
colleague's practice, 79, 108, 263; about
CT requisitions, 230; about delays, 120,
171; about difficult imaging studies, 67,
148; about film absence or disarray, 66,
171, 224

judgment, 50, 207; in assigning CT protocols, 68; clinical, 12, 296, 310 n. 18; in
CPC, 133, 142, 154; in detection, 157; exercised by courier, 174–75; exercised by
techs, 121, 129; in making comparisons,
211, 221, 326 n. 20; perception and, 328

n. 42; private psychology and, 304, 327
n. 27, 336 n. 4; viewbox intrigue and,
216, 253, 259, 273, 305

juxtaposition, 301, 333 n. 45; in AFIP
teaching files, 287; Benjamin and, 6,
308 n. 7; in conference presentations,
220–21, 303, 339 n. 35; Cubists and,
26–28, 272, 314 n. 38; of film destination
slots, 90, 173; intrigue and, 10, 272; in
MagicView, 176; in museum displays,
165; in police photo galleries, 152–56;
in radiological viewing, 21. *See also*
comparison; correlations

label, caption: 90, 101, 107, 115, 146, 165,
198, 320 n. 110, 330 n. 14; on film and
folders, 21, 65, 166, 170–73, 198, 202, 336
n. 9

learning, radiological, 7, 46, 81–82, 115–16,
271, 273, 335 nn. 3–4

lesions, 2–3, 30, 125, 137; agency of, 5, 6, 35–
37, 90–91, 148, 166, 269, 271–72, 303–5;
culprit status of, 12, 81, 133, 154–55, 161,
163, 215, 221, 223, 271, 284, 300, 303–5;
legibility of, 139, 305; location of, 211–13,
215, 218, 261, 263, 288, 290, 296; resemblance among, 154, 219, 304

library, 53, 83–84, 91, 159–61, 194–95, 218,
279, 341 n. 15; National Library of
Medicine, 332 n. 34

life, 132, 163, 165, 273, 287, 295, 302; changing concepts of, 146, 311 n. 24, 324 n. 49

liking, disliking: by administrators, 192;
by clerks, 188; by courier, 174; by nurses,
125; by radiologists, 102–3, 148, 178,
196–97, 215, 231, 237, 253, 269, 277; by
scanner, 119; by techs, 113, 121

liminality, 105, 323 n. 41

lists: in differential diagnosis, 141, 213–14,
326 n. 20; guiding perception, 34, 70–71,
82, 315 n. 53; of hospital localities, 172;
in logbook, 77–78; of old films, 56, 171,
173–74; of studies on MagicView, 60

logic, 130, 143–45, 157, 219, 258, 327 nn.
25–26

logistics, 334 n. 55; of CT biopsy, 127; of
image movement, 55–56, 60–61, 161,
171–76; of patient flow, 103–4, 182–84;
of publication, 89; of report production, 61, 192–94, 334 n. 61; of scanner
work, 119–20; of viewbox work, 75, 211,
245, 319 n. 105; of workup, 52, 223

luck, 109, 127, 250, 252

machines, 105–7, 182, 184, 193, 195, 304; as
agents in making facts, 129, 315 n. 49;
disenchantment and, 149; at the Expo,
293–94, 298–99, 314 n. 39, 340 n. 3; history of, 95–100, 128, 320 n. 4, 321 n. 8;
with interiors visible, 294–95, 298–99,
330 n. 17; shock and, 132, 167–68

magic, 285, 333 nn. 44–45; in collecting,
179; in PACS, 175–76, 318 n. 96, 332 n. 43

MagicView, 58–60, 76, 91, 124, 175–76, 243,
254–55, 332 n. 40; learning to use, 20–21,
57, 318 n. 96

magnetic resonance imaging. *See* MRI

M&M (Morbidity and Mortality conference), 201, 224, 336 n. 7

manuals, 107, 111, 114–15, 186, 197

map, mapping, 170, 332 nn. 36–37;
abduction and, 327 n. 25; of citation
networks, 335 n. 68; in court, 338 n. 23;
empire and, 25, 160, 169, 294, 332 n. 35,
332 n. 39; realism and, 26, 314 n. 37; of
visual interest, 22–23, 25, 76

matching, 68, 149–50, 171, 222, 297

measurement, metrology, 332 n. 39; in
CT image making, 44; in police work,
152; of radiological outcomes, 7; on
viewbox, 76, 127, 235, 261

media: presentational, 38, 201–2, 220, 299,
301, 336 n. 8; storage, 107, 124, 160–61,
163, 175, 178–81, 283, 342 n. 24; studies
of, 8, 10

medical humanities, 8, 309 n. 16

photography (*cont.*)
  radiography as extension of, 26, 207, 337 n. 18, 338 n. 23; for rad-path correlation, 283
physical examination, 14–15, 258–59, 311 nn. 5–6
physiognomy, 146–47, 152, 280, 328 n. 38. *See also* craniology
picture archiving and communications systems. *See* PACS
pleasure, 33, 48–49, 83, 90, 142, 149, 192, 258, 270, 296
plumbing: of hospital, 184, 202; in interventional radiology, 239–41, 265, 339 n. 39; of patients, 103, 112, 117; near scanner, 106–7
Poe, Edgar A., 10, 13, 93–94, 98, 130, 199, 275, 326 nn. 16–17, 329 n. 11; science and, 146–47, 159, 327 n. 29, 328 n. 37. *See also* detection: Poe and
pointing, 16, 18–20, 50–51, 62, 90, 154, 203, 210–12, 299, 312 n. 15. *See also* index
police, 10, 132, 152–57, 161, 164, 284, 301, 326 n. 24
population, 2, 160, 196–97
positron emission tomography, 328 n. 38
possession: of films, 166–67, 177–79, 337 n. 18; of specimens, 278
postures, 245, 287–88; of patients, 103–6, 108, 110–12, 117, 126, 167–68, 189, 223, 267, 290, 300; at viewbox, 18, 50, 62, 74, 78, 214, 300
prediction, 56, 68, 75, 91, 117, 188, 200, 227–28, 239, 258
primal scene, 145, 222, 300
privacy: of corporate innovation, 294; of examination, 14, 311 nn. 1–2; exclusivity of professional Expo and, 298–99; of hot-seat conference, 205; of offsite film archive, 163; social learning of judgment vs., 304; in stereoscopic viewing, 46–47; urban crowds vs., 132

probability, 51, 141, 213, 215, 218, 225, 335 n. 67
production, 9–10, 14, 24; of affect, desire, and imaginings, 99, 141, 160; of diagnoses/cases, 141, 145, 199; of discontinuity, 12; of images, 54, 95–97, 116, 129, 140, 168; of published research, 87–89, 196; of reality, 32, 168; of reports, 75, 91, 182, 191–92
progress: celebrated by Expos, 276, 282, 296–99, 330 n. 17; medical triumphalism and, 3–5, 9, 305; or regress of lesion over time, 221; stepwise, in inspecting images, 34; trackable in scientific literature, 335 n. 68; through workday, 80, 113
proof, 241, 327 n. 26, 336 n. 6; diagnostic, 90, 137–39, 141, 151, 200, 219, 224–27, 245, 259–62, 266, 270–72, 283. *See also* courtroom; evidence; tissue
protocols: in CT, 66–70, 91, 105, 109, 114–17, 188, 197–98, 230–31, 319 nn. 100–101, 325 nn. 54–55, 325 n. 58; in medicine, 12, 319 n. 101, 324 n. 53
proximity, 132; to specimen, 99, 166, 179, 245, 311 n. 4; in viewing, 15–16, 18, 22, 202, 205; among work sites, 89, 173–76. *See also* touching
public: bodily display and, 168, 311 n. 10; as constituency of museums, 146–47, 155, 164–65, 278–82, 329–30 n. 13, 330 nn. 17–18; as consumers of doc-at-viewbox, 303; excluded from professional Expo, 298–99; reception of x-rays and, 27; relations with technology and, 296, 314 n. 39
publication: academic, 217, 299, 320 n. 109, 334 n. 63, 339 n. 33; in books, 44, 177, 195–96; in journals, 85, 177–78, 195–97, 270, 284, 335 n. 68, 342 n. 28
purity, 8, 102–4, 106–7, 202, 232, 320 n. 3, 338 n. 31
pursuit, 2, 131–32, 155, 157, 304

research (*cont.*)
289–93, 297, 304, 312 n. 16, 334 nn.
65–66; shaped by apparatus, 315 n. 49;
subjects of, 87, 90, 257
resemblance, similitude: in appearance of
RSNA presentations, 292, 303; in cul-
prit/specimen morphologies, 152–54,
166, 226, 297, 303–4; historicity of, 330
n. 21; in homeopathic magic, 285, 333
n. 45; among imaging findings, 81, 83,
141, 153–56, 211, 219–21, 246, 249, 272,
303–4; in photographic representation,
167; sameness and, in copies, 166, 332
n. 31
residents in radiology/CT, 7, 53, 55, 61–65,
79–80, 108, 125–26, 140–42
rhetoric: of aesthetic involvement, 48;
of conjoint viewing, 29, 34–36, 90;
of dictation, 74, 232, 244, 325 n. 59; of
imaging revolution, 12, 131; of interrogation, 215; of intrigue, 12, 136, 149,
215, 217, 245; scientific, 327 n. 29; of
testimony, 12, 16–17, 205, 208, 232, 271,
298, 301, 335 n. 1
ritual, 323 n. 41; bearing witness and, 198,
207–8; of CPC, 133–34, 138–39, 146, 302,
326 n. 11; of cutting, 2, 94, 343 n. 1; in
hospital, 8–9, 11, 131, 309 n. 12, 311
n. 6; of teaching, 11, 180, 200, 205, 224,
271, 280, 336 n. 5; at viewbox, 15, 75
rogues' galleries, 152–57, 161, 167, 280, 284,
301, 305, 329 n. 10
roles, 7–8, 51, 53, 108, 149, 205, 271; of
hosts/shopkeepers, 53, 317 n. 87; role
distance and, 92, 108, 324 n. 47; in transcribed dialogue, 308 n. 8
Romanticism, 41, 157, 310 n. 22, 330 n. 20
Röntgen, Wilhelm, 149, 193, 207, 314 nn.
39–40, 315 n. 44
routines, 128; of conferences, 200–204; CT
protocols and, 68, 114; defamiliarized
by camera, 22; varying among hospi-

tals, 319 n. 105; at viewbox, 14, 34,
54–55
RSNA (Radiological Society of North
America) Meeting, 11, 47, 220, 275–76,
287–299, 317 n. 80, 343 nn. 40–41
rules, 12, 68, 70, 143–45, 150, 186, 215

sacrifice, 128, 282, 296, 343 n. 1
satisfaction, 103, 136, 151, 182, 188, 193, 231;
of search, 30, 148, 315 n. 47
scanner control room, 103, 107–8, 118, 189;
logbook in, 113, 187; radiologists in, 58,
62–63, 109, 116, 124–27, 319 n. 98, 324 n.
48. *See also* technologist, CT
scanner room, 103–7, 112, 189
scheduling: of conferences, 200, 203, 207,
224; of radiology/CT studies, 56, 75–76,
107, 113, 173, 182, 184–90, 198
science and technology studies (STS), 8,
309–10 n. 16
sciences: conjectural, 146, 280; credibility
in, 197; as disenchanting, 149, 310 n. 111;
documents of, 194, 217, 289–92, 295, 320
n. 110, 334 n. 63; fact-construction and,
297, 315 n. 49; gentlemanly society and,
41, 298; national pride and, 165, 282,
296; popular representation and, 91,
296–97, 327 n. 29, 330 n. 13; radiology as
one of, 207–8; religion and, 8, 25, 160;
trade show and, 276
screening, 100, 197, 253, 287, 322 n. 32, 329
n. 9, 334 n. 65, 339 n. 40
search, 30, 36, 130, 152, 159, 254; in literature and databases, 81, 85, 87, 174;
satisfaction of, 30, 148, 315 n. 47
secret, 130, 134–36, 299, 311 n. 1. *See also*
hiding; puzzle
section, as mode of representation, 38,
41, 91, 294, 297–98; in biology and
medicine, 41–45, 81, 95, 280–81, 283, 316
nn. 70–72, 316 n. 75, 321 n. 11, 340–41
n. 3; in CT/MR, 38, 44, 46, 49, 97, 105,

168, 273, 301–2, 323 n. 42; in geology, 38–41, 44, 91, 301, 316 nn. 62–64, 316 n. 67, 331 n. 24. *See also* slices

section, in radiology departmental organization, 102, 276, 284, 293

semiotics, 76, 91, 131, 160, 171–73, 273, 310 n. 111, 331 n. 28; of abduction, 143, 157, 327 n. 25

service personnel in CT, 104, 108, 294

shock: in mechanical reproduction, 98, 100, 105, 129, 131–32, 137–38, 157, 167–69, 301–2, 332 n. 31; in urban crowd, 132, 149

Siemens, 107–8, 110, 116–17, 123, 175–76, 293–95, 332 n. 40, 332 n. 42

signing: report verification and, 74–75, 173, 176, 191, 244; when transferring films, 171, 174

simulacrum, 168, 217, 288–89, 222, 332 n. 32

singing, humming, and whistling, 73–74, 92, 243

skepticism, 27, 99, 116, 207, 225–26, 254, 259

skill: of acumen and judgment, 273; of connoisseur, 166; of courier, 174; cumulative experience and, 287; of detective, 142–43, 325 n. 3; of image making, 16; of pathologists, 340 n. 43; of presentation, 202–3; of radiologists, 29, 36, 46, 65, 102, 130, 148–49; the tacit and, 310 n. 21

skin, 117, 126–27, 172, 288, 290, 294

slices: of cadaver, 94, 129, 168; composing Visible Humans, 44, 168; CT gantry and, 105; dynamic viewing of, 21, 37; in geology, 38–40, 91; making, 94, 332 n. 33; progressively finer, 123; reconstructing three dimensions from, 44–48, 288; summoning comparison, 245, 288, 301–2, 323 n. 42; superficiality of, 12, 27, 148, 157; tomographic devices and, 94. *See also* section, as mode of representation

sounds, 7, 14, 57, 117, 119, 126, 191, 204, 323 n. 45

space, spatiality, 1, 22, 89, 94–96, 98, 320 n. 3; bodies and, 26, 41, 97, 129, 183–84, 288, 297; CT and, 26, 38, 44, 46, 140, 170, 176, 301, 343 n. 35; disease and, 2–3, 307 n. 3; Expo and, 276, 287, 293. *See also* perspectivism; reconstruction

speaking: in case presentation, 140; observed, 320 n. 3; prerecorded instructions to patient, 110; in presentations at RSNA meeting, 290; recorded and transcribed (CT report), 190–92; recorded in fieldwork, 308 n. 8; resident to patient, in CT biopsy, 126–27; by scheduler on phone, 185–88; tech to patient, 119; twined with seeing and touching, 18, 20, 211. *See also* dictation; vision: saying and

specialties: generalists and, 201, 211, 271, 318 n. 89; in medicine, 55, 259, 279–80, 318 nn. 90–91; in nursing, 324 n. 51; in radiology, 54–55, 102–3, 196, 229, 255, 292, 323 n. 37; radiology as one of, 74, 207, 279, 318 n. 90, 337 n. 18; relations among, 7, 51–52, 92, 151, 199–201, 212, 228–29, 255–56, 259, 263–65, 268, 271, 273, 279, 296, 318 n. 92, 323 n. 34, 343 n. 39; in science, 149

specimen, 146, 157, 278–83, 289, 296–97; for AFIP case, 342 n. 26; cultic image vs., 169, 198, 303; labeling of, 165, 288, 330 n. 14; in museums, 330 nn. 16–17, 341 n. 15; radiograph as, 90, 128, 136, 164–65; in the scanner, 93, 97, 99, 106, 125, 129, 189–90, 320 n. 1. *See also* body; film: materiality of; natural history

speed, 10, 22, 79, 123, 149, 198, 211, 293, 301. *See also* time: pace of radiological work and

standardization, 99, 110, 120–22, 193–94, 214, 302, 324 n. 53; of radiological reporting, 18, 70, 168, 319 n. 103. *See also* phantom; protocols

statistics, 141, 215, 290, 328 n. 1, 334
n. 65–66, 336 n. 6; social history of, 152,
154, 197, 315 n. 51
stereoscopy, 46–48, 96, 277, 288, 301, 317
nn. 78–79
storage, 162–63, 175, 329 n. 7
stranger, 8, 14, 36, 53, 132, 205, 230, 256,
309 n. 11
students in radiology/CT, 79, 84–85, 90,
133, 177, 203
studying, 80–82, 177, 195, 205, 218, 305
substitution: of beam for blade, 100;
exhibition value and, 166, 331 n. 26;
of image for tissue/body, 10, 12, 14,
198, 273; of paper for gold, 137; scanner
bed as site of, 93; slice as locus of,
301
surgeons, 5, 8, 23, 234, 236, 259, 288, 324
n. 47; case conferences of, 224, 309,
336 n. 7; innovative displays and,
47–48, 180–81; as radiological clients,
44, 46, 60–61, 124, 172–73, 183, 226,
252–53, 261, 264; ritual studies and, 309
n. 12; "surgical anatomy" and, 38, 41,
44, 46
surprise, 273: at detective's solution,
142–43, 145; of ethnographer, 14, 294; by
imaging findings, 150–51, 204, 216–17,
255; by MagicView, 176; preempted by
CT, 138–39
suspense, 35, 136, 141, 151, 219, 228, 241, 245
suspicion, 87, 145, 161, 259, 263, 273,
294–95; of clinicians by radiologists,
226, 230; detective stories and, 132; of
doctors by clerks, 162–63, 172; of images, 333 n. 44; in viewing radiographs,
20, 22, 35–36, 76, 163, 235, 237, 248. *See
also* culprit
sympathy, 48, 66, 113, 121, 224, 257

table: of CT scanner, 5, 93, 103, 105–6,
111–12, 138, 323 n. 44, 339 n. 42; for

dissection, 2, 94, 138, 272, 300; of taxonomic categories, 157, 160, 211, 272, 303,
307 n. 3, 327 n. 36
tacit dimensions of practice, 1, 7, 10, 14,
22–24, 129, 174, 304, 310 n. 21
targeting: biopsy and, 52, 125–28, 262–63,
265–68, 290; cancer and, 169, 312 n. 16,
323 n. 34
Taylorism, 22, 182, 193, 273, 334 n. 55. *See
also* efficiency
teaching, 7, 29, 139, 228, 273, 284–85, 296;
of anatomy/dissection, 335 n. 3, 336 n. 5;
conferences, 141–42, 149, 203–5, 271–72,
276–77, 302, 336 n. 5, 337 n. 15; didactic
modes of, 204, 215, 220–21, 271, 276–77;
files, 176–82, 195, 285–86, 333 n. 50; of
judgment, 12; as radiological mission,
199, 335–36 n. 4
technique, 9, 44, 146, 160, 165, 197, 221, 261,
297; in radiography, 31, 38, 66, 70–71,
95–96, 119, 140, 319 n. 100
technologist, CT, 298; relations with
patients and, 103–5, 110–13, 117, 119, 182,
190, 324 n. 52; relations with radiologists and, 32–33, 62–64, 66–67,
109, 124–27, 268–70, 324 n. 48; skill
of, 68, 113–14, 116; tasks of, 55–57, 66,
102–6, 108, 110–21, 123–24, 177, 182,
186, 189–90, 198, 297, 324 n. 48, 325 nn.
54–56
technology: assessment, 99–100, 196–97,
334–35 nn. 65–67; effects on social relations of, 9, 12, 46, 99, 128, 157, 200, 259,
293, 301, 304, 310 n. 18, 311 n. 5; nurses
and, 324 n. 50
telemedicine, 333 n. 46, 336 n. 8
telephone, 16, 132; dictation and, 70, 190,
262; in film management, 162, 170–71,
173; in rad-clin communication, 7,
69–70, 187, 190, 230, 246, 248, 251–52,
254–55, 266; in scanner/control room,
107, 113, 117–18, 124, 126, 188; in sched-

Barry F. Saunders is an associate professor of social medicine,
a clinical associate professor of medicine and family medicine,
and an adjunct associate professor of anthropology and religious
studies at the University of North Carolina, Chapel Hill.

• • •

Library of Congress Cataloging-in-Publication Data
Saunders, Barry F., 1959–
CT suite : the work of diagnosis in the age of noninvasive cutting /
Barry F. Saunders.
   p. cm. — (Body, commodity, text)
Includes bibliographical references and index.
ISBN 978-0-8223-4104-8 (cloth : alk. paper)
ISBN 978-0-8223-4123-9 (pbk. : alk. paper)
1. Tomography. 2. Diagnostic imaging. I. Title. II. Series.
[DNLM: 1. Tomography, X-ray Computed—psychology.
2. Anthropology, Cultural. 3. Diagnosis. 4. Diagnostic Imaging—
psychology. 5. Diagnostic Services—organization & administration.
6. Interprofessional Relations. WN 206 S257c 2008]
RC78.7.T6S28 2008
616.07'57—dc22   2008032017